Civil Engineering: Beyond the Basics

Volume I

Civil Engineering: Beyond the Basics
Volume I

Edited by **Sarah Crowe**

CLANRYE
INTERNATIONAL

New Jersey

Published by Clanrye International,
55 Van Reypen Street,
Jersey City, NJ 07306, USA
www.clanryeinternational.com

Civil Engineering: Beyond the Basics
Volume I
Edited by Sarah Crowe

International Standard Book Number: 978-1-63240-103-8 (Hardback)

Printed in the United States of America.

Contents

Preface

Civil engineering has existed since Stone Age. It has formed the fundamentals of our structural existence across centuries. Environmental engineering, geotechnical engineering, geophysics, geodesy, control engineering, structural engineering, transportation engineering, earth science, atmospheric sciences, forensic engineering, municipal or urban engineering, water resources engineering, materials engineering, offshore engineering, quantity surveying, coastal engineering, surveying, and construction engineering are all sub-disciplines of Civil Engineering. It is an indispensable part of both public sector and private sector.

This book aims to aid the advanced study of the principles, processes and practices of physics, mathematics and related fields and their intricate relationship with Civil Engineering. Some major topics discussed in this book will aid the development of this discipline and further the scope of research.

Civil Engineering is seamlessly weaved in our everyday life; beginning with taking a shower to driving back home from work; every network is a part of the civil engineering. This book elucidates important issues concerning the subject and the modern concepts revolving around it. I would like to thank our enabled force of researchers, writers, industry veterans, editors, our publication house and everyone who has been a part of this project.

Editor

Engagement of Facilities Management in Design Stage through BIM: Framework and a Case Study

Ying Wang,[1] Xiangyu Wang,[1] Jun Wang,[2] Ping Yung,[1] and Guo Jun[3]

[1] School of Built Environment, Curtin University of Western Australia, Australia
[2] School of Construction Management and Real Estate, Chongqing University, China
[3] CCDI, China

Correspondence should be addressed to Xiangyu Wang; xiangyu.wang@curtin.edu.au

Academic Editor: Ghassan Chehab

Considering facilities management (FM) at the early design stage could potentially reduce the efforts for maintenance during the operational phase of facilities. Few efforts in construction industry have involved facility managers into the design phase. It was suggested that early adoption of facilities management will contribute to reducing the needs for major repairs and alternations that will otherwise occur at the operational phase. There should be an integrated data source providing information support for the building lifecycle. It is envisaged that Building Information Modelling (BIM) would fill the gap by acting as a visual model and a database throughout the building lifecycle. This paper develops a framework of how FM can be considered in design stage through BIM. Based on the framework, the paper explores how BIM will beneficially support FM in the design phase, such as space planning and energy analysis. A case study of using BIM to design facility managers' travelling path in the maintenance process is presented. The results show that early adoption of FM in design stage with BIM can significantly reduce life cycle costs.

1. Introduction

According to the International Facility Management Association (IFMA), facility management (FM) is defined as "a profession that encompasses multiple disciplines to ensure functionality of the built environment by integrating people, place, processes and technology" [1]. Industries in varieties of areas are adopting BIM for FM. Organizations such university, government, healthcare, retail, and information technology are taking a survey for the adoption of BIM-based FM [2]. Different parts of FM are adopted with BIM in these organizations. Figure 1 depicts the proportion of each function.

However, few efforts in the construction industry have involved facility FM into the design phase [3, 4]. It was suggested that early engagement of FM would contribute to reducing the needs for major repairs and alternations that will otherwise occur at the operational phase [2, 5]. There have been rare effective approaches or processes to engage FM in design stage. The proposed framework of this paper is going to integrate these FM works into early

design stage which could potentially strengthen the collaboration between design team and FM team and reduce alternations. BIM is envisaged to be an effective tool, as proposed in this paper. Considering the multidisciplinary and interoperability of this process, there must be a data source providing convenient integration and access to the relevant information. Building Information Modelling (BIM) is a conceptual approach to building design and construction that comprises all the graphic and linguistic data of building for design and detailing which facilitates exchange of building information between design, construction, and operational phase [6]. A BIM model could comprise individual 3D models of each building component with all associated properties such as weight, material, length, height, geographical information system GIS and information [7]. Beyond the inherent information, BIM also includes external association between building components. For example, the column with name Col. −093 is installed between box ceiling Cei. −52 and level 2 floor with GUID number 30836. Figure 2 depicts the column model and associated properties. The main difference

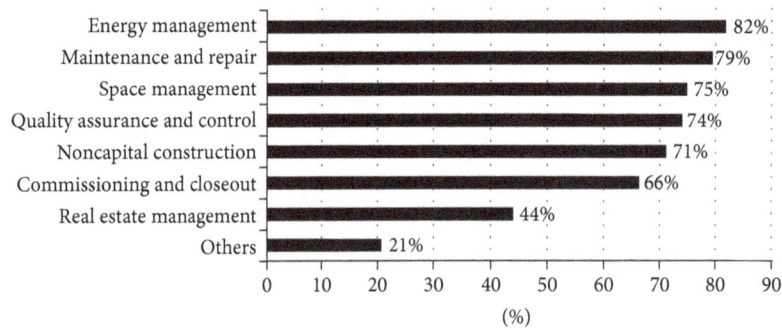

FIGURE 1: Proportion of each function [2].

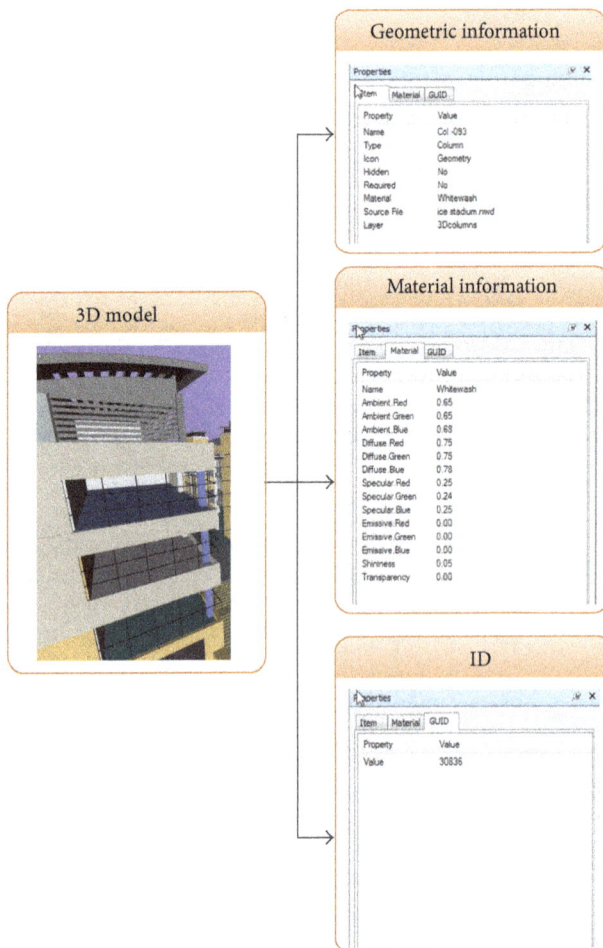

FIGURE 2: Column model and properties in BIM.

be brought to the design stage and which data should be collected are also proposed. With the ease of access to lifecycle information of all the building components BIM provided, proposed building plan could be optimized and lifecycle cost could be reduced with the FM knowledge and experience.

2. Methodology: Conceptual Framework of Integrating FM and BIM in Design Stage

Erdener [5] developed a framework linking design with FM by programming—an extension of "problem defining-solving method" which however did not classify the specific field of work in FM that should be involved into the design stage. Additionally, the backend database was not adopted as an approach to integrate the massive information such as asset portfolios, instructions, and design manuals in this multidisciplinary process. Mostly, the operation and maintenance process of a facility occupies more than 80% in its lifecycle for both cost and time [10]. During the FM process, facility managers have to acquire, integrate, edit, and update massive information related to diverse building elements such as operational costs, warranties, and specifications from varieties of systems. BIM could effectively merge these primary data and provide convenient storage and retrieval of these FM data. Based on the work of Becerik-Gerber et al. [2], three types of FM data should be incorporated into BIM: (1) equipment and systems, (2) attributes and data, (3) portfolios and documents. Figure 3 illustrates the structure of the proposed BIM database for FM. Every facility in buildings is regarded as an individual entity with two kinds of properties—attributes and portfolios. Six types of basic equipment such as HVAC, plumbing, and electrical are represented as entities in BIM. Each entity has its attributes (vender information, location information, etc.) and attached documents (specifications, warranties, manuals, etc.). Specifically, serial numbers of products specified by vendors will be collected as unique identifier for each facility. Model and part numbers will act as reference information during the maintenance. Location information is comprised of building number, floor, and room number. Description stores the status of the facility. Attributes include weight, power, and energy consumption. In order to integrate the whole information into one standardised BIM database, interoperability needs to be assured. This is partially because in different circumstances different

between BIM and 2D CAD is that the latter describes a building layer by layer [8]. Editing one layer will result in massive updating and checking work of associated floor plan. In contrast, BIM models are designed in terms of building components in 3D view. An error-prone process such as clash detection could be conducted automatically [9]. This paper aims to develop a framework of bringing facility management into design stage through BIM. Which field of FM work should

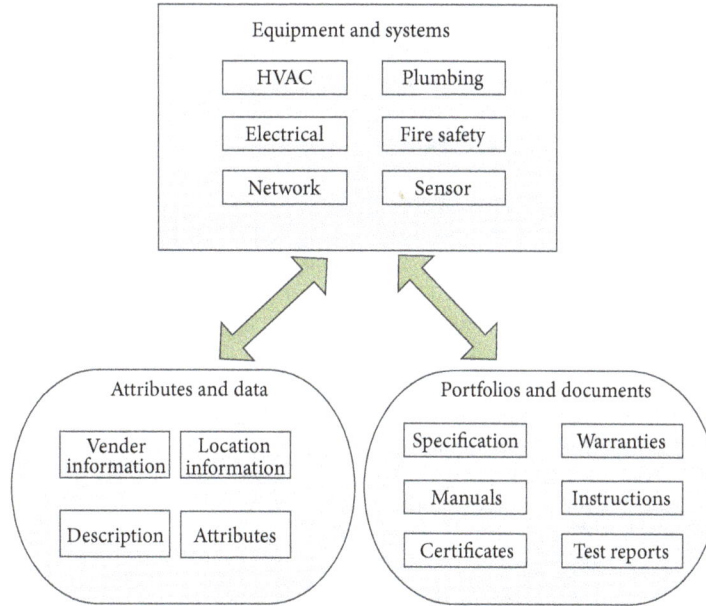

FIGURE 3: FM based BIM database.

FM software systems are adopted, that is, *Mainpac* for building maintenance; *FaPI* for monitoring building condition; *TRIM* for document management. Nevertheless, all these software have their own data structure and usually they are not compatible with each other. International Alliance for Interoperability (IAI) published the Industry Foundation Classes (IFC), a standard for BIM data structure based on an ISO standard (ISO, 1994) enabling exchange of information among heterogeneous systems [11].

BIM will provide supporting information for many categories of FM work such as maintenance and repair, energy management, commissioning, safety, and space management. Three categories of FM during the building's lifecycle are determined to be the most proper and specially discussed in this paper—(1) maintenance and repair, (2) energy management, and (3) commissioning. The decisions made in the design stage affect all aspects in maintenance stage and vice versa. The designer's relationship with the other participants in maintenance stages is very important [12]. Therefore, the maintenance team should also be involved into design stage for decision making. Additionally, different energy saving alternatives can be explored and simulated in early design stage with BIM [13]. Last but not least, commissioning stage ensures that a new building or system begins its life cycle at optimal productivity [14], in which coordination and information sharing between designer and participants essential.

This transformation will provide evaluation information for the design team and make the decision making much easier in both strategic-tactical and operational phase.

For the former, the facility manager could provide post occupancy evaluation of facilities for the design team as feedback. For the latter, bringing these FM jobs into the design stage will avoid redesign and reduce the maintenance job. The following subsection discusses the BIM role in FM engagement in design stage in detail.

2.1. Maintenance and Repair. Maintenance is defined as activities required keeping a facility in as-built condition, while continuing to maintain its original productivity [12]. During this procedure, FM personnel have to identify the components' location and get access to the relevant documents, and finally, the maintenance information. In the state-of-the-art design phase, facility management relevant information such as working space of equipment, storage condition, and weight are not considered. This directly leads to the inappropriate allocation of space and incorrect estimation of load expectations.

Location information of facility could help facility managers efficiently identify the location of specific building components, especially for those who outsource the FM tasks. The knowledge and experience of facility managers could inform the architecture designers with working condition and space of different facilities. Both interior and exterior space requirements must be considered for the normal installation and implementation. Interior space refers to the working space, storage space, and privacy of the space. Exterior space includes the spaces needed for installation and, in case of emergency, for people's escape route. All the above issues could be incorporated into BIM and shown in graphical interface for the discussion between designers and facility managers.

Additionally, FM personnel could retrieve the relevant data of task from BIM's graphical interface in real time. For example, when troubleshooting a printer, FM personnel have to check the maintenance history, get the maintenance manual, generate maintenance reports, and close the request. Conventionally, they have to log on to different electronic document management systems (EDMSs) and toggle between multiple databases to retrieve relevant information.

Preventative maintenance (PM) is defined as the care and servicing by personnel for the purpose of maintaining equipment and facilities in satisfactory operating condition by providing for systematic inspection, detection, and correction of incipient failures either before they occur or before they develop into major defects [15]. For the matter of regular inspections, a schedule will be prepared. Detailed work description is preferred for improving the overall productivity that is work order ID, facility ID, location, description of the preventive work, documents required to perform maintenance, estimated and actual labour hours, and frequency of maintenance work [12]. All these data could be incorporated into BIM database as attributes and documents. Considering the unique ID of every facility, each one could be assigned an associated barcode for the ease of access to relevant information in real time through mobile device. Additionally, after every time of maintenance, status information and working hours will be sent to BIM as feedback and reference for next turn. Figure 4 depicts the workflow of BIM-based PM. Through predesigning of maintenance information such as location information, relevant maintenance history, and schedule for PM with BIM, incorporated information could be accessed conveniently. Future maintenance will be reasonable, and redesign is avoided.

2.2. Energy Management. Statistics from the US Green Building Council [16] show that in the United States, 72% of electricity consumption, 39% of energy use, and 38% of all carbon dioxide are from buildings. However, most buildings are not optimized in terms of energy consumption or not professionally optimized with advice or knowledge of facility management teams [13]. Torcellini et al. [17] identified "designing and constructing low-energy buildings (buildings that consume 50% to 70% less energy than code-compliant buildings) require the design team to follow an energy-design process that considers how the building envelope and systems work together." Energy consumption design must be set by the design team in the predesign phase. Afterwards, virtual prototyping will be created to simulate the energy efficiency. Acquainted with knowledge on energy codes and standards, the building energy consultant in FM team can provide all the information related to energy consumption for the basic energy analyses [17]. However, traditionally, most building energy analyses have been conducted late in design. Due to the difficulty and expense of modelling the energy systems, identifying and validating energy saving alternatives with different models is not economically possible. A large portion of time will be consumed in converting floor plans to energy management system graphics [4]. BIM is envisaged to be the platform for data exchanging avoiding reentering all the building geometry, enclosure, and HVAC information. Interoperability can be overcome by the data exchange standard gbXML (Green Building Extensible Markup). BIM software such as *Bentley*, *Autodesk Revit*, *Graphisoft ArchiCAD*, and *Google sketchUp* are able to export energy analyses data in gbXML format. For overall BIM energy design of a building, three steps have to be executed based on Kim et al.'s [13] work as follows.

(1) Create BIM Model of Building. BIM model could be created based on the existing floor plan. This model comprises

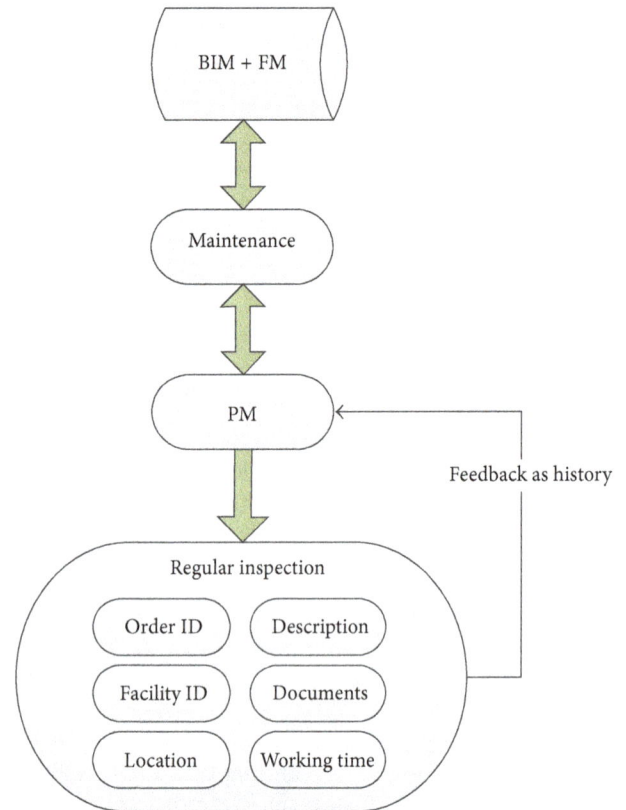

FIGURE 4: BIM-based PM workflow.

structured building components which include spatial data, texts, and databases of other properties. Based on these data, volume can be calculated, as well as energy estimates.

(2) Integrate Energy Consumption Data and Test to Identify an Alternative. After modelling, energy consumption relevant data could be exported in gbXML format and analysed by tools such as *Ecotect*, *Green Building Studio*, and *DOE-2*. gbXML data could be used to analyse the energy consumption of the whole building, estimate water usage and cost evaluation, visualize solar radiation on surfaces, and simulate daylight factors [18]. Alternative design could be simulated by changing the lighting, roof, and walls.

(3) Validation of the Proposed Design. After a design alternative is specified, validation from energy consultant is essential. Logic and assumptions of the energy model must be carefully reviewed. Figure 5 depicts the framework of energy management design using BIM.

Treating each energy-consuming object as an entity, real time and period energy consumption will be collected as one of its property. Thus, cost information of a room/zone could be calculated by adding up all the energy cost of energy-consuming objects in it. Some high-energy-consumption facilities' information could be predicted by the historical data.

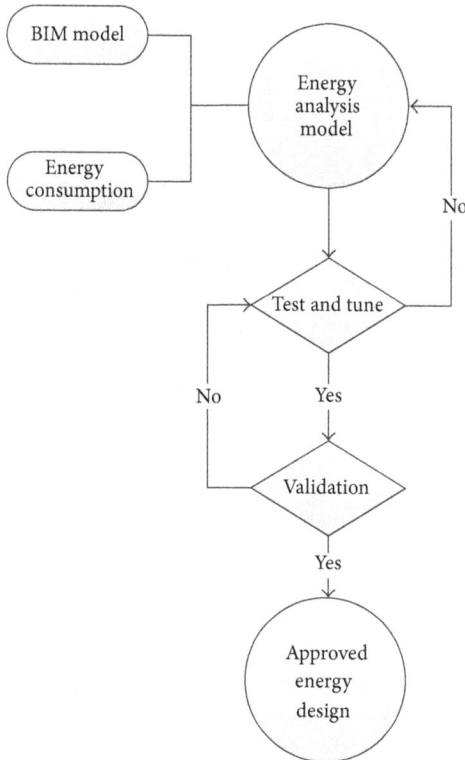

FIGURE 5: Framework of energy management design using BIM.

2.3. Commissioning and Handover. Building commissioning is defined as "a quality assurance program intended to demonstrate that the building is constructed well and performs as designed. If the building materials, equipment and systems were not installed well and are not operating as intended, the health, productivity and other benefits of high performance design will not be achieved" [19]. Building commissioning is a key process for the building operation and maintenance, as the Department of Energy (DOE) suggested that "it ensures that a new building or system begins its life cycle at optimal productivity and improves the likelihood that the equipment will maintain this level of performance throughout its life. Building commissioning is the key to quality assurance in more than one way" [20]. For evaluating the project quality and identify potential significant design defects before it is too late or expensive to make changes, building commissioning should be embedded in the following phases: predesign phase, design phase, construction phase, transition to operational sustainability, postoccupancy and warranty phase, and retrocommissioning [21]. This section is focused on the first two processes: predesign phase and design phase. In design stages, commissioning scope and commissioning team must be identified. Since different facilities have different features and budget limitations and different projects have special systems to be commissioned, commissioning team has to be involved in early design stages for the decision making. This approach will also enable knowledge sharing between different parties. However, in the process of commissioning, massive 2D documents, images, maintenance, and operation information need to be collected and accessed frequently.

For example, in the commissioning process of the Maryland General Hospital (MGH), the following systems need to be commissioned: a new 2000KVA normal power substation, a new 500 KW emergency generator and paralleling switchgear, three new automatic transfer switches and distributions, 2 new 650 ton electric centrifugal chillers and 650 ton cooling towers, temperature and humidity systems, and duct work, air handlers, dampers, and fans [21]. There has to be an easily accessed platform for data exchange and integration. BIM is envisaged to overcome the problem of interoperability and provide easy access for these massive data. When scoping which facilities need to be commissioned, similar projects' information could be retrieved from BIM database, as reference for the decision. The graphical interface of BIM could also improve the collaboration among owner, designer, and contractor. Schedules and commission facilities could be predesigned and stored in BIM models by each facility/zone/room. These plans are shown as timeline, which act as a simulation of actual commissioning practice. Thus, the logical faults and collision between activities can be easily identified. Moreover, design errors and conflicts of plumbing, HVAC, and electrical from different team could be discovered in the integrated view of BIM. In the design phase, different commissioning tasks have to be assigned to the specified experts for individual commissioning. After all these commissioning subtasks are approved, the whole system has to be commissioned together. For example, plumbing and electrical systems in a room need to be commissioned in a designed order. After both are approved, it must be ensured, these systems could work together successfully. Thus, the overall commissioning task must be done. When commissioning the plumbing or electrical system in different areas, a logical order must be specified. Simulating this schedule and identifying an optimal alternative will reduce the commissioning cost and time. Figure 6 depicts the BIM-based commissioning streamline.

Barcode system could also be incorporated into BIM for the ease of accessibility of commissioning documents. Each commissioning facility is assigned a unique ID in the BIM model for storing the properties and relevant documents when it is designed. Each ID could be associated with one barcode. The commissioning team could scan the barcode and retrieve the product data, operation data, and maintenance manuals right in the field. Commissioning critical tasks and checklists could be pushed on to a mobile device. The results will be automatically uploaded to BIM central database in a standard format after commissioning tasks.

As project proceeds, these data are handed over to the operational phase as well as the updated BIM model. Besides correcting design defects, simulating commissioning schedules, and bringing easy access for information, bringing commissioning to design phase through BIM could also improve energy saving performance in operation and maintenance phase.

3. Case Study

The project of Shanghai Disaster Tolerance Center is an ideal example of bringing FM to design stages through BIM.

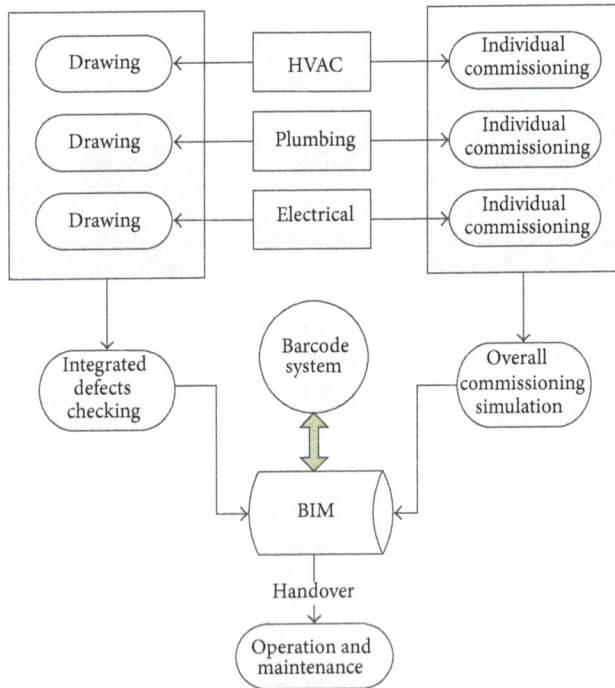

FIGURE 6: BIM-based commissioning streamline.

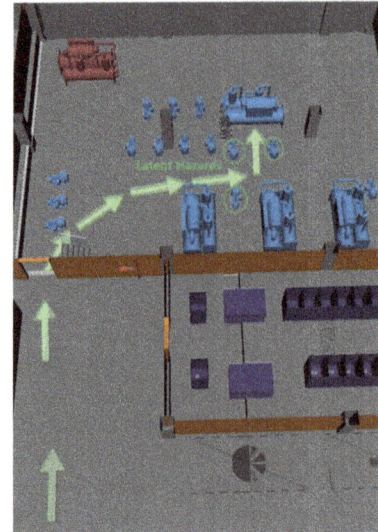

FIGURE 7: Traditional travelling path and latent hazards.

Travelling path of facility manager is predesigned in BIM thus reducing the maintenance time and providing easy access to the location information of facilities.

3.1. Project Overview.

Shanghai Disaster Tolerance Center is in the north of the North Industrial Park in Shanghai, China. It was designed in September of 2010 as a State Grid Corporation of centralized information systems data center. The construction area is 28,124 square meters with one underground layer and four floors on the ground. The diesel generator room and pump are 9.1 meters in height, with construction area of 1,703 square meters. Shanghai Municipal Electric Power Company is the construction company. Shanghai Modern Architectural Design Co., Ltd. is the design company. This project is complicated in facility systems with a tight schedule. High-standard requirements of materials and labor cost control are other characteristics of this project. BIM has been decided to be the tool to bring the FM work into design stage for predesign and simulating the maintenance work in FM.

3.2. BIM Services Content.

Accurate BIM model of the mechanical, electrical, construction, and interior decoration are created based on 2D drawings provided by the owner. Clash detection of pipelines is conducted and optimized. BIM model is used for scheduling and guiding the on-site construction work. A 4D construction simulation is also conducted based on the BIM model. Security control and quantity takeoff are based on analysis of pedestrian stream. Construction schedule needs to be incorporated into model in order to visualize the construction process in BIM model. Last but not least, a database platform is developed to read BA surveillance data and conduct real time positioning in 3D mode to get location of facilities thus facilitating maintenance work. This improves the monitoring ability and security level of the disaster tolerance center.

3.3. BIM-Based Travelling Path Optimization in Maintenance.

During the process of maintenance, FM personnel have to identify the components' location, getting access to the relevant documents, and finally, the maintenance information. Location information of facility could help facility managers efficiently identify the location of specific building components, especially for those who outsource the FM tasks. Conventionally, they have to log on to different electronic document management systems (EDMSs) and toggle between multiple databases to get the location information, relevant maintenance manuals, and warranty documents. BIM could integrate all these information together in a graphical view. By predesigning the travelling path in the maintenance job, travelling time is well scheduled and reduced and latent hazards could be avoided. Traditionally, after identifying the building number and room number, FM personnel just go to the maintenance spot through a normal path, which may be not the shortest path. Moreover, latent hazards are not identified because of lack in relevant knowledge. Figure 7 depicts the normal travelling path and latent hazards. Since there are different kinds of latent hazards in different areas, the travelling path needs to be identified with the knowledge of all departments of FM teams. After discussion between spatial experts and FM team, an optimal path is specified and incorporated into BIM database, which is safe and consumes the littlest time.

In the following scenario, using BIM to design, optimize, and simulate the path of troubleshooting, the reciprocating compressor is illustrated. Figure 8 depicts the reciprocating compressor with problem.

Firstly, FM staff receives a manual request of troubleshooting the reciprocating compressor with ID 98241620-001f-49d7-94e9-7104b0a3a93d in the underground floor of

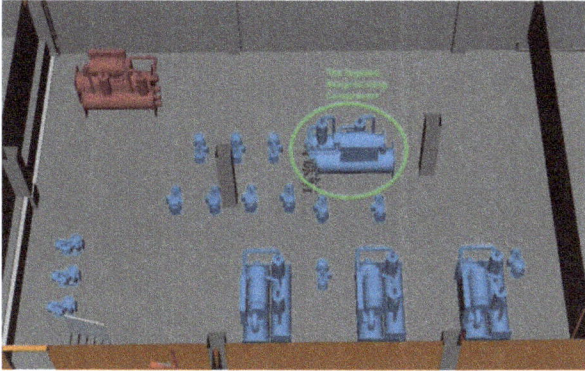

FIGURE 8: Targeted reciprocating compressor.

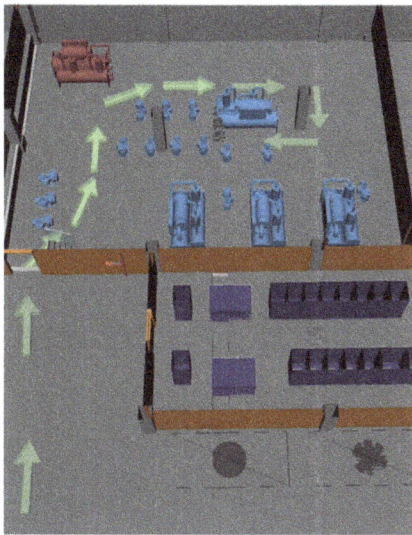

FIGURE 10: Third person view of travelling.

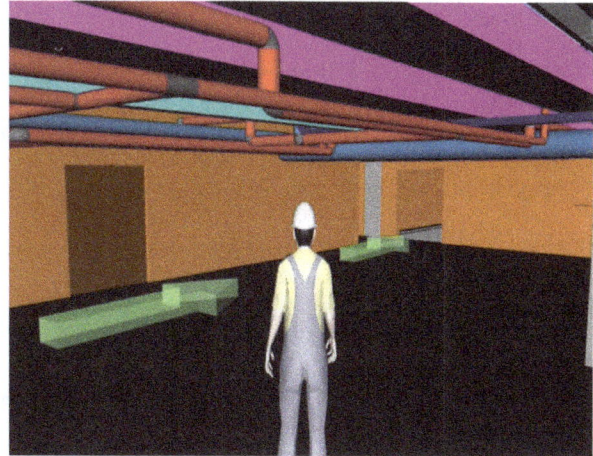

FIGURE 9: Optimal maintenance travelling path.

the Shanghai Disaster Tolerance Center. He then opens the BIM model and searches with this ID. A section view of Figure 8 is shown, and the targeted reciprocating compressor is highlighted. Access to the maintenance manuals, warranty documents, and maintenance history is also provided in the BIM model. After choosing the action of "go the maintenance site," an optimal path is visualized which is safe and timesaving as depicted in Figure 9. A third person view is also provided for the simulation of travelling. Arrow keys in keyboard can be used to control the character. Figure 10 depicts the third person view. The FM staff follows the path and troubleshoots the reciprocating compressor with maintenance manuals in a mobile device. Reports are uploaded to BIM central database as history. Status of the reciprocating compressor is updated as "Normal."

In this case, BIM is utilized as a database and visualization platform to predesign the travelling path in the maintenance job. With knowledge of FM experts, travelling time is well scheduled and reduced, and latent hazards are avoided in the design stage.

4. Conclusion and Future Work

This study developed a framework of considering FM in design stage with BIM. The contribution of this paper is the development of an innovative framework, which integrates FM work into early design stage via BIM. Furthermore, one aspect of the whole framework is validated for the proof of the concept. An innovational concept of gathering designers with the FM team through BIM is proposed for strengthening collaboration as well as information sharing and gathering. The purpose is to avoid and reduce the potential issues such as rework and inappropriate allocation of workspace in the operational phase. As little research has identified the approach and benefit of integrating FM with early design stage, this study aims at bridging this gap by providing a working pattern of providing the essential information with BIM. Due to the difficulty of altering the main structure and core service areas in the operational phase, it is practical to design for adaptability by considering operational condition and the facilities' own attributes. It is very difficult to achieve without the relevant information from FM team and appropriate integration platform. With the ease of access to lifecycle information of all the building components BIM provided, the proposed building plan could be optimized, and lifecycle cost could be reduced with the FM knowledge and experience.

Acknowledgments

CCDI of China provided support for the traveling path optimization case study, which is a large architectural consulting firm that provides integrated services for urban construction and development. Data and projects of Shanghai Modern Architectural Design Co., Ltd., Shanghai Municipal Electric Power Company, and Grid Corporation of centralized information systems data center belong to the CCDI company. The contents of this study reflect the views of the authors who are responsible for the facts and the accuracy of the data presented herein. The contents do not necessarily reflect the official views or policies of these companies.

References

[1] International Facility Management Association, http://www.ifma.org/.

[2] B. Becerik-Gerber, F. Jazizadeh, N. Li, and G. Calis, "Application areas and data requirements for BIM-enabled facilities management," *Journal of Construction Engineering and Management*, vol. 138, no. 3, pp. 431–442, 2012.

[3] B. Nutt and P. McLennan, *Facility Management: Risks and Opportunities*, Blackwell Science, 2000.

[4] D. G. Cotts, K. O. Roper, and R. P. Payant, *The Facility Management Handbook*, Amacom Books, 2010.

[5] E. Erdener, "Linking programming and design with facilities management," *Journal of Performance of Constructed Facilities*, vol. 17, no. 1, pp. 4–8, 2003.

[6] R. Sacks, I. Kaner, C. M. Eastman, and Y. S. Jeong, "The Rosewood experiment—building information modeling and interoperability for architectural precast facades," *Automation in Construction*, vol. 19, no. 4, pp. 419–432, 2010.

[7] V. Singh, N. Gu, and X. Wang, "A theoretical framework of a BIM-based multi-disciplinary collaboration platform," *Automation in Construction*, vol. 20, no. 2, pp. 134–144, 2011.

[8] J. P. Duarte, G. Celani, R. Pupo et al., "Inserting computational technologies in architectural curricula," in *Computational Design Methods and Technologies: Applications in CAD, CAM and CAE Education*, N. Gu and X. Wang, Eds., IGI Global, Hershey, Pa, USA, 2010.

[9] X. Wang and P. S. Dunston, "Comparative effectiveness of mixed reality-based virtual environments in collaborative design," *IEEE Transactions on Systems, Man and Cybernetics C*, vol. 41, no. 3, pp. 284–296, 2011.

[10] M. R. Devetakovic and M. Radojevic, "Facility Mangement: a paradigm for expanding the scope of architectural practice," *International Journal of Architectural Research*, vol. 1, no. 3, pp. 127–139, 2007.

[11] International home of openBIM. 1994, http://buildingsmart.com.

[12] D. Arditi and M. Nawakorawit, "Designing buildings for maintenance: designers' perspective," *Journal of Architectural Engineering*, vol. 5, no. 4, pp. 107–116, 1999.

[13] H. Kim, E. Jenicek, and A. Stumpf, "Early design energy analysis using bims (building information models)," in *Proceedings of the Construction Research Congress*, pp. 426–436, April 2009.

[14] ASHE, *Healthcare Facility Commissioning Guideline*, ASHE, Chicago, Ill, USA, 2010.

[15] J. W. Korka, A. A. Oloufa, and H. R. Thomas, "Facilities computerized maintenance management systems," *Journal of Architectural Engineering*, vol. 3, no. 3, pp. 118–123, 1997.

[16] U. S. G. B. Concil, *Green Building Facts*, 2009.

[17] P. A. Torcellini, S. J. Hayter, and R. Judkoff, "Low-energy building design—the process and a case study," in *Proceedings of the ASHRAE Annual Meeting*, pp. 802–810, June 1999.

[18] B. Dong, K. P. Lam, Y. C. Huang, and G. M. Dobbs, "A comparative study of the IFC and gbXML informational infrastructures for data exchange in computational design support environments," in *Proceedings of the Building Simulation*, 2007.

[19] US Environmental Protection Agency, http://www.epa.gov/.

[20] DOE, *BUILDING COMMISSIONING: The Key to Quality Assurance*, Rebuild America Guide Series, DOE, Washington, DC, USA, 1998.

[21] G. C. Lasker, H. Y. Dib, and C. Chen, "Benefits of implementing building information modeling for healthcare facility commissioning," in *Computing in Civil Engineering*, vol. 2011, pp. 578–585, 2011.

2

Pounding Effects in Simply Supported Bridges Accounting for Spatial Variability of Ground Motion: A Case Study

G. Tecchio, M. Grendene, and C. Modena

Department of Structural and Transportation Engineering, University of Padova, Via Marzolo 9, 35131 Padova, Italy

Correspondence should be addressed to G. Tecchio, tecchio@dic.unipd.it

Academic Editor: Sami W. Tabsh

This study carries out a parametrical analysis of the seismic response to asynchronous earthquake ground motion of a long multispan rc bridge, the Fener bridge, located on a high seismicity area in the north-east of Italy. A parametrical analysis has been performed investigating the influence of the seismic input correlation level on the structural response: a series of nonlinear time history analyses have been executed, in which the variation of the frequency content in the accelerograms at the pier bases has been described by considering the power spectral density function (PSD) and the coherency function (CF). In order to include the effects due to the main nonlinear behaviours of the bridge components, a 3D finite element model has been developed, in which the pounding of decks at cap-beams, the friction of beams at bearings, and the hysteretic behaviour of piers have been accounted for. The sensitivity analysis has shown that the asynchronism of ground motion greatly influences pounding forces and deck-pier differential displacements, and these effects have to be accurately taken into account for the design and the vulnerability assessment of long multispan simply supported bridges.

1. Introduction

Earthquake ground motion is usually assumed as a spatially uniform dynamic input in seismic analysis; this assumption is correct for structures standing on a reasonably restricted area, in which the soil characteristics are presumed to be homogeneous and the seismic wave propagating velocity can be neglected, but becomes inadequate for spatial structures standing on large sites such as extended foundations or dams, and long-estending structures such as bridges, viaducts, tunnels, and pipelines. In these cases the spatial variability of ground motion should be considered to avoid gross evaluation errors or at least underestimation of the dynamic response, since the phenomenon affects the response considerably and, hence, the level of protection of these structures (Lupoi et al. [1]). In particular for long multispan simply supported bridges, a spatial variation in the input acting at supports (pier and abutment foundations) should be considered since it can induce pounding effects and deck unseating. It has been observed during the recent major seismic events that this kind of bridge structure very often experiences pounding phenomena between adjacent structural segments (between neighbouring decks or cap-beams and decks, with

a component of impact force transferred to the piers), which can amplify differential movements between adjacent spans and determine cracks or brittle fractures at beam endings. These amplified differential displacements can induce pull-off and drop collapses of spans when the displacement capacity of the bearing devices is exceeded or the seating length of girders is not sufficient for them to rest on their supports during strong ground motions.

For this type of bridges are required quite complex numerical models to represent with acceptable approximation the global structural response taking into account the inchoerency of the seismic excitation at the supports, the impact phenomena between neighbouring structural segments, and the nonlinear behaviour of the substructural components (piers and decks).

In the present study, the acceleration and displacement time histories at the several prescribed locations on the ground surface corresponding to the bridge supports, are generated using the spectral representation method [2–4]. In order to generate the stochastic field, three basic components are required: (i) power spectral density (PSD) which gives the frequency content of the random process, (ii) coherency function (CF) which gives an analytical representation of spatial variation of the ground motion in the frequency domain, and (iii) shape function (SF) for determining a nonstationary random process in the time domain. Some expressions have been proposed for the target spectral density (i.e., the Clough-Penzien form [5], and the expressions given by seismic codes [6]), for the coherency function [7–10] for the shape function [9]. The generated time histories are compatible with prescribed response spectra and duration of strong ground motions for the considered seismic area and reflect the wave passage and loss of coherence effects.

As regards evaluation of the pounding effects, it has to be said that the interest of researchers is quite recent; the problem was first investigated by [11] who studied the pounding phenomenon between two adjacent buildings, modelling the collision through impact elements which connected simple single-degree of freedom structures. In 1992 the same problem was examined also by [12]; in 1998 the study made by [11] was taken up again and applied to bridge structures in [13]. Further investigations were developed in [14] on the numerical simulation of the pounding process with the aim of calibrating the impact element between neighbouring structures by comparing the numerical results with the exact solution based on the wave propagation theory. In recent years more complex finite elements models have been developed: a numerical 3D simulation applied to a multispan simply supported bridge is described in [15].

From the aforementioned studies interesting conclusions can be drown for an improved modelling of the pounding effect:

(1) pounding between adjacent segments can be described with fair accuracy through an impact element characterised by stiffness and damping (which accounts for energy dissipation);

(2) it has been noted in [13] that there is no need of modelling the entire structure for long bridges in order to assess the middle span response with fair accuracy; it is enough to study the seven central spans since there are no relevant differences in the numerical results between a model with an infinite number of span and a seven-span model;

(3) in the finite element model the stiffness k_1 of the impact element should be calibrated considering the number n of finite elements which compose the deck ([14]);

(4) it is important to define opportunely the time step used for the integration in the time-domain to avoid that colliding adjacent segments of neighbouring decks may behave like rigid bodies, since they influence the dynamic response with their axial deformation.

In the present study state-of-the-art models have been used to simulate the asynchronous ground motion as a multisupport seismic excitation and describe the pounding effects, as described in Section 2.

FIGURE 1: Fener bridge: (a) lateral view of the bridge on the Piave river.

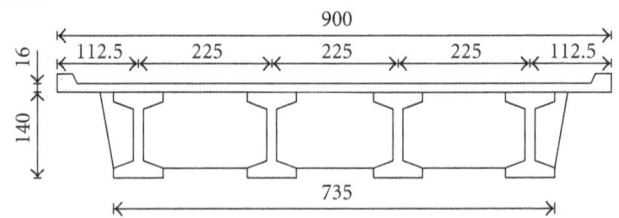

FIGURE 2: Typical transverse section of the superstructure.

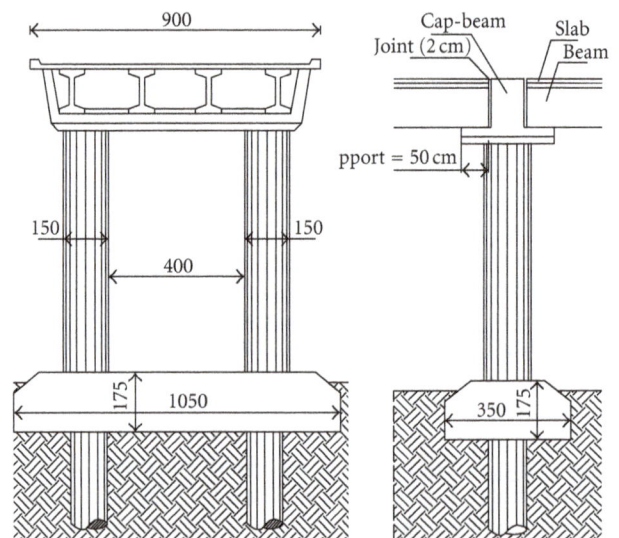

FIGURE 3: Pier elevation and longitudinal section.

2. Seismic Response Accounting for Spatial Variability of Ground Motion: A Sensitivity Analysis

2.1. The Fener Bridge. This study carries out a parametrical analysis of the seismic response to asynchronous earthquake ground motion of a long multispan rc bridge, the Fener bridge (see Figures 1, 2, 3 and 4), located in the Veneto region, in the Treviso province.

It represents an important overcrossing of the Piave river for the region road network. It was built in the mid nineteen seventies, and it consists of 24 regular spans having the same length of 24.75 m, except for the lateral spans near abutments, which are shorter (in particular at one end there

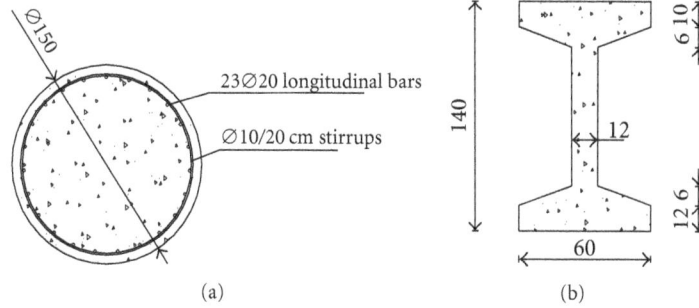

FIGURE 4: Cross sections of (a) typical column with reinforcement and (b) typical longitudinal precast concrete beam.

are two spans with reduced lengths of 18 m and 17.5 m, respectively, whilst at the other end only the last span has a slightly shorter lenght of 23.75 m). The overall structure is about 579 m long.

The deck lodging two lanes has an overall width of 9 metres; the deck structure is made up by four I-shaped precast beams with a constant height of 1.4 m and by a 16 cm high rc slab. The transverse distribution of traffic loads is obtained through 3 orthogonal rc girders positioned in the middle and at both ends of each span. Piers have a portal-shaped structure with circular rc columns, whose height varies gradually along the plan from a minimum of 5 to 8 m roughly, since the deck slope in the longitudinal direction is about 2%, while the extrados levels of plinths at the base remain constant. Piers raise on deep foundations as illustrated in Figure 3.

The pier section is shown in Figure 4(a): reinforcement for each of the two columns is provided by 23 longitudinal bars of 20 mm diameter and transverse stirrups of 10 mm diameter (pitch = 20 cm).

The materials used for piers can be classified as follows:

(i) concrete: grade C25/30;

(ii) reinforcing steel: smooth bars, characteristic yield stress f_{yk} = 315 MPa.

According to the national seismic zonation map, Fener bridge is located in an area characterised by PGA = 0.25 g, on a soil of medium stiffness (type B soil, according to the national zonation map [16]).

2.2. FE Model of the Multisupported Structure. In the numerical model of the bridge elements with linear and nonlinear behaviour have been adopted in order to represent effectively the global structural response: main beams, cap-beams, and transverse girders have been modelled with linear beam elements, the rc slab has been modelled with plate elements, whilst a nonlinear behaviour has been adopted for columns to simulate their hysteretic behaviour (using the Takeda-model [17] see Figure 5), for gap elements simulating impact between adjacent structural segments, and for frictional connections between longitudinal beams and cap-beams.

For the pier-element it has been necessary to assign in input a nonlinear force-displacement law, which has been

obtained through a push-over analysis both in the longitudinal direction, where the column has a cantilever deflection, and in the transverse direction where the piers behaviour is that of a portal frame. A lumped plasticity element has been employed for modelling the piers; the derived force-displacement curves are plotted below (see Figures 7 and 8).

Girders sit on cap-beams without any bearing devices therefore restraint of superstructure segments from longitudinal displacement is given only by friction; the force-displacement law for frictional bearings is assumed as an idealized rigid-plastic behaviour, with a friction coefficient taken as μ = 0.60 in the analysis (see Figure 6). In the transverse direction a rigid restraint between deck and cap beam is assumed: the cap-beam lateral sides, being in direct contact with beams and acting as shear keys, do not allow any differential displacement.

Pier-deck pounding has been modelled through nonlinear gap elements which react only under compression, after the initial gap closure corresponding to the joint width (2 cm).

The gap element stiffness k_1 has been determined normalising to 1 parameter γ in the following expression [14]:

$$\gamma = \frac{k_1 L}{nEA}, \tag{1}$$

where A is the deck cross section, E its elastic modulus, L the span length, and n the number of finite elements into which the span length has been divided, taken as n = 10 in this study. In particular for the impact element a damping equivalent to energy dissipation has not been considered and a perfectly elastic collision has been modelled since impact energy dissipation does not influence the global structure response significantly [13].

As to external restraints, they have been considered fixed both in translation and rotation, because foundations are plinths on piles and in a first approximation the soil-structure interaction can be neglected. The superstructure segments not considered in the model (which represent a boundary condition to it) have been substituted with gap elements as illustrated in [13].

The FE model of the bridge and the related nonlinear dynamic analyses have been performed using *CSI SAP2000 release 9* software [18]. The model represents only the 7 central spans of the bridge (see Figure 9), with adequate

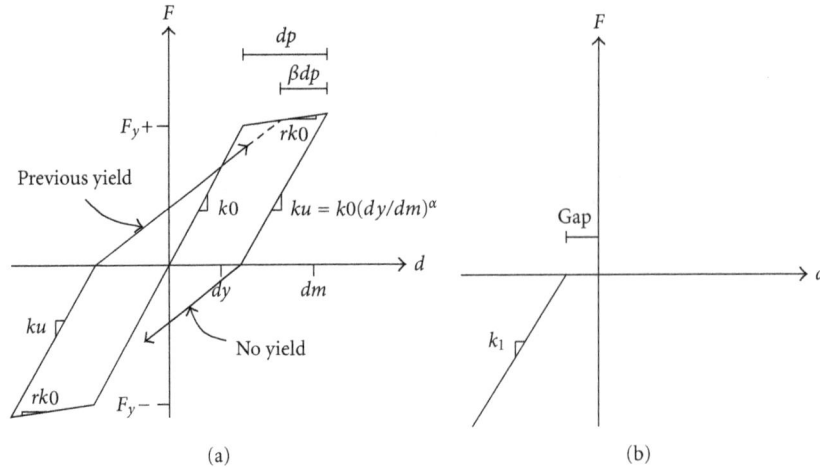

FIGURE 5: Models for (a) hysteretic behaviour of piers (Takeda model) and (b) gap element between adjacent structural segments.

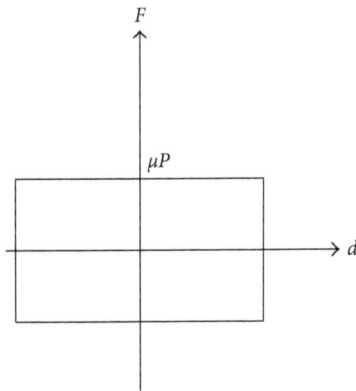

FIGURE 6: Connection between girder and cap-beam (frictional behaviour).

boundary conditions, instead of all the 24 spans; this has allowed to reduce substantially the computational effort due to the nonlinear effects included, without influencing the numerical accuracy of results because, as reported in [13], the seismic response of the central span in a model with a number of spans not less than five is a good approximation of the response obtained modelling the complete structure.

2.3. Charactherisation of Spatial Variability.

In the present study, the acceleration and displacement time histories at several prescribed locations on the ground surface corresponding to the bridge supports are generated using the spectral representation method. A uniform soil type is considered. As mentioned before, in order to generate the stochastic field, three basic components are required: (i) power spectral density function, (ii) coherency function, and (iii) shape function.

2.3.1. Power Spectral Density Function.

Different analytical models for PSD are advanced by some authors; in this study the expressions given in EC8 [6] have been used, which are approximate relations for power spectra corresponding to the site-dependent response spectrum proposed in the code. The expressions are derived as follows:

$$S_a = 0.2\xi' A^2 T^{1.4} \quad \text{for } T < T_B,$$

$$S_a = 6\xi' V^2 T^{-0.74} \quad \text{for } T_B < T < T_c, \qquad (2)$$

$$S_a = 300\xi' D^2 T^{-3.1} \quad \text{for } T > T_c,$$

where S_a is the acceleration power spectrum, ξ' is the value of the damping ratio, A, V, and D are the values of spectral acceleration, velocity and displacement, and T_B and T_C are the response spectrum parameters.

2.3.2. Coherency Function.

Assuming that the seismic wave field can be completely described by a single plane vawe, its spatial variation can be quantified by means of the coherency function, which expresses the dependence in the frequency domain between the PSD of time histories ground motions occuring at two different stations k and l (with relative distance given by d_{kl}) [15]. It is generally defined as follows:

$$\gamma_{kl}(\omega) = \begin{cases} \dfrac{S_{kl}(\omega)}{\sqrt{S_{kk}(\omega) \cdot S_{ll}(\omega)}} & \text{for } S_{kk} \neq S_{ll}, \\ 0 & \text{for } S_{kk} \cdot S_{ll} = 0, \end{cases} \qquad (3)$$

where ω is the circular frequency, $S_{kk}(\omega)$ and $S_{ll}(\omega)$ denote the autopower spectral density of the time histories at the stations k and l and $S_{kl}(\omega)$ is the cross-spectral density function of the considered pair of processes.

In general $\gamma_{kl}(\omega)$ is complex valued; its bounded modulus $0 \leq |\gamma_{kl}(\omega)| \leq 1$ measures the linear statistical dependence between the two time-histories: in particular $\gamma_{kl} = 1$ represents perfect correlation between the two motions, whereas $\gamma_{kl} = 0$ denotes complete lack of linear dependence, which means totally uncorrelated signals.

There are several models available in literature for the coherency function; in the present study the formulation

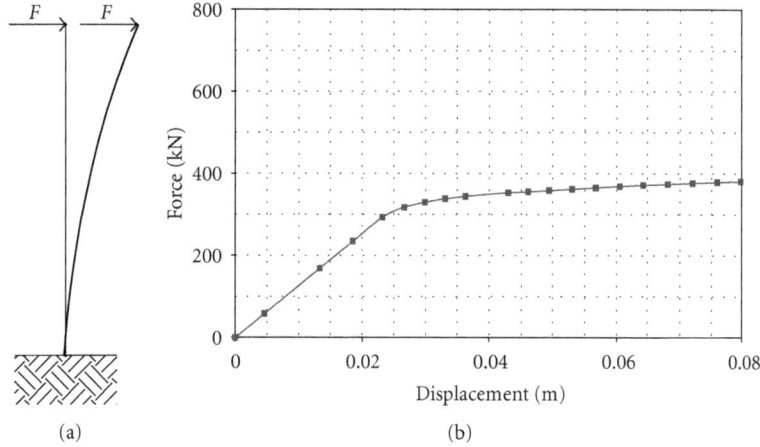

FIGURE 7: Piers behaviour in the longitudinal direction: (a) deflection shape; (b) force-displacement curve.

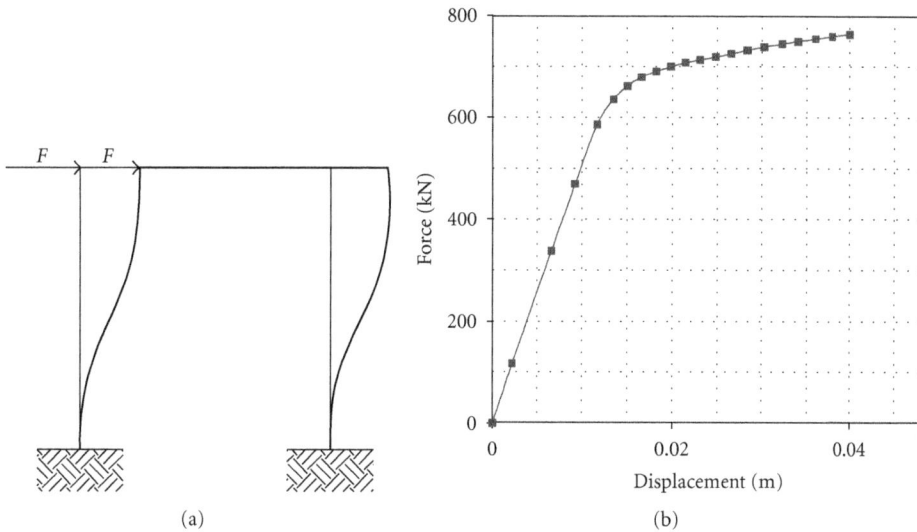

FIGURE 8: (a) Piers behaviour in the transverse direction: deflection shape; (b) force-displacement curve.

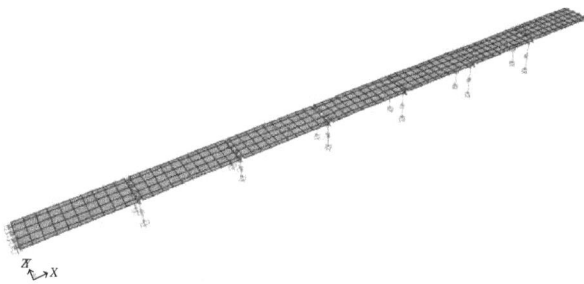

FIGURE 9: Three-dimensional FE model of the seven central spans.

given in [10] has been adopted (see Figure 10), and its general expression is

$$\gamma_{kl}(d_{kl}, \omega) = \exp\left\{-\left(\frac{\alpha \cdot \omega \cdot d_{kl}}{v_s}\right)^2\right\} \cdot \exp\left\{i \frac{\omega \cdot d_{kl}}{v_{\text{app}}}\right\}, \quad (4)$$

where the first term represents the geometrical incoherence, which arises from the scattering of waves in the heterogeneous soil medium, while the second term accounts for the velocity of seismic waves and the difference in the times of arrival at different stations (vawe-passage effect). The parameters describing these phenomena are, respectively (v_s/α), in which v_s is the shear wave velocity in the medium, α a measure of loss of the coherency rate with distance and frequency, v_{app}, is the value of the apparent horizontal velocity of the surface wave. The relative distance between the two different stations k and l is given by the span length d_{kl}, while ω is the circular frequency. Both parameters (v_s/α) and v_{app} usually vary in the range $[300\,\text{m/s}, \infty[$; if $(v_s/\alpha) \rightarrow \infty$ and $v_{\text{app}} \rightarrow \infty$, the modulus of coherency function tends to be 1: the two signals are then totally correlated (identical and in-phase ground motions).

2.3.3. Shape Function. The shape function of the oscillatory process is defined in a general exponential form as suggested in [15]; it is the normalised envelope function of the time history and is governed by the parameters t_1 and t_2 which define the ramp duration and the decay starting time; t_{max}

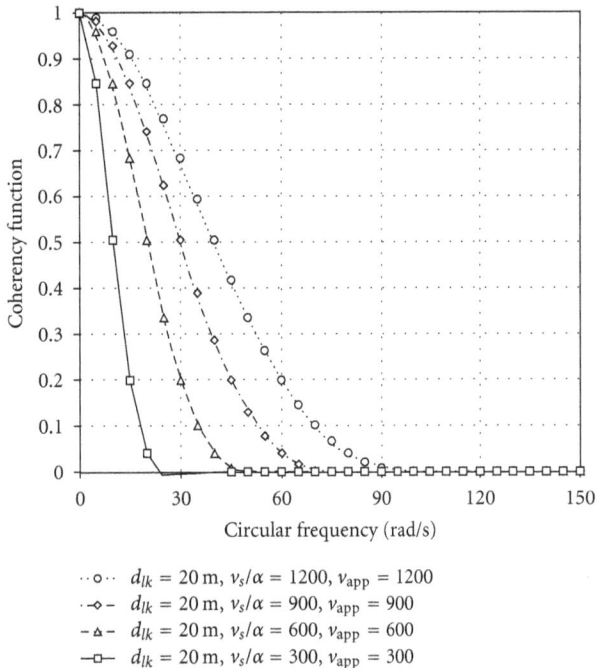

FIGURE 10: Coherency function modulus obtained for different correlation levels corresponding to a set of 4 values of parameters v_s/α e v_{app} (formulation by Luco and Wong).

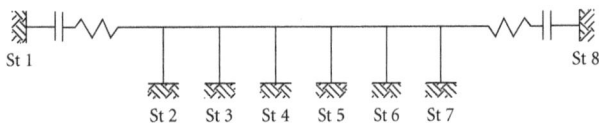

FIGURE 11: Ray of the 8 stations implemented in the FE model of the structure.

is the time history duration and $\nu = 0.2$ is the ratio of the amplitude envelope. The analytical formulation is as follows:

$$\xi(t) = \begin{cases} (t/t_1)^{\eta}, & t \leq t_1, \\ 1, & t_1 \leq t \leq t_2, \\ \exp\left\{\dfrac{t - t_2}{t_{max} - t_2} \ln \nu\right\}, & t_2 \leq t \leq t_{max}. \end{cases} \quad (5)$$

The parameters values in this work are taken as follows: $t_1 = 6$ sec, $t_2 = 16$ sec, $t_{max} = 40$ sec, $\eta = 2.0$, and $\nu = 0.02$.

2.4. Generation of Compatible Time History Sets.
In this study the formulation proposed in [6] for target spectral density function has been used to generate time history sets compatible with the response spectrum given by the code for a soil of medium stiffness with PGA = 0,25 g. The accelerograms, based on the coherency function previously described, have been generated trough the implementation of opportune alghoritms [2]; the nonstationarity has been impressed to the stationary simulated motions by means of the shape function. In order to use the generated sets of response spectrum and coherency compatible time histories

as multisupport seismic inputs at the stations numbered from 1 to 8 (see Figure 11), the acceleration time histories have been doubly integrated to obtain the corresponding displacement time histories [15].

Different patterns of coherency have been selected in the parametric study, in order to represent the intermediate levels between the full correlation and the total uncorrelation of the time histories: 16 combinations of parameters v_s/α and v_{app} varying in the interval 300–1200 m/s have been considered (see Table 1), and for each combination five sets of generated time histories associated with a linear array of 8 stations (corresponding to locations of piers in the 7 central modelled spans of the bridge) have been applied, for a total of 80 sets examined.

A superimposition of the displacement time-histories generated in the simulation is reported as an example in Figure 12; the two extreme cases of strongly correlated ground motions ($v_s/\alpha = 1200$ m/s, $v_{app} = 1200$ m/s) and weakly correlated motions at supports ($v_s/\alpha = 300$ m/s, $v_{app} = 300$ m/s) are presented for the 8 stations considered in the analysis.

2.5. Analysis in the Time Domain.
In order to determine the nonlinear response of the structure to a large set of earthquake ground excitations, it is necessary to use an efficient and not much time-consuming time-integration algorithm; in the present study the mode superimposition procedure based on load-dependent Ritz vectors [19] has been employed instead of direct integration methods in the time domain to reduce the computational effort and maintain an accurate solution. Further, the duration of the time step used for the integration has been limited by the follows condition:

$$\Delta t < T_1, \quad (6)$$

where T_1 is the expected impact duration. Thus it has been possible to capture in the model the effect that colliding adjacent segments of neighbouring decks produce by behaving not like rigid bodies, but influencing the dynamic response with their axial deformation [20]. The impact duration is calculated as follows:

$$T_1 = \frac{2L}{C_0}, \quad (7)$$

where L is the span length of the deck subjected to the pounding effect and C_0 is the propagation velocity of the impact wave travelling in a continuous medium, defined as follows:

$$C_0 = \sqrt{\frac{E}{\rho}}, \quad (8)$$

with E representing the elastic modulus of the superstucture and ρ its density

The values calculated for T_1 and the corresponding Δt adopted in this study are listed in Table 2.

It should be noticed that in fact a superstructure segment does not hit the neighbouring deck directly, due to the

TABLE 1: Combinations of v_s/α and v_{app} values considered in the parametric study.

Vawe-passage effect	Geometrical incoherence v_s/α (m/s)			
v_{app} (m/s)	300	600	900	1200
300	x	x	X	x
600	x	x	X	x
900	x	x	X	x
1200	x	x	X	x

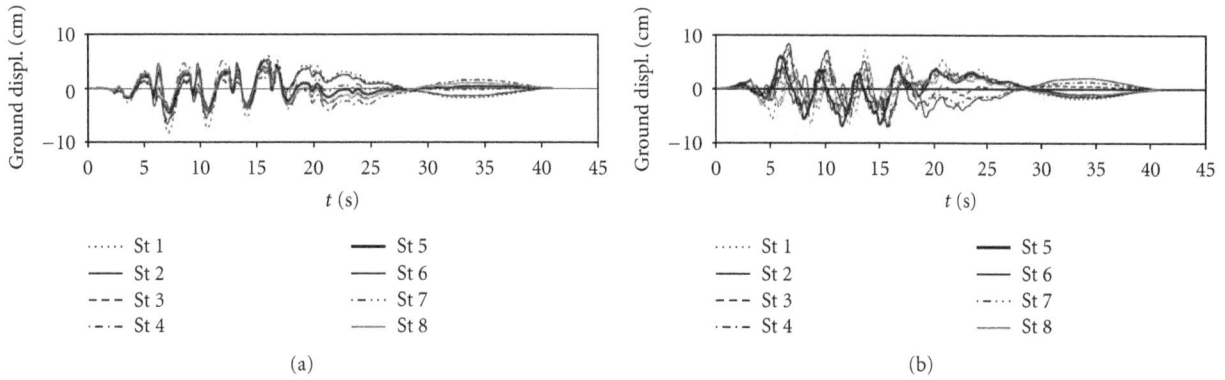

(a)

(b)

FIGURE 12: Displacement time histories for stations 1 to 8: (a) highly correlated time-histories (set 1/5 v_s/α = 1200 m/s.

TABLE 2: Integration time step adopted Δt.

Elastic modulus E (MPa)	24821
Density ρ (Kg/m^3)	2500
Deck span lenght L (m)	24.75
Impact duration T_1 (s)	0.016
Integration time step Δt (s)	0.01
$\Delta t/T_1$	0.625

presence of the cap beam, but this element has been assumed as transmitting the impact rigidly and not influencing the wave propagation.

3. Results of the Numerical Analyses

A sensitivity analysis of the structure dynamic behaviour due to different spatially varying ground motion sets has been carried out, evaluating the influence of the seismic input correlation on the structural response, in terms of the following:

(i) differential displacement between piers and deck segments;

(ii) pounding forces between cap-beams and decks;

(iii) effects on piers: shear forces at the bases and maximum displacements at the tops.

The response analysis focuses on the central span of the FE model, in order to provide results unaffected by the boundary coditions; as previously said, for each prefixed level of ground motion correlation (16 in total, each one determined by a couple of the parameters v_s/α and v_{app}) 5 nonlinear dynamic analyses have been performed, using

compatible time history sets. The mean value of the five results has been adopted.

3.1. Differential Displacements. Differential displacements between piers and decks are represented in Figure 13: it can be observed that in all cases the calculated values are relatively small and remain under the threshold of 5 cm; the maximum differential displacement (d_d = 4.4 cm) is obtained, as expected, for the extreme case of maximum coherency loss (v_s/α = 300 m/s, v_{app} = 300 m/s).

The limited amplitude of differential displacements prevents pull-off-and-drop collapse of deck segments and can be explained considering that joint gaps at span ends are small (2 cm) and do not allow the development of high inertia forces at the deck level; consequently the displacements cannot be considerably amplified. These results are in accordance with the observations reported in [20].

3.2. Pounding Forces. Impact forces between cap beams and decks are highly influenced by the correlation level of ground motions at the structural supports: as Figure 14(a) shows, there is a trend in pounding forces, which rapidly increase with the loss of coherency of seismic inputs: the magnitude of impact force F_I, obtained in the extreme case of weakly correlated time histories (v_s/α = 300 m/s, v_{app} = 300 m/s), assumes a value 3 times larger (F_I = 1293 kN) than that derived by analysis with uniform inputs, F_I = 428 kN (represented by the case v_s/α = 1200 m/s, v_{app} = 1200 m/s). Consequently, even though in the case of uniform seismic excitation no damage should occur to the cap beams and decks, with weakly correlated input time histories, the pounding effects could determine considerable damage to the local area of bridge decks.

(a)

(b)

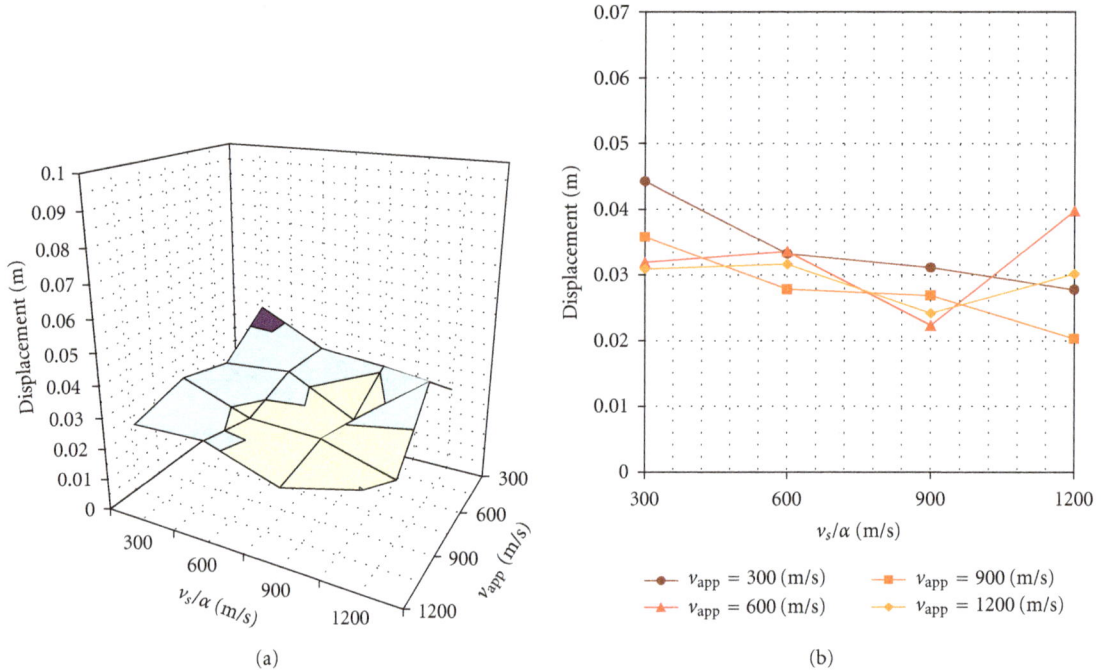

Figure 13: Pier-deck differential displacement varying with seismic input correlation level (represented by parameters v_s/α and v_{app}): (a) 3D view and (b) 2D view.

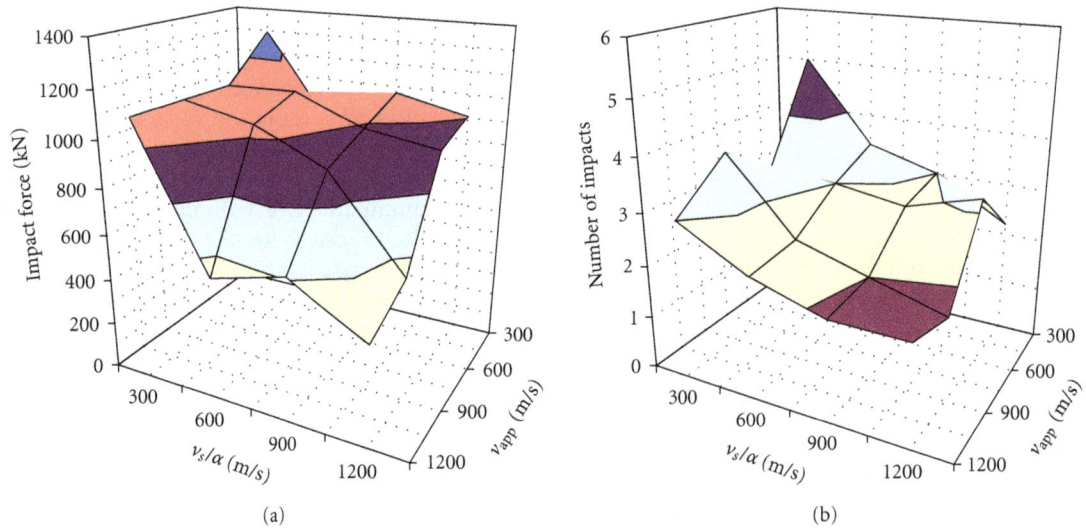

(a)

(b)

Figure 14: (a) Pounding forces varying with correlation level of input ground motions; (b) Total number of registered impacts (mean value).

Registered impacts follow a similar tendency (see Figure 14(b)): the numerical results show that collisions occur more frequently as the correlation level of time history inputs decreases. However in all cases, the total number of impacts (mean value of the 5 non linear dynamic analysis performed) is found to be relatively small (less than 5).

3.3. Effects on Piers. As regards the effects on piers in the longitudinal direction, they are represented in Figure 15 in terms of shear forces and displacements (maximum values at the top of the pier) obtained as functions of the correlation

level between the time histories. It can be observed that the maximum value of shear force $V = 346$ kN is obtained in the case of highly correlated time histories, and the minimum $V = 274$ kN is derived using inputs with the weakest correlation ($v_s/\alpha = 300$ m/s, $v_{app} = 300$ m/s). Similarly it can be said that there is a general trend for displacements at the pier top (see Figure 15(b)) that become larger as the correlation increases, with the maximum value $D = 3,7$ cm calculated in the case of the highest correlation level of input ground motions.

These effects can be explained considering that when ground excitations are weakly correlated or uncorrelated, the movement of deck segments can be in opposite direction

(a)

(b)

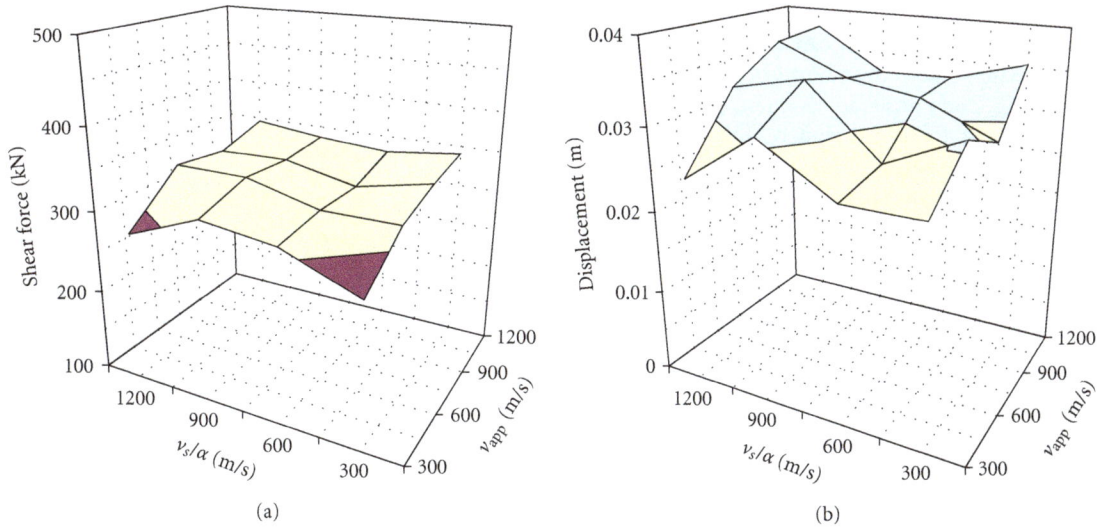

FIGURE 15: Longitudinal direction: (a) shear forces at the pier base: maximum values obtained for each correlation level of input ground motions; (b) maxima values of displacement at pier top.

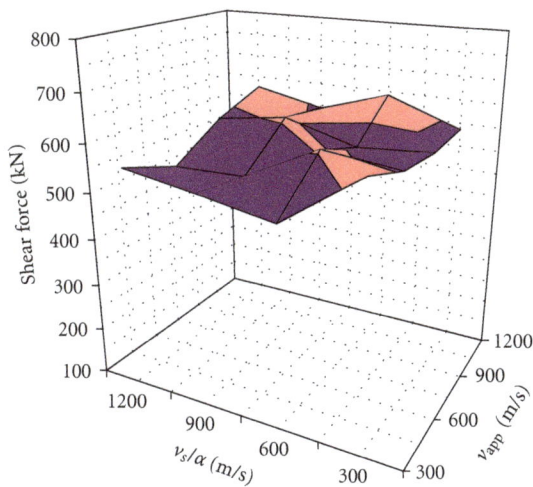

FIGURE 16: Transverse direction: maximum shear forces.

due to out-of-phase vibrations, and this fact determines collisions that reduce displacements at the top of the pier (and consequently the shear and bending moment at the base induced by the deformation of the pier itself). When the ground excitation is highly correlated, the responses of the bridge spans are in phase, the inertial forces at the pier tops are maximised, and in consequence displacements at the top and shear forces increase.

Regarding the response in the transverse direction, it should be noted that structural behaviour is not clearly affected by seismic input correlation (see Figure 16); one can observe that there is a slight tendency for shear forces at the pier base to increase with higher correlation levels, but the values are weakly affected by impacts between deck segments. This is consistent with the results presented in [19].

4. Conclusions

A parametrical analysis has been performed with the aim of investigating the influence of the seismic input correlation level on the structural response of a long multispan girder bridge. A series of nonlinear time history analyses have been performed, in which the main nonlinear behaviours of the bridge components, have been included: (i) the pounding of decks at cap-beams, (ii) the friction of beams at bearings, and (iii) the hysteretic behaviour of piers. The following conclusions can be drawn:

(i) differential displacements between decks and pier-tops are affected by input correlation level but remain within a limited range (under the threshold of 5 cm) with the maximum value obtained for the extreme case of maximum coherency loss. The fact that they are relatively small prevents decks from unseating and can be explained by the limited width of the bridge joints;

(ii) asynchronous ground motion influences greatly the pounding forces between decks and pier-tops, which can assume values 3 times larger than those calculated by an analysis with uniform input (represented by the case with the highest correlation level between the time-histories). The amplified pounding effects might determine considerable damage to the local area of bridge decks;

(iii) as regards the effects on piers, it can be observed that in the longitudinal direction there is a general trend for displacements and shear forces, which increase with higher correlation levels of input ground motions. In the transverse direction the seismic response is not clearly influenced by the correlation level of ground excitations.

The results highlight that the spatial variation properties of the earthquake ground motion can significantly change the structural response especially in terms of pounding forces and deck unseating, and consequently these effects have to be taken into account for the design or the vulnerability assessment of long multispan simply supported bridges.

References

[1] A. Lupoi, P. Franchin, P. E. Pinto, and G. Monti, "Seismic design of bridges accounting for spatial variability of ground motion," *Earthquake Engineering and Structural Dynamics*, vol. 34, no. 4-5, pp. 327–348, 2005.

[2] G. Deodatis, "Non-stationary stochastic vector processes: seismic ground motion applications," *Probabilistic Engineering Mechanics*, vol. 11, no. 3, pp. 149–167, 1996.

[3] M. Shinozuka and G. I. Schueller, "Stocastic fields and their digital simulation," in *Stochastic Methods in Structural Dynamics*, G. I. Schueller and M. Shinozuka, Eds., pp. 93–133, Martinus Nijhoff, Dordrecht, The Netherlands, 1987.

[4] M. Shinozuka and G. Deodatis, "Simulation of multi-dimensional Gaussian stochastic fields by spectral representation," *Applied Mechanics Reviews*, vol. 44, no. 4, pp. 191–204, 1991.

[5] R. W. Clough and J. Penzien, *Dynamics of Structures*, McGraw Hill, New York, NY, USA, 1993, International Edition.

[6] UNI ENV 1998-2, "Eurocode 8—design provisions for earthquake resistance of structures—part 2," Bridges, 1998.

[7] A. Der Kiureghian, "A coherency model for spatially varying ground motions," *Earthquake Engineering and Structural Dynamics*, vol. 25, no. 1, pp. 99–111, 1996.

[8] H. Hao, C. S. Oliveira, and J. Penzien, "Multiple-station ground motion processing and simulation based on smart-1 array data," *Nuclear Engineering and Design*, vol. 111, no. 3, pp. 293–310, 1989.

[9] R. S. Harichandran and E. H. Vanmarcke, "Stochastic variation of earthquake ground motion in space and time," *ASCE Journal of Engineering Mechanics*, vol. 112, no. 2, pp. 154–174, 1986.

[10] J. E. Luco and H. L. Wong, "Response of a rigid foundation to a spatially random round motion," *Earthquake Engineering and Structural Dynamics*, vol. 14, no. 6, pp. 891–908, 1986.

[11] S. A. Anagnostopoulos, "Pounding of buildings in series during earthquakes," *Earthquake Engineering and Structural Dynamics*, vol. 16, no. 3, pp. 443–456, 1988.

[12] B. F. Maison and K. Kasai, "Dynamics of pounding when two buildings collide," *Earthquake Engineering and Structural Dynamics*, vol. 21, no. 9, pp. 771–786, 1992.

[13] R. Jankowski, K. Wilde, and Y. Fujino, "Pounding of superstructure segments in isolated elevated bridge during earthquakes," *Earthquake Engineering and Structural Dynamics*, vol. 27, no. 5, pp. 487–502, 1998.

[14] K. Kawashima and Watanabe G., "Numerical simulation of pounding of bridge decks," in *Proceedings of the 13th World Conference on Earthquake Engineering Vancouver*, British Columbia, Canada, August 2004, Paper no.884.

[15] G. Zanardo, H. Hao, and C. Modena, "Seismic response of multi-span simply supported bridges to a spatially varying earthquake ground motion," *Earthquake Engineering and Structural Dynamics*, vol. 31, no. 6, pp. 1325–1345, 2002.

[16] Ordinance of the Presidency of the Council of Ministers 3431, "Initial elements on the general criteria for classifying national seismic zones and technical standards for construction," *Official Gazette of the Italian Republic, 2003*.

[17] T. Takeda, A. M. Sozen, and N. N. Nielsen, "Reinforced concrete response to simulate earthquakes," *ASCE Journal of Structural Engineering Division*, vol. 96, no. 12, pp. 2257–2273, 1970.

[18] Computers & Structures Inc., "*CSI SAP2000 release 9*," Berkeley Calif, USA, 2004.

[19] E. L. Wilson, M. W. Yuan, and J. M. Dickens, "Dynamic Analysis by direct superposition of Ritz vectors," *Earthquake Engineering and Structural Dynamics*, vol. 10, no. 6, pp. 813–821, 1982.

[20] R. Jankowski, K. Wilde, and Y. Fujino, "Reduction of effects in elevated bridges during earthquakes," *Earthquake Engineering & Strucutral Dynamics*, vol. 29, pp. 195–212, 2000.

3

An Investigation into the Response of GFRP-Reinforced Glue-Laminated Tudor Arches

S. Alshurafa,[1] H. Alhayek,[1] and F. Taheri[2]

[1] *Department of Civil Engineering, University of Manitoba, 15 Gillson Street Winnipeg, MB, Canada MB R3T 2N2*
[2] *Department of Civil and Resource Engineering, Dalhousie University, 1360 Barrington Street, Halifax, NS, Canada B3J 1Z1*

Correspondence should be addressed to F. Taheri, farid.taheri@dal.ca

Academic Editor: Muhammad Hadi

This paper presents the results of an experimental and computational investigation tailored to examine the response of glass fiber-reinforced-plastic-(GFRP-) reinforced glue-laminated curved beams and arches. The main objective was to ascertain the viability of GFRP as an effective reinforcement for enhancing the load carrying capacity and stiffness of such curved structures. The study included optimization of the length and thickness of the GFRP reinforcement. In doing so, first a parametric finite element study was conducted to evaluate the influence of unidirectional GFRP reinforcement applied onto the arch using eleven possible configurations and different thicknesses. Subsequently, an experimental investigation was conducted to verify the results established by the finite element method as well as the integrity of actual GFRP-reinforced glue-laminated curved structures. The results indicate that GFRP can be considered as an effective and economically viable solution for strengthening and stiffening glulam arches, without adding any appreciable weight to the structure.

1. Introduction

1.1. Background. Over the past few decades, many studies have been performed on studying the response of fiber-reinforced plastic (FRP) laminates when combined with other structural materials such as concrete and wood. An important application of FRP in recent years has been in retrofitting of existing wood, concrete, and steel structural members, such as those used in bridge and other civil structures. Retrofitting structures with FRP is nowadays considered as an effective and economical alternative to the replacement of the structural components, since members are rehabilitated instead of being replaced.

Another advantageous application of FRP has been in increasing the strength of wood beams. Wood is a resilient material, but its relatively low stiffness and statistically varying strength impede its use in long span applications. Glue-laminated technology (hereafter reinforced to as glulam) partially resolves the varying strength issue, but the relatively low stiffness of glulam structural components still impedes their use in moderately large span applications, even when reinforced with composites. This is because while the addition of FRP can significantly increase the strength, nevertheless, the improvement in stiffness would be marginal.

To alleviate the issue, in one of the earliest studies recorded regarding reinforced wood members, Mark [1] used aluminum strips bonded to the compression and tension faces of wood beams. Failure of the beams occurred mainly by separating and buckling of the aluminum facings. Bohannan [2] reinforced glulam beams of low-grade Douglas-fir using pretensioned steel wire strands in the tension zone. In a similar study, Peterson [3] reinforced low grade Douglas-fir glulam beams with a prestressed flat steel strip bonded in the tension zone. Both studies reported an increase in strength and stiffness for the prestressed beam. Lantos [4] performed an experimental research on glulam beams reinforced with steel bars and found a substantial reduction in the coefficient of variation for bending strength as well as an increase in strength directly proportional to the reinforcement ratio. Krueger and Sandberg [5] studied laminated timber reinforced in the tension zone with a composite of high-strength bronze-coated woven steel wire that was bonded with epoxy.

The use of steel plates for reinforcing glulam beams was studied by Bulleit et al. [6] and was found to be effective, achieving remarkable stiffness increase as high as 32 percent and moment capacity increase as high as 30 percent. Although the use of metallic reinforcement has shown promising results with respect to increases in the overall performance of wood products, the issue of inadequate bond between wood and steel reinforcement has been identified as significant problem in such reinforced structures.

Consequently, several researchers considered the use of FRP an effective reinforcing agent for wood structural components. Wangard [7] and Biblis [8] studied the effect of bonding unidirectional fiberglass/epoxy-reinforced plastic to the compression and tension faces of wood cores of various species. Increases in modulus of elasticity (MOE) ranging from 20 to 50 percent using only 10 percent reinforcement by volume were reported. Theakston [9] studied the feasibility of strengthening both laminated and solid wood beams with a fiberglass cloth and woven roving. Increases in load-carrying capacity ranged from 30 to 60 percent. Theakston observed that even after failure occurred in the wood core, the fiberglass reinforcement retained enough strength to support the load. From a safety standpoint, this observation was an important discovery.

There has also been a renewed interest in past two decades in the application of FRP for reinforcing timber beams. Some examples are Dagher et al. [10], who studied the effect of FRP reinforcements on low-grade eastern Hemlock glulam by fabricating nine glulam beams reinforced with FRP on the tension side and three unreinforced controls. The FRP reinforcement ratios used in his research varied from 0.3% to 3.1%. All of the glulam beams were tested to failure in four-point bending and the results showed a substantial increase in strength up to 56% and an increase in stiffness up to 37% compared to the control beams. Johns and Lacroix [11] investigated the length effect of CFRP (Epoxy) bonded onto the tension side of timber beams (CFRP layer on the full length or on the constant moment area only). Strength increases from 40 to 100% were reported when compared to the unreinforced control beam. The effect of reinforcement length has also been studied, which indicate that longer reinforcement length shows higher strength for CFRP-reinforced timbers. More failures occurred in the compression side, which indicates a more ductile behaviour. Olsson [12] studied glulam timber arches and developed a method for reliability-based optimization of glulam arches. He then linked his developed method to a commercial software package for optimization of glulam arches.

Radial reinforcement of curved glue-laminated wood beams with composite materials was investigated by Kasal and Heiduschke [13]. The objective of this research was to study the application of fiber-reinforced composite materials in reinforcing laminated wood arches subjected to radial tension. The application of composite materials in radial reinforcement was found to be feasible and possibly advantageous over the glued-in steel rods approach, because of greater flexibility of sizes and properties of reinforcing elements, lower mass, and potential ease of installation.

More recently, Buell and Saadatmanesh [14] also investigated special reinforcement configuration. It consisted of placing CFRP reinforcement at the bottom of the timber beam in the tension side, far from the neutral axis to maximize the bending resistance. The shift of the CFRP was achieved by positioning long pieces of wood to the bottom of the beam. An additional carbon fabric was wrapped around the beam in the side and the tension area. A 69% increase of the bending strength was reported when compared to the control beam and a compression failure mode. An increase of the stiffness by 18% was also reported. The increase in mean load capacity was between 44% and 63%. Moreover, ductile failure mode in compression side was observed in reinforced glulam beams.

As can be seen, there have been several studies conducted in consideration of the performance of GFRP reinforced glulam beams, including those conducted by Yahyaei-Moayyed and Taheri [15, 16]. In the latter investigations, the creep parameters of aramid fiber-reinforced epoxy (AFRP) and plain wood were determined experimentally. Moreover, a nonlinear finite element model was developed to predict the creep response of the AFRP-reinforced timber beams based on the creep characteristics of the individual components. The results of the finite element analysis showed a good agreement with the experimental results conducted on the AFRP-reinforced samples. Moreover, the influence of reinforcement on the flexural strength of reinforced beams was also examined.

As can be seen from the previous brief survey, the application of FRP in reinforcing other types of glulam structural components is relatively very scarce. An example of such studies is that conducted by Taheri et al. [17], who investigated the response of glue-laminated columns reinforced with GFRP. In that study, the authors conducted a complimentary experimental and computational investigation to characterize the response of axially loaded glulam timber columns strengthened with GFRP. Several parameters such as slenderness ratio, boundary conditions, FRP reinforcement length, and relative cost were considered in that study.

1.2. Motivation. The present work therefore aims at studying the response of GFRP-reinforced glulam three-hinge Tudor arches. The three-hinge Tudor arch is one of the most commonly used curved glulam structural members. They are used in large open structures such as churches, school gyms, warehouses shelters, and barns because of their excellent structural performance and pleasing appearance. The effect of different parameters such as number of layers of laminae and location of the reinforcements on the performance of the arches is investigated in this study.

1.3. Objective. The main objective of this work was to examine the viability of the use of GFRP, a relatively inexpensive composite, in conjunction to inexpensive and relatively low grade wood to produce an effective structural material. The application of the resulting composite into a three-hinge Tudor arch produces a cost-effective and efficient structure. The ultimate goal of this study was to enhance the current engineering database and to provide practical and valuable insight to the designers of wood structures and practicing engineers. The result of this study will partially address the

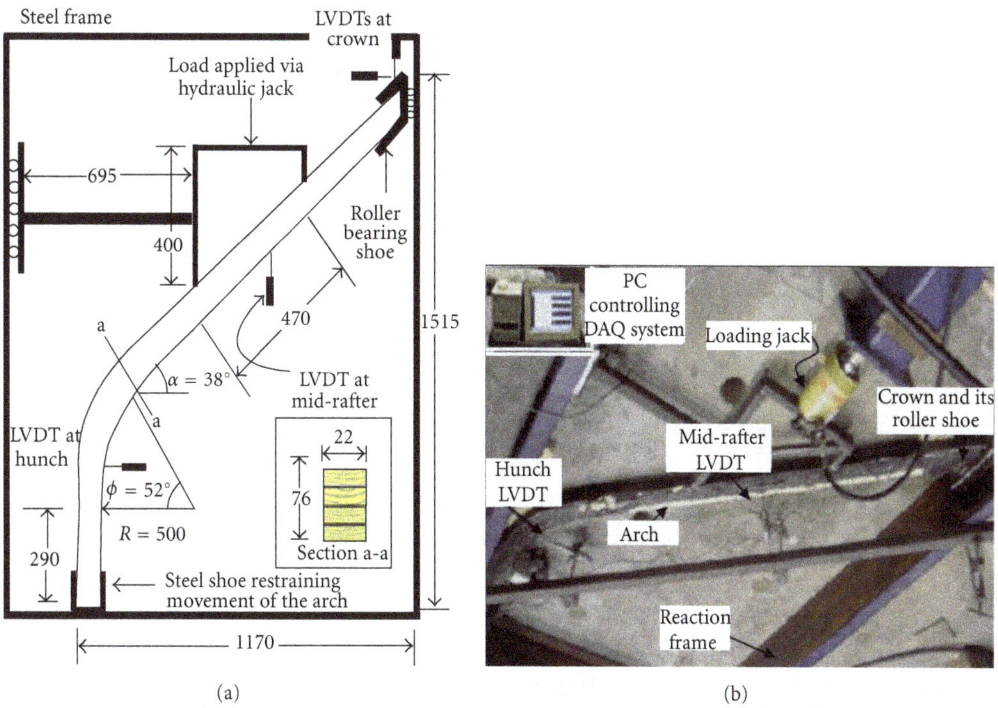

FIGURE 1: (a) Schematics and dimensions of the test arch specimen and the experimental setup. (b) photo of the actual test setup (drawing not to scale; all dimensions in mm).

lack of engineering database in regards to the more diverse applications of FRP when considering wood structures.

2. Glulam Arch Test Specimens

2.1. Arch Configuration. In an effort to establish the optimum configuration of GFRP for reinforcing our test arches, the combined loading of dead and snow loads was selected as the loading condition, since this loading condition is one of the most critical loading conditions governing the design of such arches. This paper assumed symmetrical snow load on arch; this indeed is a special case since snow load distribution on arch is not always symmetric. It should be reemphasized that the main objective of this study was to investigate the influence of GFRP reinforcement in enhancing the overall moment capacity and stiffness of Tudor arches, rather than the design aspects of such structures. This loading condition was also selected, because of the resultant symmetry, in that it facilitates easier test setup in laboratory setting as well as optimizes the onerous effort required for the fabrication of the arches tested in this investigation, since only one-half of the arch could be tested to produce the full-arch response. Therefore, the selected loading and arch configuration would facilitate the necessary means to achieve the goal of our study.

2.2. Lumber. The type of wood utilized to construct the glulam arches was eastern white pine, a relatively inexpensive lumber. Both "clear" and "common" grades of the lumber were used in this study. Clear grade is the finest architectural heartwood, which is carefully selected and manufactured to be more or less free of knots and other flaws, and contains sapwood in limited amounts. Common grade is a combination of heartwood and sapwood, containing knots of varying sizes and other slight imperfections. The higher-grade (clear) pine was used to form the other layers (laminations) of the cross section of the arch, where bending stresses are the greatest. The lower-grade (common) pine was used to form the core laminations. The cross section of the glulam arch was 76×22 mm^2, with the other dimensions and configuration being illustrated in Figure 1. Since the arch has a hinge on its crown, and the loading combination is symmetric, with the aim of simplifying the investigation, especially arches fabrication, only one-half symmetry of the arch was fabricated and subsequently loaded.

2.3. GFRP Reinforcement Configurations. In order to establish the optimum reinforcement configuration (i.e., the optimized location and length of the GFRP reinforcement along the arch), a total of eleven feasible combinations of GFRP reinforcement (as well as the virgin arch (i.e., with no reinforcement)) were considered in this study. The finite element method was utilized to analyze the various cases and to establish the most optimum reinforcement configuration. Figure 2 shows the various configurations of reinforcements considered in the study. These combinations of reinforcement were selected mainly based on the expected bending moment and shear force variations along the arch. Subsequently, the most promising reinforcement configurations (i.e., cases (a) and (k)) were considered in our experimental investigation.

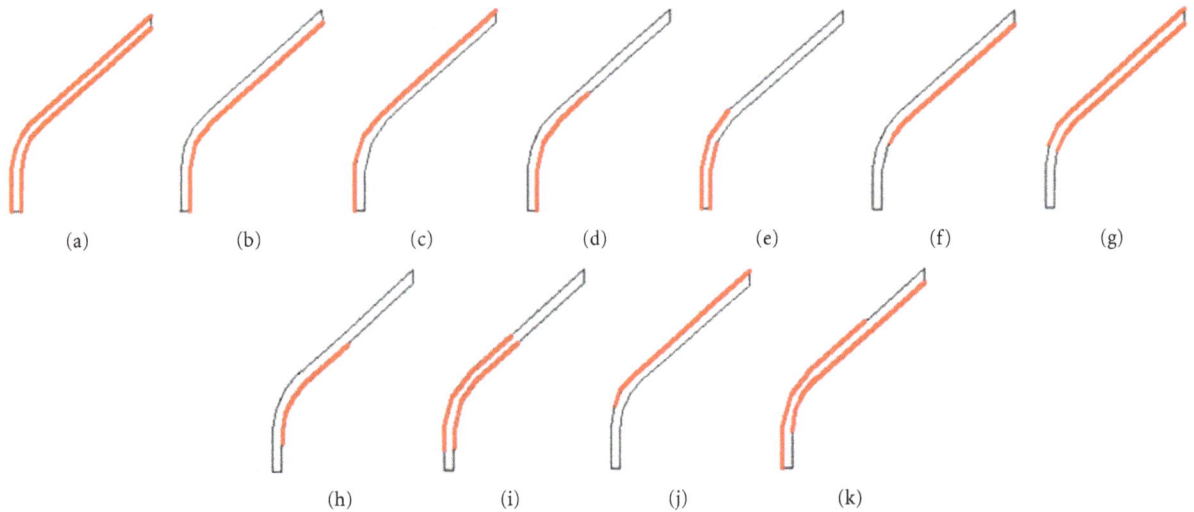

FIGURE 2: Various GFRP reinforcement configurations considered.

The NISA (Numerically Integrated System Analysis), a commercially available finite element package, was employed for this investigation. First-order plane stress element (NKTP1) was used in constructing the models. Some of the mechanical properties of E-glass/vinyl ester composite used to reinforce the arches were obtained experimentally, while the other values were obtained from the available literature. The values are reported in Table 1. The modulus of elasticity of the glulam section was estimated from the experimental load-displacement curves obtained from testing of nonreinforced glulam arches, reported in Table 1; the other wood properties were obtained through the literature.

Resorcinol (more specifically, CASCOPHEN LT-5210, produced by the Momentive Specialty Chemicals Inc. of Columbus, OH), a standard adhesive used in the industry, was used to adhere the lumber pieces as well as adhering the FRP sheets to the arch. The material properties of the Resorcinol used in the analysis were obtained from the supplier's technical sheet and are also summarized in Table 1.

Results of these analyses, along with the lengths of the reinforcement applied on the top and bottom surfaces of the arch, in each case, have been summarized in Table 2. The locations of the reinforcement configurations are shown in Figure 2. The locations of the combination of FRP reinforcements considered by the finite element method are shown in Figure 2. The results revealed that as expected, the fully reinforced arch, with reinforcement applied on the upper and lower surfaces along the entire arch length (represented by case (a)), produced the highest moment capacity and stiffness. However, this configuration is not the most cost-effective alternative. In turn, one can attain relatively very high improvement in load capacity and stiffness by considering the partially reinforced configuration identified as case (k) in Figure 2, while economizing the improvement as well. Consequently, these two configurations were selected for further experimental examination.

2.4. Manufacturing Process. The pine wood material was cut to the required lengths for preparation of the test specimens.

TABLE 1: Material properties used in the analysis.

Property	Easter White PineGlulam Lumber*	E-Glass/Vinylester Composite**	Resorcinol LT-5210 adhesive ***
E_{11} (MPa)	10500	31470	3500
E_{22} (MPa)	—	8807	3500
G_{12} (MPa)	2952	2000	1500
v_{12}	0.25	0.27	0.10
Shear strength-τ_u (MPa)			16.80
Peel strength, σ_u (MPa)			8.50
Longitudinal tensile strength (MPa)		550*	

* Values obtained from laboratory tests (Load versus deflection curve).
** Obtained from coupons testing [18].
*** Values obtained from Resorcinol Technical Data Sheet.

The arch cross section was constructed using four layers of these pine strips. As stated, two strips of high-grade clear pine laminates were used to form the upper and lower layers of arch's cross section, while two strips of common (knotty) pine were sandwiched in between the two clear strips to form the core of the cross section. These strips were completely immersed in a pool of clean water for a period of 24 hours to make them malleable, so that they could be steamed (to further soften the wood) and bent to shape. The water-saturated lumber strips were subsequently steamed in a chamber for an average of 2.75 hrs per 25 mm of thickness of the lumber to make them malleable, so they could be bent to the required tight radius of 500 mm. After the completion of the steaming process, the strips were placed into a special jig to facilitate their bending to the specified curvature. The arches were therefore clamped in the jig and let dry for a 24-hour period (see Figure 3). Once dried, the strips were then

TABLE 2: Summary of the finite element results of the twelve models.

Cases	Length of reinforcement at top surface (mm)	Length of reinforcement at the bottom surface (mm)	Vertical displacement at the peak (mm)	Horizontal displacement at the haunch (mm)	Vertical displacement at the mid-rafter (mm)	von Mises stress at the haunch (MPa)
Control			−5.34	−4.32	−8.96	8.44
Case (A)	**2034**	**1973.26**	**−3.57**	**−2.75**	**−5.98**	**4.89**
Case (B)		1973.26	−4.27	−3.36	−6.64	7.80
Case (C)	2034		−4.40	−3.48	−6.74	4.98
Case (D)		770	−3.99	−3.17	−7.22	7.98
Case (E)	873	582	−3.63	−2.75	−7.06	5.36
Case (F)		1390	−5.03	−4.01	−7.23	8.21
Case (G)	1450	1390	−4.93	−3.94	−6.45	6.66
Case (H)		1000	−4.04	−3.18	−7.06	7.92
Case (I)	1060	1000	−4.05	−3.16	−6.77	5.64
Case (J)	1452		−5.22	−4.21	−7.25	6.81
Case (K)	**1500**	**1600**	**−4.09**	**−3.16**	**−6.34**	**5.13**

FIGURE 3: Special jig used for bending and gluing the glulam strips.

glued together using liquid Resorcinol mixed with a powder catalyst and clamped and let cured for another 24 hours.

For the reinforcement, two GFRP composite panels with dimensions of 1600 mm × 600 mm with two unidirectional layup sequences of $[0]_2$ and $[0]_4$ were manufactured using a vacuum-bagged hand lay-up method. E-glass/vinylester prepreg, supplied by Simex Technologies Inc. (Montreal, Canada), was used to form the composite panels. The laminates were cured in an oven at 145°C for 2 hours, per the supplier's specifications. The processed laminate sheets were then carefully cut to 22 mm strips, using a diamond coated saw. The thickness of the two unidirectional layup sequences of $[0]_2$ and $[0]_4$ was 0.4 mm and 0.8 mm, respectively. The tensile properties of the composite were evaluated using appropriate size test coupons according to the method outlined in ASTM D3039-08.

3. Experimental Investigation

3.1. Glulam Arch Test Setup. As explained earlier, since the test arch was a three-hinge arch and symmetric with respect to geometry and loading condition with respect to a plane passing through the crown of the arch, the advantage of symmetry was used to test only one-half of the arch. The test program therefore consisted of testing of 15 glulam wood half-arches (hereafter referred to as arches). Three of these 15 half-arches were the control specimens (i.e., nonreinforced glulam arches), while the other 12 specimens were divided into two groups; one group had GFRP reinforcement as per configuration (a), shown in Figure 2, while the other group was reinforced per configuration (k). Each group of 6 arches was further divided into two subgroups, based on the layup of the unidirectional GFRP reinforcements (i.e., $[0]_2$ and $[0]_4$).

The surface areas of the arch, to which the GFRP reinforcement was to be applied to, were sanded and cleaned with compressed air prior to application of the GFRP. The arches were tested in pseudo-four-point bending configuration (see Figure 1). This loading scenario is different from that considered in earlier FE analysis; however, it produces more or less similar deflection and stress patterns on the arch, and is experimentally feasible.

As such, the arch specimen was placed in a reaction steel frame. The load was applied via two rollers attached to stiff subframe, as shown in the figure. This subframe assembly was free to travel vertically on a set of roller-bearings attached to the exterior vertical member of the frame. The load was applied through a hydraulic jack, and a calibrated load cell, with a maximum capacity of 10 kN, was placed in between the jack and loading frame to record the exact magnitude of the applied load. Four linear variable displacement transducers (LVDTs) were magnet mounted to the steel frame to measure the displacement of the Glulam arches at various locations, and with the locations of the applied load (see Figure 1). The arches were tested according to the procedures outlined in ASTM D198-84. The test was conducted as displacement controlled, at a rate of 2.0 mm/min, with data from all transducers being collected simultaneously via a data acquisition system at a sampling rate of 2 Hz, and stored on a personal computer (PC).

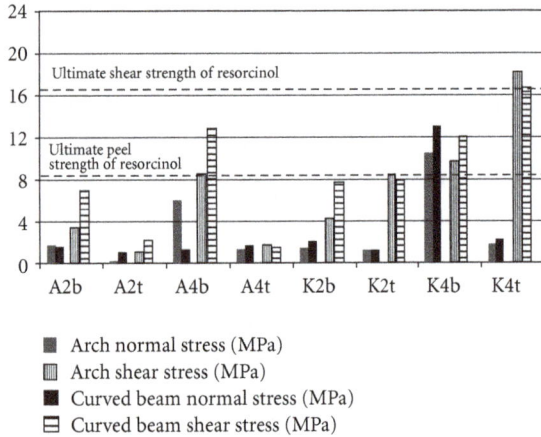

FIGURE 4: Maximum values of the peel and shear stresses within the adhesive along various interface bond lines obtained by FE analyses.

Legend for Figure 4:
- ■ Arch normal stress (MPa)
- ▥ Arch shear stress (MPa)
- ■ Curved beam normal stress (MPa)
- ▤ Curved beam shear stress (MPa)

ID	Location of interface bond	Arch/beam configuration	GFRP lay-up
A2b	Bottom layer interface bond	a	$[0]_2$
A2t	Top layer interface bond	a	$[0]_2$
A4b	Bottom layer interface bond	a	$[0]_4$
A4t	Top layer interface bond	a	$[0]_4$
K2b	Bottom layer interface bond	k	$[0]_2$
K2t	Top layer interface bond	k	$[0]_2$
K4b	Bottom layer interface bond	k	$[0]_4$
K4t	Top layer interface bond	k	$[0]_4$

4. Finite Element Analysis of the Tested Arches

The NISA finite element package was employed to simulate the response of the tested arches. In total, five sets of analyses were conducted (one for the unreinforced arch, and four for the subgroups of the reinforced arches). The thickness of each layer of GFRP as measured from the manufactured GFRP composite panel was taken as 0.20 mm, and the thickness of the adhesive used to adhere the GFRP to the arch was measured as 0.15 mm. The total depth of wood lamination, interface (glue), and the GFRP reinforcements was measured using a caliper. The thickness of the interface was calculated as the difference of the total depth minus the thickness of GFRP. The glue between wood strips was not significant in the analysis because as soon as the steaming process of strips was finished, the warm strips were taken out from the steam box and were immediately glued and bent by placing them on a special setup, therefore the grain surface at this stage of the wood was easily able to absorb the liquid resinol glue between these strips. Also these strips were closely clamped and left under load for one day and as a result the interface between wood lamination was not visually seen. Moreover, it was observed during testing that there was no delamination occurred between wood lamination and therefore the interface between wood lamination was not indeed considered in this analysis.

The models were constructed using the NKTP1 4-node element. A mesh convergence study was conducted to establish a suitable number of elements and mesh topography.

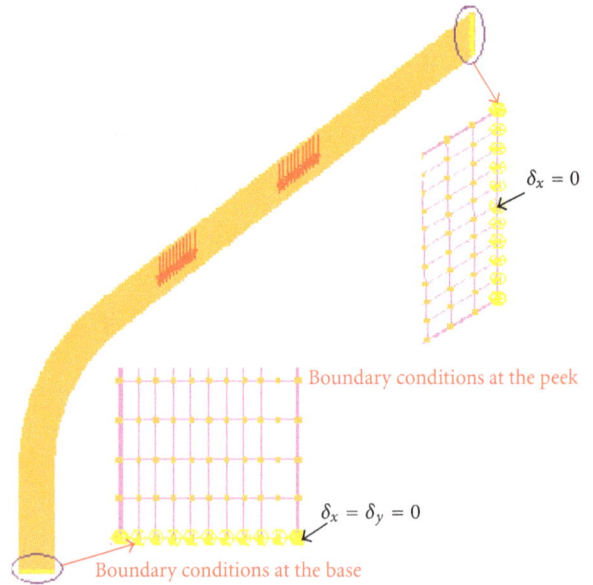

FIGURE 5: Glulam arch finite element boundary conditions.

It should be noted that, as will be explained in the subsequent section, our experimental observation indicated that during the experiments, the half-arches rotated slightly at their crown location, thus violating the assumption of absolute half-symmetry. As a result, two modelling schemes were tried to consider the response of an actual GFRP-reinforced Tudor arch as well as mimic the actual arched specimen's response under the laboratory condition. In the first scheme (hereafter referred to as the arch), all the nodes along the vertical plane (at the arch crown) were restrained in the horizontal direction (see Figure 5), thereby modelling a half-symmetry condition. In the other modelling scheme, only the lower node on the plane of symmetry at the crown was restrained in the horizontal direction (see Figure 6). In this configuration, arch's crown allowed to rotate about the restrained node, therefore mimicking the actual rotation of the crown observed during the experiments. This configuration of the arch is referred to as the curved beam configuration hereafter.

In the FE analysis, the failure in the wood was established based on the specified strength for glue-laminated timber for pine, as stipulated by CAN/CSA O86-09. As such, the value of the ultimate compressive strength parallel to grain was taken as 25.20 MPa, while the tensile strength parallel to grain value was taken as 13.40 MPa. To assess the failure in the GFRP/wood interface, a commonly used second-order criterion was used, represented by the following equation [19]:

$$\left(\frac{\sigma_n}{\sigma_u}\right)^2 + \left(\frac{\tau_{nt}}{\tau_u}\right)^2 \leq 1, \quad (1)$$

where σ_n represents the normal stress in the adhesive (obtained by FEA). τ_{nt} represents the shear stress in the adhesive (obtained by FEA). σ_u represents the peel strength of the adhesive (taken as 8.50 MPa, based on supplier's technical sheet). τ_u represents the shear strength of the adhesive (taken as 16.80 MPa, based on supplier's technical sheet).

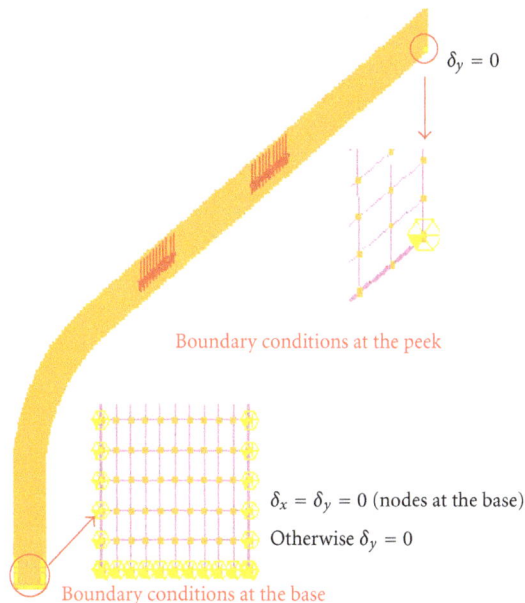

FIGURE 6: Glulam curved beam finite element boundary conditions.

The failure of the GFRP was assessed using the maximum stress theory equation (i.e., $\sigma_{11}/S_{11(\text{tensile})} \leq 1$), as noted in most composite text book (see e.g., Mallick [19]), using the experimentally obtained value of 550 MPa as the longitudinal tensile strength of GFRP.

The design was based on the lower limit of the previous two criteria. The summary of the results (peel and shear stresses) obtained from the FE analyses has been presented in Figure 4 for the two modelling schemes considered in this study (the arch and curved beam configurations discussed above). Furthermore, the values of the failure limits established by the interface criterion and maximum stress criterion (for the GFRP), as also explained previous, are tabulated in Table 3. As seen, the rotation of the crown significantly influences of the variation in the magnitude of stresses from one configuration to another.

5. Experimental Results

5.1. Group 1: Control Glulam Wood Arches. Three unreinforced glulam arches (also referred to as the control arches) were tested to failure, and the resulting ultimate loads, moments, and deflection can be seen in Table 4. The average ultimate load and the maximum deflection were determined to be 5120.33 N and 13.30 mm, respectively.

The load versus mid-rafter deflection curves are illustrated in Figure 7. It can be seen that the three control arches performed similarly as seen by the initial slope of the load- deflection curves as well as the ultimate capacities. Both controles arches 1 and 2 sustained loads at approximately 5,197 N and 5,014 N with the maximum mid-rafter displacements of 13.60 mm and 13.90 mm, respectively. The third control arch sustained 5,150 N of load with the least displacement of 12.50 mm compared to the arches 1 and 2. The maximum displacement of the three control arches

occurred at the mid-rafter point before the final failure of the specimens. All beams failed in Tension. After the glulam arches were tested to failure, the modulus of elasticity of each one was calculated by using the linear portion from its load deflection curve. Because of the inherent heterogeneity of wood, it would be impractical to attempt to predict the ultimate load of such structures using the linearly elastic finite element method. The average ultimate load of each group obtained from the experimental testing was used in the finite element program as an indicator requesting the program to quit as soon as the load applied surpassed that given load. The deflections data obtained from the FE analysis were compared with those obtained from the experiment. Finite element analysis showed good agreement to the experimental results. The mode of failure in all three specimens was initiated by development of a crack at the haunch of the arch, where the bending moment was the maximum.

Finite element prediction showed good agreement to the experimental results obtained. The mode of failure in all three specimens was initiated by development of a crack at the haunch of the arch, where the bending moment was the maximum.

5.2. Partial Reinforcement of Glulam Wood Arch (Configuration (K)). As stated earlier, this group contained of six arches, which were further subdivided into two sets of three half-arches. The first set comprised of the partially reinforced (configuration (K)), with the $[0]_2$ unidirectional GFRP being applied on the upper and lower surfaces of the half-arches. The second set of arches was also partially reinforced, but with four layers of GFRP (i.e., $[0]_4$). The load versus mid-rafter deflection of two of the arches and the FE results are illustrated in Figure 8. The third arch was not tested, since there was a visible manufacturing flaw, which had occurred inadvertently during its fabrication.

The failure mode of this set of GFRP-reinforced arches was different from that observed for the unreinforced (control) set. Brittle mode cracking was heard and observed during various loading stages, and the failure occurred rather catastrophically and without any warning, when the load reached to the highest recorded load value. At that stage, a delamination in the interface of the GFRP and glulam arch became evident at the midpoint of the rafter, which was followed by a sudden failure of the lower most layer of wood.

As it can be seen from Table 4, the comparison of the average ultimate capacity of the partial reinforcement of set 1 showed 30.78 percent improvement in strength. The stiffness, EI, was calculated from the initial linear elastic portion of each curved beam's load-deflection curve, using the conventional beam deflection equation, established based on the virtual-work method. The stiffness was found to have increased by 9.7 percent in comparison to that of the control arch.

The partially reinforced arches with four layers of GFRP (set 2 of group 2) were tested as well, and all of them were delaminated before the load reached an average of 3,600 N. In all arches, delamination of GFRP from wood occurred at the upper surface of the arch, at the free edges near to the location of the applied load. Another distinct delamination

TABLE 3: Comparison of the maximum values obtained by the use of failure criteria for the adhesive interface and GFRP obtained by FE analysis.

| Location of stresses | Adhesive Interface Crit. $((\sigma_n/\sigma_u)^2 + (\tau_{nt}/\tau_u)^2 \leq 1)$ | | GFRP failure Crit. $(\sigma_{11}/S_{LT} < 1)$ | |
| | FRP layup sequence | | FRP layup sequence | |
	$[0]_2$	$[0]_4$	$[0]_2$	$[0]_4$
	GFRP configuration (a)			
Arch-lower	0.80	8.81	0.21	0.2
Beam-lower	0.96	1.86	0.21	0.2
Arch-upper	0.014	0.47	12.30	13.79
Beam-upper	0.34	0.65	12.82	12.29
	GFRP configuration (k)			
Arch-lower	0.66	25.65	0.15	0.16
Beam-lower	1.36	39.64	0.22	0.23
Arch-upper	0.99	3.70	0.15	0.16
Beam-upper	0.91	3.61	0.22	0.2

Arch/beam-lower: FRP applied onto the lower face of the arch/Beam.
Arch/beam-upper: FRP applied onto the upper face of the arch/Beam.

TABLE 4: Experimental and finite element analysis results of the control glulaminated curved beams and arches (average values of the test results described in the text).

	Average ultimate Load (N)	Horizontal displacement at hunch (mm)	Vertical displacement at mid-rafter (mm)	Horizontal displacement at crown (mm)	Vertical displacement at crown (mm)
		Control arches			
Experimental	5,120	1.13	−13.3	−5.01	−0.41
FEA		0.25	−12.01	−3.41	−2.10
		Fully GFRP-reinforced arches—configuration (a)			
Experimental	8,100	0.42	−20.14	−4.80	−3.62
FEA		1.50	−15.40	−3.70	−3.21
		Partially GFRP-reinforce arches—configuration (k)			
Experimental	6,696	0.60	−18.12	−1.41	−1.23
FEA		0.35	−15.10	−3.81	−2.21

The noted horizontal displacement at crown was measured at the highest point on the arch/beam (see Figure 1).

occurred at the end of the same reinforcement. The delamination in the interface between the GFRP and the wood occurred when the applied load reached 3,600 N, as seen from the typical graph presented in Figure 8.

5.3. Full-Length Reinforcement of Glulam Wood Arch (Configuration (A)).
Group 3 also contained six arches divided into two sets of three arches each. The first set was fully reinforcement with $[0]_2$ layup of unidirectional GFRP being applied on the upper and lower surfaces of the arches along the entire length. The second set was reinforced in the same way as the first set, except with $[0]_4$ layup of GFRP.

Figure 9 illustrates the Load versus mid-rafter deflection response of the two arches in set 1 of group 3. The third arch prematurely failed due to delamination of its reinforcement,

due to manufacturing cause anomaly (lack of adhesive penetration in some sections).

As can be seen from Table 4, this set of curved beams exhibited 42 percent increase in the ultimate load carrying capacity as well as 27 percent increase in their stiffness, compared to the unreinforced (control) arches. The mode of failure in these arches was different from that observed in the partially reinforced cases. No visible cracks could be observed during the experiments. The response of the arch was elastic up to approximately 90% of its ultimate load carrying capacity, after which the response was relatively fairly ductile, with a rather plasticity-like response, up to the ultimate load. After reaching the ultimate load carrying plateau, the arches could no longer sustain further loading, and the load gradually decreased. This type of ductile response is highly desirable, since it provides warning before the final failure.

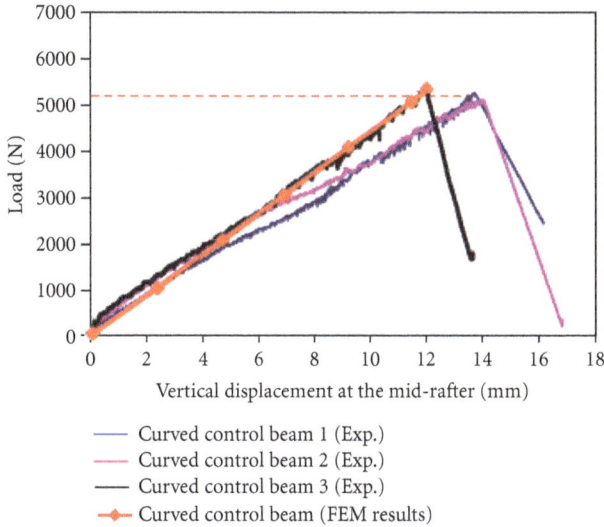

FIGURE 7: Load versus displacement of the three control arches and that obtained from FEA.

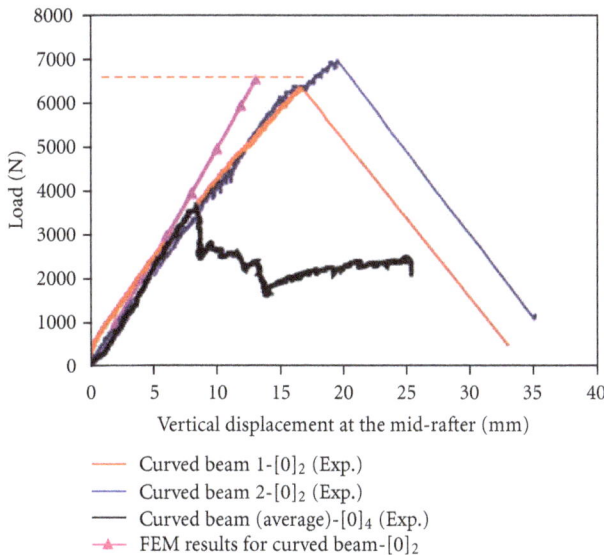

FIGURE 8: Typical load-displacement curves of the half-arch (case K) with two and four layers of partial reinforcement compared to the FEM results.

The fully reinforced arches with four layers of GFRP in set 2 of group 3 were also tested. Similar to the partially reinforced arches, they all failed prematurely due to delamination of the GFRP from the arch, as well as within the wood layers, which occurred at an average load of 4000 N, as can be seen from Figure 9. The failure mode occurred in twofold: (i) by delamination of wood layers at region close to the support at the foot of the arches, followed by (ii) delamination of GFRP from arches in the region in between the two concentrated applied loads.

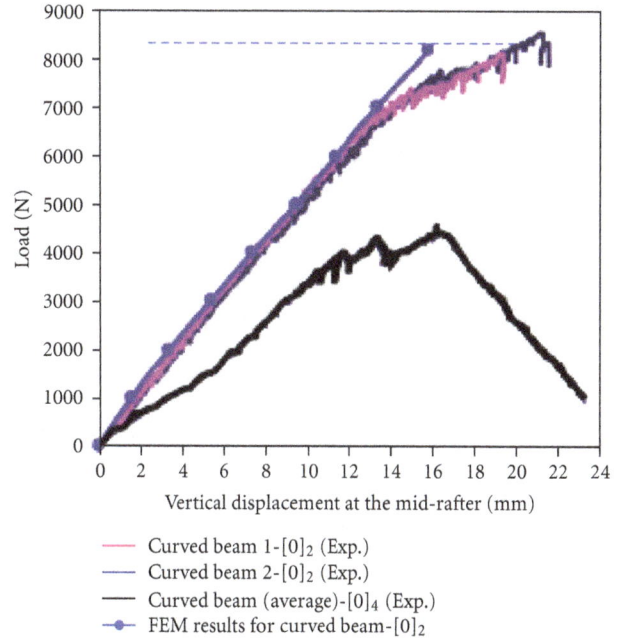

FIGURE 9: Typical load-displacement curves of the half-arch (case A) with two and four layers of full-length reinforcement compared to the FEM results.

6. Discussion on the Results

The results of the vertical and horizontal displacement obtained from the finite element analyses of the curved beam and arch models compared to the experimental values are tabulated in Table 4. The computational results agree well with the experimental results. The comparison of the computational results for the two modeled boundary conditions confirmed that, contrary to what was initially anticipated, the half-arches reacted like a curved beam. This was primarily due to the excessive rotation of the crown caused by the application of nonuniform loading (despite the fact that the resulting bending moment distribution due to the uniformly distributed load and the concentrated loading scheme were established to be more or less similar).

The finite element analyses results presented in Table 3 also indicated that delamination would occur due to the presence of very high shear stress in some portions of GFRP-wood interface, especially when the arch was reinforced with a laminate comprised of four layers of GFRP (i.e., $[0]_4$), regardless whether the GFRP reinforcement was applied along the entire length of the arch or along the partial length. This shortfall could be better addressed by using a suitable epoxy adhesive, as opposed to the resorcinol used in this project (which is commonly used by the glulam industry). Resorcinol is a relatively strong, but very brittle adhesive. The glulam arches tested in this study underwent relatively large magnitude of deformation; as a result, large shearing and peel stresses were developed within the adhesive interface of the arch and GFRP reinforcement. The thicker GFRP (i.e., $[0]_4$) had significantly higher stiffness than the thinner GFRP ($[0]_2$), thus, producing larger strain incompatibility within the resorcinol interface. This caused premature failure of the resorcinol adhesive layer in those cases.

7. Conclusions

Through a computational investigation, the two most effective configurations of GFRP reinforcement for strengthening and stiffening glue-laminated Tudor arch were established; one configuration was to apply the GFRP reinforcement on the upper and lower surfaces of the cross-section along the entire length of the arch (configuration (a) in Figure 2); the other was the application of the reinforcement along the partial length of the arch (configuration (k) in Figure 2). It was found that the thickness of the GFRP was the most significant factor in limiting the performance of the GFRP reinforcement. Interestingly, thicker reinforcement caused premature delamination of GFRP from the wood.

The following summarizes the findings of this study.

(i) The average ultimate load for a glulaminated curved beam with $[0]_2$ layup (i.e., GFRP reinforcement ratio of 1.05% by volume) applied on the upper and lower surfaces of the arch/curved beam, along the full-length of the arches, showed an increase of 42 percent in strength and a 27-percent increase in stiffness compared to those of the control curved beam, respectively. This remarkable enhancement was achieved by using only 1% by volume of reinforcement. There were no clear cracks or delamination in the structure with this reinforcement configuration up to the stage when the applied load approached the ultimate load. When the load exceeded the values corresponding to the yield strength of the materials, the specimens reinforced with GFRP along their full-length (configuration (a)) exhibited gross plastic-like deformation. The resulting deformation was permanent upon release of the load. This type of failure is the preferred mode, since it provides warning before the final failure.

(ii) All GFRP glulam-curved beams reinforced with $[0]_4$ layup GFRP (with a reinforcement ratio of 2.10% by volume) also applied along the full-length of the arches delaminated before the load reaching approximately 80 percent of the ultimate load carried by the control glulam-curved beams. This was prompted due to excessive shear and peel stresses in the adhesive layer joining wood to GFRP.

(iii) The ultimate load for the glulaminated curved beam reinforced with $[0]_2$ layup of GFRP applied along the partial length exhibited a 30.80-percent increase in strength and a 9.70-percent increase in stiffness, compared to the control glulam-curved beam. Cracks occurred without any warning, immediately after the curved beam reached its ultimate load carrying capacity, at which stage, a very strong blast occurred at the lower side of the cross section, followed immediately by delamination of the GFRP from the wood.

(iv) In the partially reinforced curved beams reinforced with $[0]_4$ layup GFRP, the failure occurred by at the interface between the wood and the GFRP at a load 27 percent less than the ultimate load of the control curved beams. The delamination occurred as a result of large combined shear and peel stresses developed at the free edges of the reinforcement that was adhered to the bottom surface of the curve beams.

References

[1] R. Mark, "Wood-aluminium beams within and beyond the elastic rangepart1: rectangular sections," *Forest Products Journal*, vol. 10, no. 11, pp. 477–484, 1961.

[2] B. Bohannan, "Pre stressed wood members," *Forest Products Journal*, vol. 12, no. 12, pp. 596–602, 1962.

[3] J. Peterson, "Wood beams prestressed with bonded tension elements," *Journal of Structural Engineering*, vol. 91, no. 1, pp. 103–119, 1965.

[4] G. Lantos, "The flexural behaviour of steel reinforced laminated timber beams," *Wood Science*, vol. 2, no. 3, pp. 136–143, 1970.

[5] G. P. Krueger and L. B. Sandberg, "Ultimate strength design of reinforcing timber "evaluation of design parameters"," *Wood Science*, vol. 6, no. 4, pp. 316–330, 1974.

[6] W. Bulleit, L. Sandberg, and G. Woods, "Steel-reinforced glued laminated timber," *Journal of Structural Engineering*, vol. 115, no. 2, pp. 433–444, 1989.

[7] F. Wangard, "Elastic deflection of wood-fiber-glass composite beams," *Forest Products Journal*, no. 14, pp. 256–260, 1964.

[8] E. J. Biblis, "Analysis of wood fibre glass composite beams within and beyond the elastic region," *Forest Products Journal*, vol. 15, no. 2, pp. 81–88, 1965.

[9] F. H. Theakston, "A feasibility study for strengthening timber beams with fibreglass," *Canadian Agricultural Engineering*, vol. 7, no. 1, pp. 17–19, 1965.

[10] H. J. Dagher, T. E. Kimball, and S. Shaler, "Effect of FRP reinforcement on low grade Eastern Hemlock," in *Proceedings of the National Conference on Wood-Transportation Structures*, United States Department of Agriculture, Forest Service, Forest Products Laboratory, Madison, Wis, USA, 1996.

[11] K. C. Johns and S. Lacroix, "Composite reinforcement of timber in bending," *Canadian Journal of Civil Engineering*, vol. 27, no. 5, pp. 899–906, 2000.

[12] N. Olsson, *Glulam timber arches strength of splices and reliability-based optimization*, Ph.D. thesis, Lulea University, Luleå, Sweden, 2001.

[13] B. Kasal and A. Heiduschke, "Radial reinforcement of curved glue laminated wood beams with composite materials," *Forest Products Journal*, vol. 54, no. 1, pp. 74–79, 2004.

[14] T. W. Buell and H. Saadatmanesh, "Strengthening timber bridge beams using carbon fiber," *Journal of Structural Engineering*, vol. 131, no. 1, pp. 173–187, 2005.

[15] M. Yahyaei-Moayyed and F. Taheri, "Creep response of glued-laminated beam reinforced with pre-stressed sub-laminated composite," *Construction and Building Materials*, vol. 25, no. 5, pp. 2495–2506, 2011.

[16] M. Yahyaei-Moayyed and F. Taheri, "Experimental and computational investigations into creep response of AFRP reinforced timber beams," *Composite Structures*, vol. 93, no. 2, pp. 616–628, 2011.

[17] F. Taheri, M. Nagaraj, and P. Khosravi, "Buckling response of glue-laminated columns reinforced with fiber-reinforced plastic sheets," *Composite Structures*, vol. 88, no. 3, pp. 481–490, 2009.

[18] ASTM, "Standard test method for the tensile properties of polymer matrix composite materials," Tech. Rep. ASTM D3039M-08, American Society for Testing and Materials, Conshohocken, Pa, USA, 2008.

[19] P. K. Mallick, *Fiber-Reinforecd Composites: Materials, Manufacturing, and Design*, Marcel Dekker, New York, NY, USA, 1993.

4

Long-Term Field Performance of Pervious Concrete Pavement

Aleksandra Radlińska, Andrea Welker, Kathryn Greising, Blake Campbell, and David Littlewood

Department of Civil and Environmental Engineering, Villanova University, 800 Lancaster Avenue, Villanova, PA 19085, USA

Correspondence should be addressed to Aleksandra Radlińska, aleksandra.radlinska@villanova.edu

Academic Editor: Serji N. Amirkhanian

The work described in this paper provides an evaluation of an aged pervious concrete pavement in the Northeastern United States to provide a better understanding of the long-lasting effects of placement techniques as well as the long-term field performance of porous pavement, specifically in areas susceptible to freezing and thawing. Multiple samples were taken from the existing pavement and were examined in terms of porosity and unit weight, compressive and splitting tensile strength, and the depth and degree of clogging. It was concluded that improper placement and curing led to uneven pavement thickness, irregular pore distribution within the pervious concrete, and highly variable strength values across the site, as well as sealed surfaces that prevented infiltration.

1. Introduction

The Stormwater Research and Demonstration Park at Villanova University (Greater Philadelphia area) is comprised of more than one dozen Stormwater Control Measures (SCMs), which are also known as Best Management Practices (BMPs). The pervious concrete site, which is the focus of this paper, is located between two dormitories in the middle of Villanova University campus. Originally, this site consisted of an asphalt paved area and a conventional storm sewer system which channeled runoff directly to the headwaters of Mill Creek, a high-priority stream on Pennsylvania's Clean Water Act (303d) list. The site serves primarily as a pedestrian walkway that occasionally experiences light maintenance vehicle and automobile traffic. In August 2002, the conventional materials were demolished and three infiltration beds (sized to accept approximately 5 cm of rainfall) were constructed with a pervious concrete surface. Originally, the entire pedestrian area was pervious; however, the surface failed and the site was reconstructed in May 2003 with a combination of traditional and pervious concrete [1]. In October 2004, several sections of the pervious concrete were again removed and repaired due to localized deterioration and cracking. Since that time, significant deterioration of the concrete has occurred.

Several Key Factors: environmental effects, material inconsistencies, and inadequate installation methods led to the deterioration of the original pervious concrete surface. Excessively high temperatures during installation resulted in inconsistent mixing and difficulties with finishing. Additionally, the curing process was improperly accelerated by the excessive heat, reducing strength and affecting aggregate bonding. The mixing time in the trucks varied with inconsistent travel times to the site. Workability of the concrete was low as there was a short window of time to finish the concrete surface. The finishing technique used was also inadequate. At least two different finishing methods were used, including a vibratory screed and modified plate tamper. Neither of the methods proved to be functional, however, and several areas were left in an unfinished state [2, 3].

In May 2003, the original pervious concrete surface, which covered the entire pedestrian area, was replaced. The pervious area was significantly decreased from the original pervious design. It was determined through site runoff data and visual observations of the first design during storm events that the pervious area could be decreased if it was strategically located to collect the same amount of runoff. A new layout (Figure 1) was designed which consisted of a pervious concrete perimeter (marked red) around three crowned standard concrete sections (marked purple) [1–3].

(a)   (b)

FIGURE 1: (a) Aerial view of Villanova Campus Pervious Concrete Site in 2003. (b) pervious concrete (red), traditional concrete (purple), and paved brick (green) surfaces identified.

In October 2004, approximately half of the sections of the pervious concrete area were showing signs of deterioration and these sections were removed and replaced with new pervious concrete (marked green "repaired" in Figure 2). Subsequent evaluation of the pavement condition in June 2006 revealed that some sections of the pavement were completely clogged (marked red "sealed" in Figure 2), while other sections experienced cracking (marked blue "cracked" in Figure 2). It should be mentioned here that the pervious concrete placed in 2003 and 2004 was a standard Florida Mix [4].

The construction process of the present system was intended to avoid the problems encountered during construction of the first surface. For example, the construction was completed during cooler months and the water was added at the site to create a more consistent and workable mix. On-site mixing allowed for better control of the mixing and curing process. A weighted drum was rolled over the concrete after it was placed for compaction and finishing. A modified version of this method is still used in practice today and includes a weighted vibratory hand roller [2, 3].

Despite the second placement and repairs, the pervious concrete surface installed in May 2003 and October 2004 has degraded significantly and is slated for demolition in 2012. The impending removal of this concrete presented a unique opportunity to perform a forensic evaluation of one of the earliest pervious concrete installations in the northeastern portion of the United States.

The objectives of this research were to evaluate the following properties of pervious concrete:

(i) extent of clogging and the resulting decrease in infiltration capacity,

(ii) extent of phosphorous adsorption onto the concrete,

(iii) mechanical properties evaluation.

The results of this work are placed into the context of the current state of understanding of pervious concrete properties. This can provide researchers and practitioners with insights into mechanisms of failure as well as an understanding of

the ability of pervious concrete to adsorb phosphorous in the long term. A thorough examination of failures, as well as successes, is required to enable the widespread acceptance of this material.

2. Literature Review

Pervious surfaces are attractive alternatives to traditional concrete pavements because of their role in stormwater management and mitigating Urban Heat Island (UHI) effects [5, 6]. The concrete industry's understanding of pervious concrete is evolving with the ongoing research of the materials mechanical and chemical properties [7]. The existing literature on the properties of pervious concrete, both mechanical and hydrological, are examined in this section to provide a better understanding of the behavior of pervious pavements.

2.1. Mechanical Properties. Adequate strength of pervious concrete is critical to ensuring that it provides a long-term, durable solution to storm water management. There is a tradeoff between strength and porosity (related to permeability), and it is typically up to the designer to determine which parameter controls the design [8]. Pervious concrete can develop compressive strengths upwards of 28 MPa, with typical values around 17 MPa. As with traditional concrete, the properties and combinations of ingredient materials, as well as batching, placement techniques, and environmental conditions, contribute to the actual in-place strength [9]. Destructive testing of drilled cores is the best measure of in-place strengths, as compaction differences make cast cylinders less representative of field concrete. Flexural strength in pervious concrete generally ranges between about 1 MPa and 3.8 MPa. Many factors influence the flexural strength, particularly degree of compaction, porosity, and the aggregate-to-cement (A/C) ratio. However, for typical applications utilizing pervious concrete, measurement of flexural strength is not required [10].

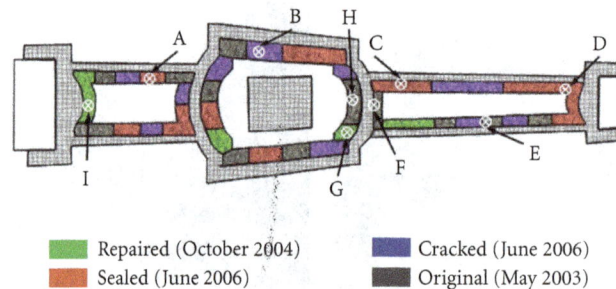

FIGURE 2: April 2011 core sample locations and delineation of physical conditions of concrete. Note: during the April 2011 inspection, all pervious concrete areas were found to be completely sealed.

Drying shrinkage develops earlier in pervious concrete, but is significantly less than conventional concrete [11]. Specific values vary with different mixtures and materials used, but typically the values are approximately half of that expected for conventional concrete mixtures. The material's low paste and mortar content is a possible explanation. Approximately 50% to 80% of shrinkage occurs within the first 10 days, compared to 20% to 30% in the same period for conventional concrete. Often pervious concrete is made without control joints and allowed to crack randomly [10].

Aggressive chemicals in soils or water, such as acids and sulfates, are a concern to conventional concrete and pervious concrete alike, and the mechanisms for chemical attack are similar. However, the open structure of pervious concrete may make it more susceptible to attack over a larger area. Pervious concrete can be used in areas of high-sulfate soils and ground water if isolated from direct contact. Placing the pervious concrete over a 150 mm layer of 25 mm maximum top size aggregate provides a pavement base, storm water storage, and isolation for the pervious concrete. Unless these precautions are taken in aggressive environments, recommendations from ACI 201 on water-to-cement ratio and material types/proportions should be followed [10]. Similarly, care should be taken to minimize the degradation caused by deicing materials by minimizing their use soon after placement [12].

The rougher surface texture and open structure of pervious concrete can lead to the raveling of aggregate particles when the pavements are plowed during snow events. This abrasion and raveling is one of the major reasons why applications such as road ways are generally not suitable for pervious concrete in seasonal climates with harsh winters. However, the open structure does prevent refreezing of melted snow, which decreases the amount of deicing materials required to clear the pavement. Most pervious concrete pavements will have a few loose aggregates on the surface in the early weeks after opening to traffic. These rocks were typically loosely bound to the surface initially and popped out because of traffic loading. After the first few weeks, the rate of surface raveling is usually reduced considerably and the pavement surface becomes much more stable [13]. Proper compaction and curing techniques reduce the occurrence of surface raveling.

2.2. Hydrological and Chemical Properties. Pervious pavements allow stormwater to infiltrate into the ground as a method to meet both volume and pollutant reduction goals [1, 14, 15]. Typical values for the infiltration rate of pervious concrete vary between 0.20 cm/s and 0.54 cm/s, although values in excess of 1.2 cm/s have been measured in the laboratory [10, 16]. The porosity of pervious concrete usually varies between 15% and 25% [10]. It is necessary for pervious pavements to retain their porosity and high infiltration rates to continue to meet the storm water control goals for which they were designed. In addition to the total porosity, the interconnectivity of the pores plays an important role as it is directly related to the infiltration rate [17]. Recently developed computer models allow virtual characterization of pervious concrete microstructure [18].

The large pore sizes, which allow for increased infiltration, can also allow smaller sediment particles to accumulate within these spaces. Clogging of the pores can greatly limit the performance of these systems from a hydrological perspective. Proper installation and routine maintenance are required to minimize clogging. Vacuum street sweeping, pressure washing, or a combination of the two are the most commonly used methods to dislodge materials clogging the pores of pervious pavements [19–21].

Pervious concrete improves the quality of water in three ways. The first way is that the pervious concrete is used in conjunction with an infiltration or retention bed to prevent polluted water from travelling downstream. The second way is that pollutants within the storm water runoff are adsorbed onto the pavement particles as the water is infiltrated through the system. The third way is that the pavement traps sediments, to which pollutants (nutrients and metals) adsorb, in the pore spaces [10, 22–26].

3. Evaluation of the Properties of the Pervious Concrete Site

The properties of the pervious concrete investigated in this study have been evaluated at several points in time. When the concrete was placed in May 2003, two cylinder samples were cast to determine the concrete's compressive strength. In June 2006, a thorough inspection of the site was performed to

FIGURE 3: Concrete core samples used for compressive strength evaluation. Note that the length differences are due to uneven pavement thickness.

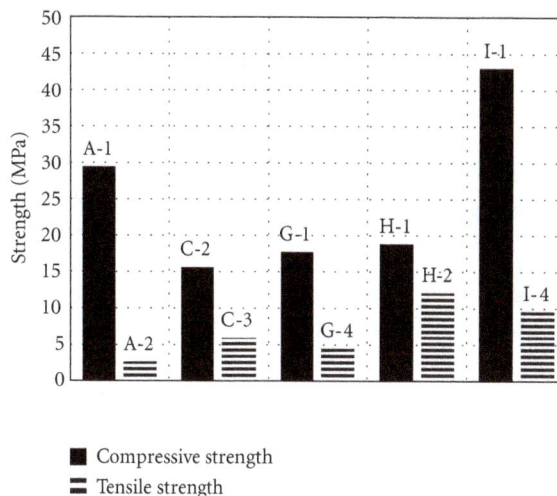

- ■ Compressive strength
- ≡ Tensile strength

FIGURE 4: Compressive and splitting tensile strength results for individual cores.

determine the extent of deterioration of the concrete as well as its infiltration capabilities (Figure 2). In April 2011, a total of 28 core samples were taken from nine different locations throughout the site. Each location was assigned a letter and number starting with location "A" and ending with location "I".

The following properties were evaluated for the 28 cores obtained in 2011: compressive and splitting tensile strength, unit weight, porosity (using both image analysis and volumetric method), infiltration testing, and phosphorous adsorption to both the concrete and the sediments found within the pores. The methods used and the results obtained are discussed for each property evaluated.

3.1. Compressive and Splitting Tensile Strength and Unit Weight. Of the 28 core samples taken from the Pervious Concrete Site on Villanova's campus, 10 were used for strength and unit weight evaluation (Figure 3). The top and bottom of each sample was saw-cut to create smooth surfaces. The core samples were measured and weighed prior to testing. The diameter of both the top and bottom as well as the length of each core sample was measured in three different locations and the average for each cylinder was reported (Table 1). The measured unit weights, with an average of 134 pcf (2147 kg/m^3), are greater than the values typically expected for pervious concrete. Typically, the in-place density of pervious concrete is between 100 pcf (1602 kg/m^3) and 125 pcf (2002 kg/m^3) [10]. These higher than expected unit weights indicate that some sections of pervious concrete were "overworked" during installation, resulting in significant reduction of porosity.

Compressive strength tests were performed following ASTM C39-05 [27]. Samples were capped with neoprene pads and loaded at a continuous rate at 1.78 kN/s to 2.22 kN/s until failure. Correction factors were applied to the strength results as per ASTM C42-04 [28], as the cores had a length-to-diameter ratio (L/D) less than 1.75. The compressive strength was measured between 15.6 and 43.0 MPa, as presented in Figure 4. Despite the fact that the design strength for pervious surfaces was achieved, significant variation in strength across the sample locations was observed. This variability indicates that there were inconsistencies in the placement and curing of the pervious concrete as well as in the intrinsic material properties of the concrete.

The splitting tensile strength was evaluated as per ASTM C496-04 [29]. Thin balsa wood bearing strips were used to distribute the load applied along the cylinder length. The samples were loaded continuously at a rate of 0.7 MPa/min to 1.4 kPa/min until failure. The maximum load sustained by the specimen was divided by appropriate geometrical factors to determine splitting tensile strength. The results of splitting tensile strength (Figure 4) show significant variability, as observed in case of compressive strength results.

As previously mentioned, two cylindrical samples of the pervious concrete were cast in May 2003, during site installation. Despite the low reliability and consistency associated with the sample casting of pervious concrete [10], the results of these tests can be used for comparison with compressive strength results obtained from the 2011 cores. The samples obtained in 2003 were placed in 10 cm diameter concrete cylinder molds and cured in a moist room for 28 days. The measured compressive strengths for the two samples were 23.9 and 23.5 MPa. These results compare favorably to the compressive strengths obtained for the samples cored in 2011 (Figure 4). The higher strength of laboratory prepared samples can be explained by the extended moist curing period. It is also important to note that the compressive strength for the 2011 samples represents the value for intact cores and therefore is not representative of the locations where the pavement has completely deteriorated.

3.2. Porosity. Pervious pavement is generally designed to contain approximately 15% to 25% voids. The void space or porosity of a pervious pavement is directly related to the rate of infiltration, and thus its effectiveness as a storm water management system [30]. Since its initial construction in 2002, care has been taken at the pervious concrete site to minimize clogging. During construction in 2002, 2003, and 2004, all surrounding vegetated areas were stabilized to ensure that sediment was not carried onto the pervious pavements during storm events. The site was swept with

TABLE 1: Pervious concrete cores sample data.

Specimen	Average diameter (cm)		Average length (cm)	Unit weight (kg/m^3)
	Top	Bottom		
A-1	9.50	9.50	9.50	2195
C-2	9.50	9.50	14.17	2082
G-1	9.52	9.50	13.91	2114
H-1	9.52	9.50	17.20	2066
I-1	9.52	9.52	16.46	2243
A-2	9.50	9.52	14.86	2210
C-3	9.50	9.50	14.76	2082
G-4	9.52	9.52	12.01	2098
H-2	9.50	9.50	16.51	2130
I-4	9.50	9.50	14.07	2275
Average	**9.50**	**9.50**	**14.86**	**2145**
STDEV	**0.01**	**0.007**	**1.52**	**64**

a vacuum street sweeper quarterly to remove any sediment trapped in the pore spaces. Despite these efforts, the pervious concrete exhibited visible clogging due to raveling, possibly caused by inadequate placing and finishing. A thorough inspection of the site performed in June 2006 revealed that approximately one-half of the pavement area was unable to infiltrate water (Figure 2). Although great care was taken to minimize clogging from sediments, the concrete was subjected to intense application of de-icing materials within several months of installation. The site under investigation is a major pedestrian thoroughfare in the university and, as such, was regularly treated with calcium chloride to prevent injuries.

The porosity of the pervious concrete specimens obtained in 2011 was examined to provide insight into the clogging that had occurred. Digital image analysis was used to determine clogging as a function of depth [31] using two-dimensional images [32]. Nine core samples were cut into 3.8 cm thick slices using a wet saw to achieve a clean, level surface at either end of the sample (Figure 5(a)). As such, six faces were created for a single core. In some cases, localized changes in pavement depth resulted in only four representative faces (samples D1, D2, E3) that were subject to further analysis. It should be noted that the top surfaces of the cores were not included in the image analysis due to significant surface damage and corresponding loose aggregate particles. Additionally, the coring and cutting process involved a wet saw that might have altered the existing pore structure.

After the samples were prepared, a digital image was taken of each face (Figure 5(b)). Once the image was obtained, it was processed using Adobe Photoshop software. All large pores visible in the image were manually filled with black color to distinguish between dark aggregate particles and pores (Figure 5(c)). The image was converted into a binary image (Figure 5(d)), and the area occupied by the pores was calculated as a percentage of the total sample cross-sectional area.

Additionally, the volumetric method was used to measure the total volume of pores in each of the analyzed samples. Each sample was measured to determine the total volume (V). The dry mass of the sample was recorded (m_{dry}). Next, samples were submerged in water and the buoyant mass was measured (m_b). Knowing the density of water (ρ_w), the porosity (ϕ) was determined using the following equation:

$$\phi = \left[1 - \frac{m_{\mathrm{dry}} - m_w}{\rho_w \times V} \right] \times 100\%. \qquad (1)$$

The porosity versus core depth for nine separate locations, as obtained using image analysis, are summarized in Figure 6. Significant variations in the porosity exist between different locations and core depths. The values of porosity recorded were between 1% and 26%, with only 19% of results larger than 15% (typically the lower threshold expected for pervious pavements). Porosity in the upper part of the pavement was consistently lower than that of the deeper sections of the pervious concrete. The extremely low porosity values observed in the top section indicate that at the time of evaluation, the pervious surface had lost its infiltration capacity. This loss of porosity is confirmed by visual inspection of the core A-3 presented in Figure 5(a). Additionally, sediment clogging of the void space of the pervious concrete system resembled hardened concrete paste rather than loose sediment, which would indicate that freezing and thawing contributed to concrete spalling, and loose pieces clogged the pores. Based on visual observations of the samples from locations D and I, these sections exhibited reduced void content due to excessive compaction during placement rather than passive clogging.

The summary of the porosity values obtained using the volumetric method is shown in Figure 7. Only 12% of the samples tested had porosity greater than 15%. Similar to the image analysis results, a trend of increasing porosity with depth was also observed.

3.3. Infiltration. Field infiltration tests were performed on the pervious concrete within 10 cm of where the cores were obtained. For this test, water was placed into a 10 cm

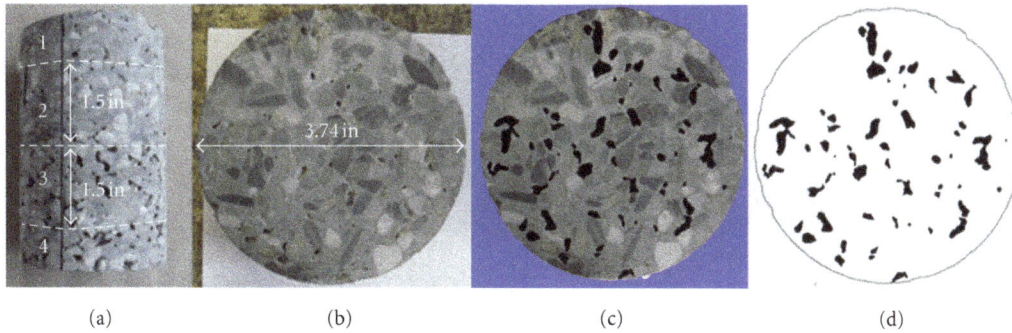

(a)   (b)   (c)   (d)

FIGURE 5: Image analysis of pervious core A-3 (porosity = 7.4%).

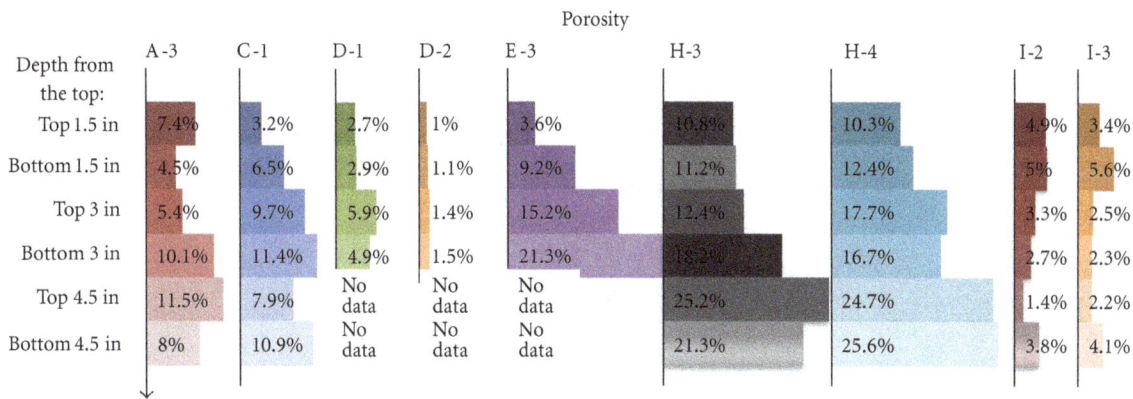

FIGURE 6: Porosity of nine concrete samples with respect to core depth evaluated using image analysis of sliced cores.

diameter concrete cylinder mold with a 22 mm hole drilled in the bottom and the length of time it took for the water to drain was recorded [33, 34]. This cylinder has a gasket around the bottom to ensure that side leakage does not occur and that the water is forced into the pavement. The time, t (seconds), was then used to calculate the infiltration rate, k (in/hr), using the following equation [35, 36]:

$$k = 2533e^{(-0.062*t)}. \qquad (2)$$

This method was used to determine the infiltration rate as it was the same method used to evaluate the infiltration rate in 2006. It is important to note that ASTM C1701-09 is a more rigorous method to determine the infiltration rate of pervious concrete [37].

All of the field tests, with the exception of one, revealed that the top surface was sealed with no infiltration. The only location that exhibited any infiltration (0.0003 cm/s) was near where sample series F was obtained. These results are consistent with the porosity values measured (Figure 6) and indicate that clogging of the site, which was approximately 50% in 2006, was complete by 2011.

3.4. *Phosphorous Adsorption.* The pervious concrete site had been extensively monitored hydrologically as well as for water quality. Metals were detected in the pore water of the natural soils underlying the infiltration bed and in the water stored

in the infiltration bed. The source of the copper was the downspouts and gutters from the roof runoff, which were directly piped into the infiltration beds; thus, this water did not infiltrate through the pavements. Once the pervious concrete storm water control measure was constructed, the university ceased using fertilizers near the site; therefore, one would expect the concentration of phosphorous adsorbed onto the sediments and to the concrete to be fairly low [1, 22].

To evaluate this hypothesis, the amount of phosphorus adsorbed to the sediments contained within the pores of the pervious concrete and the concrete cores was determined. Twenty specimens, four from each location A, C, G, H, and I, were analyzed. To determine the concentration of phosphorous adsorbed to the sediments within the pores, the samples were placed in shaker for one hour to dislodge the sediments. The samples were then removed from the shaker and placed on a flat surface until the sediment settled to the bottom. The extract solution was then removed from the sample and was filtered through a 0.45 μm filter. The filtered extract was then analyzed for phosphorous concentration using a Systea Scientific EasyChem machine [38]. The measured concentrations were averaged for each location and related to the mass of the sediment and the mass of the concrete to place the results into context (Figure 8).

As expected, the phosphorous concentrations were low for both the sediment and the concrete. In addition, the

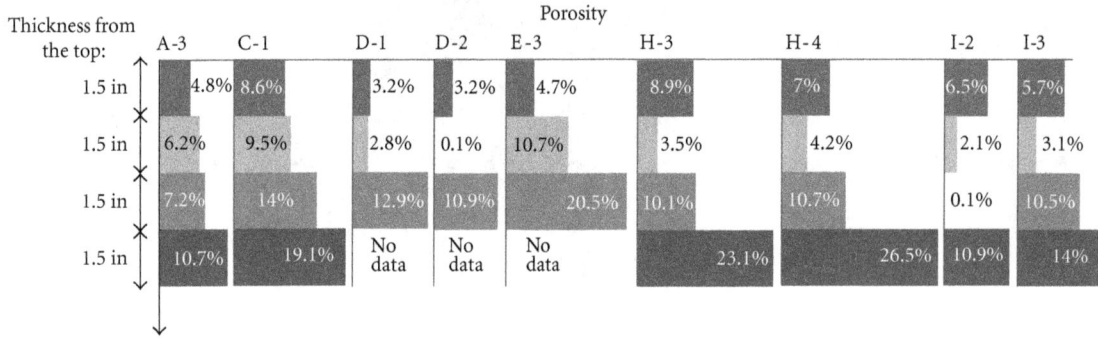

Thickness from the top:	A-3	C-1	D-1	D-2	E-3	H-3	H-4	I-2	I-3
1.5 in	4.8%	8.6%	3.2%	3.2%	4.7%	8.9%	7%	6.5%	5.7%
1.5 in	6.2%	9.5%	2.8%	0.1%	10.7%	3.5%	4.2%	2.1%	3.1%
1.5 in	7.2%	14%	12.9%	10.9%	20.5%	10.1%	10.7%	0.1%	10.5%
1.5 in	10.7%	19.1%	No data	No data	No data	23.1%	26.5%	10.9%	14%

FIGURE 7: Porosity of nine concrete samples with respect to core depth determined by volumetric method.

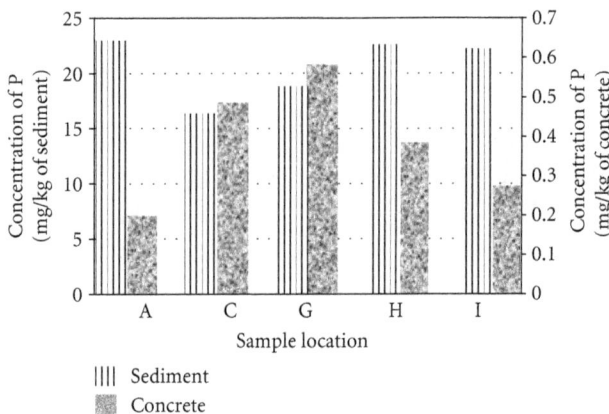

FIGURE 8: Concentration of phosphorus adsorbed to sediments within pores (primary y-axis) and to the concrete (secondary y-axis).

concentrations on the sediment were much higher than those for the concrete, due to the surface area of smaller particles being greater than that of larger particles. Water quality monitoring revealed that the pervious concrete infiltration SCM removed 94.3% of phosphorous [22]. The inflow concentrations were typically around 1 mg/L [39]. When the sample locations are compared to the site inspection that was performed in 2006 (Figure 2), some trends emerge. The phosphorous concentration of the sediment is the lowest for sample C, which is from an area identified as sealed in 2006. The other samples, A, G, H, and I, received more runoff, and thus more phosphorous, than sample C. Sample A was obtained from the border of two areas identified as sealed and cracked, and this area probably received the runoff from the sealed section as well as from the conventional concrete. Samples G and H were taken from an area of the 2003 concrete pour, while sample I was taken from an area of the 2004 pour.

4. Summary and Conclusions

The pervious concrete site at Villanova University was analyzed in terms of performance and durability after eight years of use. The strength of the concrete proved to be adequate for its intended use; however, significant variabilities in the measured porosity, unit weight, and strength values indicate inconsistencies in construction practices and material properties. Several factors that contributed to the decreased effectiveness of the pervious concrete were identified. Improper construction methods altered the desired pore distribution and significantly reduced the permeability of the sections, gradually leading to impervious surfaces. The reduced permeability decreased the effectiveness of the site to collect storm water, which further lead to reduced capacity to adsorb phosphorous. While the pervious surface allowed water ingress of 50% of its capacity in 2006, it became completely sealed by 2011. Additionally, raveling from freeze-thaw cycles played a key role in the eventual ineffectiveness of the concrete to allow infiltration into the ground, as verified by the inspection of particles locked in concrete pores. The varying porosity across the depth of the pavement was the result of improper installation procedures.

Although the design and installation practices of pervious concrete have improved tremendously over the past eight years, great care must be taken during installation to ensure that the concrete is properly placed, finished, and cured. This involves adequate training procedures in the finishing of the pervious surface. New methods that involve single-pass weighted rollers reduce the consolidation of aggregates near the surface and allow for better pore distribution. Proper mix design for the intended use of the pervious concrete with attention to freeze-thaw characteristics is also critical to the long-term durability and performance of porous concrete.

Acknowledgments

The authors gratefully acknowledge support received from Villanova University. This work was conducted in the Structural Engineering Teaching and Research Laboratory (SETRL), the Soils Laboratory, and the Water Resources Teaching and Research Laboratory at Villanova University; as such, the authors acknowledge the support that has made these laboratories and its operation possible. The authors would also like to thank Jennifer Gilbert Jenkins for her help with the phosphorous testing.

References

[1] M. Kwiatkowski, A. L. Welker, R. G. Traver, M. Vanacore, and T. Ladd, "Evaluation of an infiltration best management practice utilizing pervious concrete," *Journal of the American Water Resources Association*, vol. 43, no. 5, pp. 1208–1222, 2007.

[2] R. G. Traver, A. L. Welker, M. Horst, M. Vanacore, A. Braga, and L. Kob, "Lessons in porous concrete," *Stormwater*, pp. 130–136, 2005.

[3] R. G. Traver, A. L. Welker, C. Emerson, M. Kwiatkowski, T. Ladd, and L. Kob, "Villanova urban stormwater partnership: porous concrete," *Stormwater*, pp. 30–45, 2004.

[4] Florida Concrete and Products Association, Recommended Specifications for Pervious Pavement, publication 650.

[5] L. Frazer, "Paving paradise: the peril of impervious surfaces," *Environmental Health Perspectives*, vol. 113, no. 7, pp. A456–A462, 2005.

[6] N. Delatte and S. S. Schwartz, "Sustainability benefits of pervious concrete pavement," in *Proceedings of the 2nd International Conference on Sustainable Construction Materials and Technologies*, Universita Politecnicadelle Marche, Ancona, Italy, 2010.

[7] J. Kevern, K. Wang, M. T. Suleiman, and V. R. Schaefer, "Pervious concrete construction: methods and quality control," NRMCA Concrete Technology Forum: Focus on Pervious Concrete, Nashville, Tenn, USA, May 2006.

[8] N. Delatte, M. Miller, and A. Mrkajic, *Portland Cement Pervious Concrete Pavement: Field Performance Investigation on Parking Lot and Roadway Pavements*, Cleveland State University, Cleveland, Ohio, USA, 2007.

[9] M. Offenberg, "Producing pervious pavements," *Concrete International*, vol. 27, no. 3, pp. 50–54, 2005.

[10] P. Tennis, M. Leming, and D. Akers, *Pervious Concrete Pavements*, Portland Cement Association, Skokie, Ill, USA, National Ready Mixed Concrete Association, Silver Spring, Md, USA, 2004.

[11] V. M. Malhotra, "No-fines concrete—its properties and applications," *Journal of American Concrete Institute*, vol. 73, no. 11, pp. 628–644, 1976.

[12] K. Wang, D. E. Nelsen, and W. A. Nixon, "Damaging effects of deicing chemicals on concrete materials," *Cement and Concrete Composites*, vol. 28, no. 2, pp. 173–188, 2006.

[13] J. T. Kevern, K. Wang, and V. R. Schaefer, *The Effect of Coarse Aggregate on the Freeze-Thaw Durability of Pervious Concrete*, Portland Cement Association, 2008.

[14] Stormwater Technology Fact Sheet, Porous Pavement Publication. EPA 832-F-99-023, US Environmental Protection Agency, Office of Water, Washington, DC, USA, 1999.

[15] C. Dierks, P. Gobel, and W. Benze, "Next generation water sensitive stormwater management techniques," in *Proceedings of the 2nd National Conference on Water Sensitive Urban Design*, Brisbane, Australia, 2002.

[16] M. Legret, V. Colandini, and C. LeMarc, "Effects of porous pavement with reservoir structure on the quality of runoff water and soil," *Science of the Total Environment*, vol. 190, special issue, pp. 335–340, 1996.

[17] M. S. Sumanasooriya and N. Neithalath, "Pore structure features of pervious concretes proportioned for desired porosities and their performance prediction," *Cement and Concrete Composites*, vol. 33, no. 8, pp. 778–787, 2011.

[18] D. R. Bentz, "Virtual pervious concrete: microstructure, percolation, and permeability," *ACI Materials Journal*, vol. 105, no. 3, pp. 297–301, 2008.

[19] L. M. Haselbach, S. Valavala, and F. Montes, "Permeability predictions for sand-clogged Portland cement pervious concrete pavement systems," *Journal of Environmental Management*, vol. 81, no. 1, pp. 42–49, 2006.

[20] L. M. Haselbach, "Potential for clay clogging of pervious concrete under extreme conditions," *Journal of Hydrologic Engineering*, vol. 15, no. 1, pp. 67–69, 2010.

[21] M. Chopra, S. Kakuturu, C. Ballock, J. Spence, and M. Wanielista, "Effect of rejuvenation methods on the infiltration rates of pervious concrete pavements," *Journal of Hydrologic Engineering*, vol. 15, no. 6, Article ID 009006QHE, pp. 426–433, 2010.

[22] M. Horst, A. Welker, and R. Traver, "Multiyear performance of a pervious concrete infiltration basin BMP," *Journal of Irrigation and Drainage*, vol. 137, no. 6, pp. 352–358, 2011.

[23] K. F. Bruce, *Porous Pavements*, Taylor and Francis, Boca Raton, Fla, USA, 2005.

[24] Z. Teng and J. J. Sansalone, "In-situ storm water treatment and recharge through infiltration: particle transport and separation mechanisms," *Journal for Environmental Engineering*, vol. 130, no. 9, pp. 1008–1020, 2004.

[25] M. Scholz and P. Grabowiecki, "Review of permeable pavement systems," *Building and Environment*, vol. 42, no. 11, pp. 3830–3836, 2007.

[26] B. T. Rushton, "Low-impact parking lot design reduces runoff and pollutant loads," *Journal of Water Resources Planning and Management*, vol. 127, no. 3, pp. 172–179, 2001.

[27] ASTM Standard C39-05, *Standard Test Method for Compressive Strength of Cylindrical Concrete Specimens*, vol. 4 of *Annual Book of ASTM Standards*, ASTM International, West Conshohocken, Pa, USA, 2nd edition.

[28] ASTM Standard C42-04, *Standard Test Method for Obtaining and Testing Drilled Cores and Sawed Beams of Concrete*, vol. 4 of *Annual Book of ASTM Standards*, ASTM International, West Conshohocken, Pa, USA, 2nd edition.

[29] ASTM Standard C496-04, *Standard Test Method for Splitting Tensile Strength of Cylindrical Concrete Specimens*, vol. 4 of *Annual Book of ASTM Standards*, ASTM International, West Conshohocken Pa, 2nd edition.

[30] D. Akers, M. Leming, and P. Tennis, *Pervious Concrete Pavements*, Portland Cement Association, Skokie, Ill, USA, National Ready Mixed Concrete Association, Silver Spring, Md, USA, 2004.

[31] E. Masad, B. Muhunthan, N. Shashidhar, and T. Harman, "Internal structure characterization of asphalt concrete using image analysis," *Journal of Computing in Civil Engineering*, vol. 13, no. 2, pp. 88–95, 1999.

[32] M. S. Sumanasooriya, D. P. Bentz, and N. Neithalath, "Predicting the permeability of pervious concretes from planar images," NRMCA 2009 Concrete Technology Forum: Focus on Performance Prediction, pp. 11, 2009.

[33] N. Delatte, D. Miller, and A. Mrkajic, "Portland cement pervious concrete pavement: field performance investigation on parking lot and roadway pavements," RMC Research and Education Foundation, Cleveland State University, Cleveland, Ohio, USA, http://rmc-foundation.org/images/Long%20Term%20Field%20Performance%20of%20Pervious%20Final%20Report.pdf.

[34] P. Jeffers, *Water quality comparison of pervious concrete and porous asphalt products for infiltration best management practices*, M.S. thesis, Villanova University, 2009, http://www3.villanova.edu/vusp/Outreach/theses.htm.

[35] D. Miller, *Field performance of PCPC in severe freeze-thaw environments*, M.S. thesis, Cleveland State University, 2007.

[36] A. Mrkajic, *Investigation and evaluation of PCPC using nondestructive testing and laboratory evaluation of field samples*, M.S. thesis, Cleveland State University, 2007.

[37] ASTM Standard C1701-09, *Standard Test Method for Infiltration Rate of In Place Pervious Concrete*, vol. 4 of *Annual Book of ASTM Standards*, ASTM International, West Conshohocken, Pa, USA, 2nd edition.

[38] J. O'Dell, *Method 365.1 Determination of Phosphorus by Semi-Automated Colorimetry*, U.S. Environmental Protection Agency, 1993.

[39] M. Kwiatkowski, *Water quality study of a porous concrete infiltration best management practice*, M.S. thesis, Villanova University, 2004.

5

A Prediction Method of Tensile Young's Modulus of Concrete at Early Age

Isamu Yoshitake,[1] Farshad Rajabipour,[2] Yoichi Mimura,[3] and Andrew Scanlon[2]

[1] Department of Civil and Environmental Engineering, Yamaguchi University, Ube, Yamaguchi 755-8611, Japan
[2] Department of Civil and Environmental Engineering, Pennsylvania State University, University Park, PA 16802, USA
[3] Department of Civil and Environmental Engineering, Kure National College of Technology, Kure, Hiroshima 737-8506, Japan

Correspondence should be addressed to Isamu Yoshitake, yositake@yamaguchi-u.ac.jp

Academic Editor: Kent A. Harries

Knowledge of the tensile Young's modulus of concrete at early ages is important for estimating the risk of cracking due to restrained shrinkage and thermal contraction. However, most often, the tensile modulus is considered equal to the compressive modulus and is estimated empirically based on the measurements of compressive strength. To evaluate the validity of this approach, the tensile Young's moduli of 6 concrete and mortar mixtures are measured using a direct tension test. The results show that the tensile moduli are approximately 1.0–1.3-times larger than the compressive moduli within the material's first week of age. To enable a direct estimation of the tensile modulus of concrete, a simple three-phase composite model is developed based on random distributions of coarse aggregate, mortar, and air void phases. The model predictions show good agreement with experimental measurements of tensile modulus at early age.

1. Introduction

An accurate estimation of the Young's modulus is important for proper structural design of concrete members, and ensuring their serviceability, such as controlling deflections and crack widths. In particular, the time-dependent development of the tensile Young's modulus at early ages is needed for estimation of the tensile stresses that are generated due to restrained thermal and hygral shrinkage. These tensile stresses may lead to premature cracking of concrete members. Currently, the tensile modulus is assumed to be equal in value to the compressive modulus and is estimated using empirical correlations based on the compressive strength of concrete [1, 2]. The Architectural Institute of Japan (AIJ) [3] points out that employing the tensile modulus is more appropriate for estimation of the risk of early-age cracking; however, the specification indicates that the compressive modulus may be used instead of the tensile modulus because investigations dealing with the tensile modulus are currently insufficient. Since the tensile behavior of concrete is more significantly affected by the presence of

flaws (e.g., microcracks or large capillary pores common in early-age concrete), it is important to develop tools to predict or measure the tensile properties more accurately.

Direct tension tests have been conducted in earlier studies to investigate the tensile strength and the tensile strain capacity of concrete. Although the tensile moduli can be obtained from the linear portion of the stress-strain diagram in these reports, the focus of these earlier studies has been primarily on mature concrete. As such, little information is available on the early-age (i.e., less than 28 days) tensile modulus and its development with time. In addition, a reliable model to aid design engineers in estimating the tensile modulus based on concrete's proportions and age does not currently exist.

Xie and Liu [4] conducted a direct tension test using small and large specimens of mature concrete with various aggregate sizes, and measured the tensile strength, strain capacity, and Young's modulus. They observed that increasing the maximum aggregate size does not have a proportional impact on tensile strength and tensile modulus of concrete. Oluokun et al. [5] researched the compressive

Young's modulus and Poisson's ratio of early-age concrete. They concluded that the compressive modulus is proportional to the 0.5 power of the compressive strength, and the ACI 318 formula for estimation of the compressive modulus is valid after the age of 12 hours. Hagihara et al. [6] investigated the tensile creep of high-strength concrete at early age, and reported that the tensile Young's moduli are approximately 15% higher than the compressive moduli. This is an important conclusion and should be evaluated for other concretes with normal strength. Swaddiwudhipong et al. [7] investigated the mechanical properties of concrete containing ground granulated blast furnace slag and pulverized fuel ash. They reported that the tensile moduli of all concretes tested had correlated well with the tensile strength; although no predictive formula is presented for the tensile modulus. Aoki et al. [2] also conducted direct tension test in order to obtain tensile strength and Young's modulus of mature concrete and found that the tensile modulus is 9–12% higher than the compressive modulus for these concretes. Bissonnette et al. [8] researched the tensile creep of concrete at early age, and presented some measurements of the tensile Young's modulus at the age of 7 and 28 days. In an earlier work [9], the authors investigated the tensile Young's moduli by using a direct tension test, and presented a composite model derived from the Hirsh model [10] to predict the tensile modulus. The model showed good agreement with the experimental data as well as other composite models offered by Counto [11] and Hashin [12]. Recently, Mihashi and Leite [13] presented a state-of-the-art report on early age cracking of concrete and its mitigation techniques that includes some information on the mechanical properties of concrete at early ages.

As described above, the investigations focusing on the tensile Young's modulus of early-age concrete are few, and more experimental data will be needed to establish reliable predictive correlations for estimation of the tensile modulus. The present paper reports laboratory measurements of the tensile modulus within the first 7 days of hydration. Three concrete mixtures (with different *W/C*) were tested. In addition, to evaluate the effect of aggregate size, duplicate concrete mixtures were prepared and sieved before setting using a 5 mm mesh sieve. The resulting mortars were tested to determine their tensile Young's modulus. Using the measurements results, a composite model was developed and calibrated which can serve as a simple method for estimating the tensile Young's modulus of concrete.

2. Experimental Program

2.1. Materials and Mix Proportions of Concrete. This study employed ordinary Portland cement with a density of $3.14 \, \text{g/cm}^3$. Tables 1 and 2 provide the details of the cement and aggregate used. Proportions of the six concrete and mortar mixtures tested in this study are given in Table 3. The proportions of the concrete were designed by referring to mixture proportions used in a ready mixed concrete plant in Japan. Mortars were obtained by sieving plastic concrete mixtures as discussed above. This was done to duplicate the mechanical properties (i.e., tensile modulus)

TABLE 1: Physical and chemical compositions of cement.

	Ordinary Portland cement
Density	$3.14 \, \text{g/cm}^3$
Blaine fineness	$3340 \, \text{cm}^2/\text{g}$
Setting time start-end	2 h 26 m–3 h 34 m
Comp. strength at 3 days	30.8 MPa
at 7 days	46.3 MPa
at 28 days	63.6 MPa
Chemical compositions	
CaO	64.5%
SiO_2	20.5%
Al_2O_3	5.7%
Fe_2O_3	2.9%
MgO	1.27%
SO_3	2.15%
Cl^-	0.009%
Loss of ignition	1.89%

TABLE 2: Properties of aggregate.

	Fine agg. *S*	Crushed rock *G*
Materials	Sea sand	Andesite
Density	$2.56 \, \text{g/cm}^3$	$2.73 \, \text{g/cm}^3$
Fineness modulus	3.36	6.66
Absorption	1.3%	1.3%
Size (max. – min.)	5 mm	20–5 mm

of the mortar portion of the concretes as closely as possible. These results are needed to develop the composite model as discussed in Section 3. In all mixtures, proper dosages of air entraining and water reducing admixtures were used to ensure consistency and workability of the concrete (Table 3).

2.2. Test Methods and Specimens. Figure 1 shows the direct tension apparatus used in this study. Figure 2 shows the geometry of the dog-bone specimens tested. To reduce bending moment during test, the ends of the dog-bone specimens were not fully fixed but were allowed rotational freedom. The direct tension apparatus manually provides a tensile force to a specimen using a lever. Overall, an approximately constant strain rate of 2 to 3×10^{-6}/sec was applied. To measure tensile strain of specimens, an embedded strain sensor was used which includes an electrical resistance wire strain gage of 60 mm long coated by epoxy resin with tensile modulus of 2.8 kPa as shown in Figure 2. The overall sensor size was $120 \times 10 \times 3 \, \text{mm}$; as such the area ratio of the sensor to concrete was smaller than 1.4% to ensure that the sensor has little influence on the behavior of specimens when subjected to tensile force. While this setup was used to measure the tensile modulus, it may not be suitable for measurement of the tensile strength since a number of breaks occurred within the end zones of the dog-bone specimens (Figure 3).

TABLE 3: Mixture proportions of concrete.

ID	W/C (%)	Water (kg/m³)	Cement (kg/m³)	S (kg/m³)	G (kg/m³)	WRA (kg/m³)	Air (%)
O57	57	165	290	812	1030	2.9	4.5
O57m[#1]	57	265	466	1304	—	4.7	—[#2]
O39	39	169	434	790	933	4.3	3.8
O39m[#1]	39	257	659	1200	—	6.5	—[#2]
O25	25	170	680	694	818	6.8	3.6
O25m[#1]	25	243	971	991	—	9.7	—[#2]

[#1] Mortar is made from wet screening of concrete (maximum mesh size: 5 mm).
[#2] Air content of mortar was not measured because it is difficult to obtain wet-screened mortar volume required for the test.

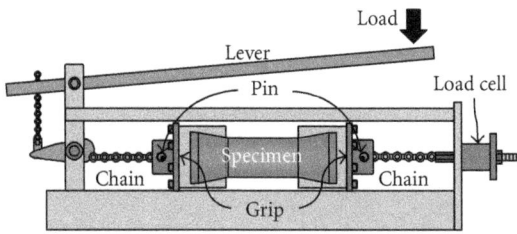

FIGURE 1: Direct tension test using a dog-bone-shaped specimen.

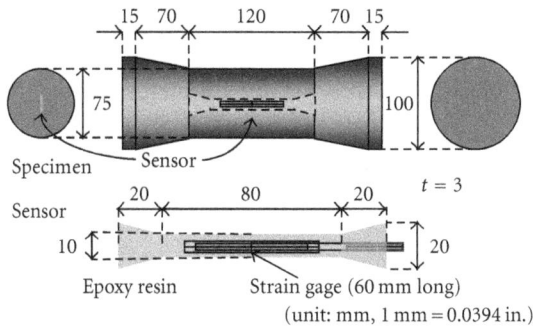

FIGURE 2: A dog-bone-shaped specimen and an embedded sensor.

(unit: mm, 1 mm = 0.0394 in.)

FIGURE 3: Typical failure within the end zone of the dog-bone-shaped specimen.

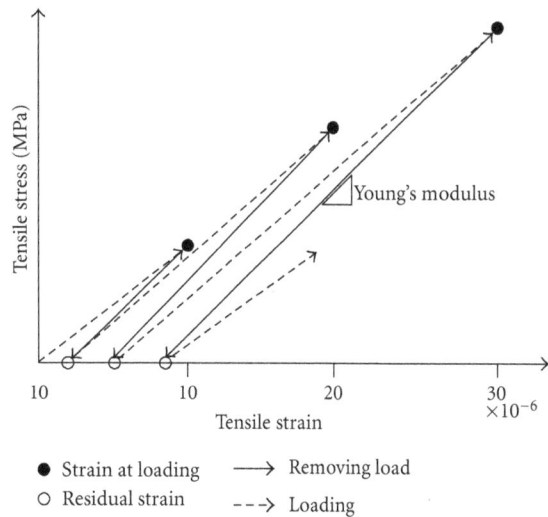

FIGURE 4: Evaluation method for tensile Young's modulus.

3. Prediction of the Young's Modulus Using a Composite Model

In addition to the tensile modulus, the compression and indirect tension tests were conducted using cylindrical specimens. The compressive modulus of concrete was obtained using an extensometer equipped with 2 displacement gages, and the modulus of mortar was measured using 2 wire strain gages 30 mm long. The cylindrical specimens (diameter × height) tested were 100×200 mm for concrete, and 50×100 mm for mortar. Three cylindrical specimens were used for each test per each mixture, and the average of the three measurements was used. The tests were performed at ages 1, 2, 3, and 7 days.

3.1. Determination of the Tensile Young's Modulus. To measure the tensile modulus, the tensile force is applied to the specimen at strain intervals of 10×10^{-6}. In order to prevent failure of specimen, the maximum strain during the test is set at 60×10^{-6}. Since the plastic strain of concrete at early age

may comprise a high percentage of the total strain measured, the tensile modulus is obtained from the unloading branches of the stress-strain relation as shown in Figure 4. The slope of the unloading branches after the specimen was loaded to 10×10^{-6}, 20×10^{-6}, and so forth, is determined and averaged to obtain the tensile modulus of the specimen. The force is measured twice at each strain level using a load cell

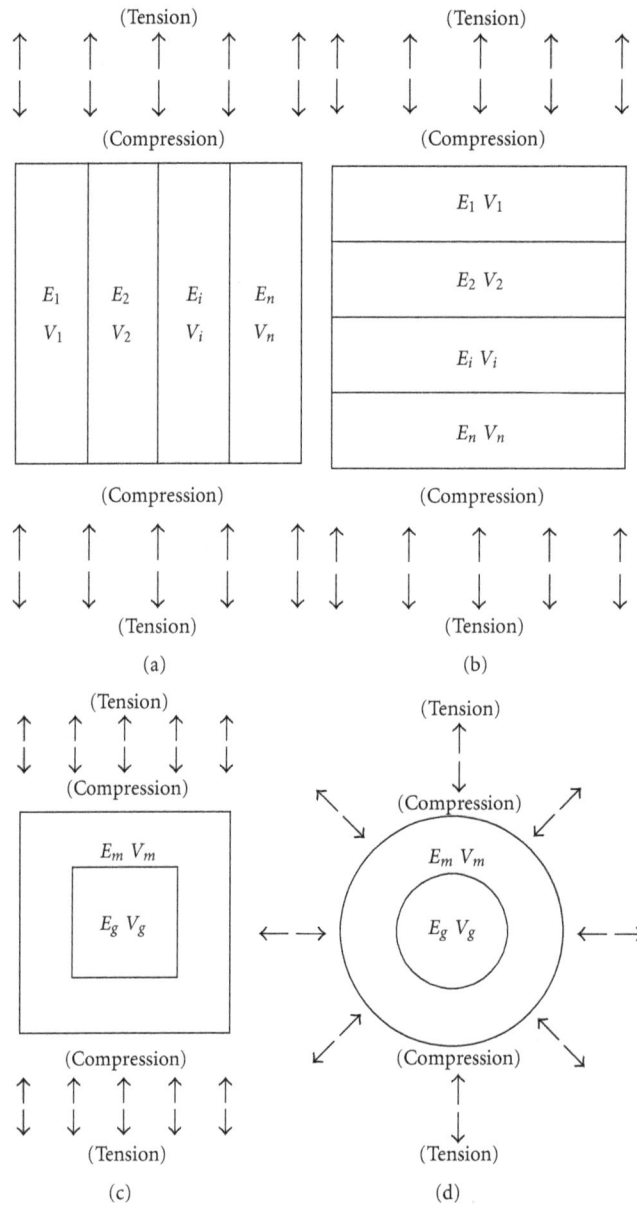

FIGURE 5: Typical composite models for predicting Young's modulus: (a) Parallel model, (b) Series model, (c) Counto model, and (d) Hashin model.

with an accuracy of 0.1 kN and capacity of 200 kN. As will be discussed later, the maximum residual strain after unloading of each specimen was measured as 3×10^{-6}. In each case, concrete and mortar specimens were made from the same batch using the sieving procedure mentioned above.

As mentioned earlier, the tensile Young's modulus of concrete is often assumed to be equal in value to the compressive modulus. In addition, the compressive modulus of concrete is frequently estimated based on empirical correlations with concrete compressive strength. The compressive modulus of concrete has also been related to the volume and mechanical properties of concrete's constituents (aggregates, paste, etc.) using some classical composite models such as those presented by Zhou et al. [14], Topçu [15], and

Yoshitake et al. [9]. A brief overview of these models is provided below.

The typical composite models for estimation of the elastic modulus are illustrated in Figure 5. These include (a) the Parallel model, (b) the Series model, (c) the Counto model [11], and (d) the Hashin model [12]; as represented by the following:

$$\text{Parallel model: } E = \sum_{i=1}^{n} E_i \cdot V_i, \tag{1}$$

$$\text{Series model: } \frac{1}{E} = \sum_{i=1}^{n} \frac{V_i}{E_i}, \tag{2}$$

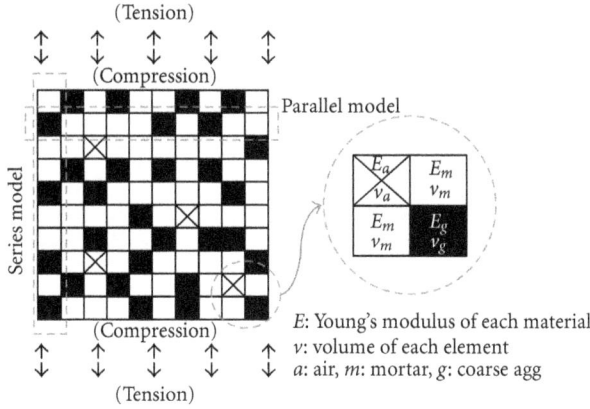

FIGURE 6: An example of a simple composite model using parallel and series models.

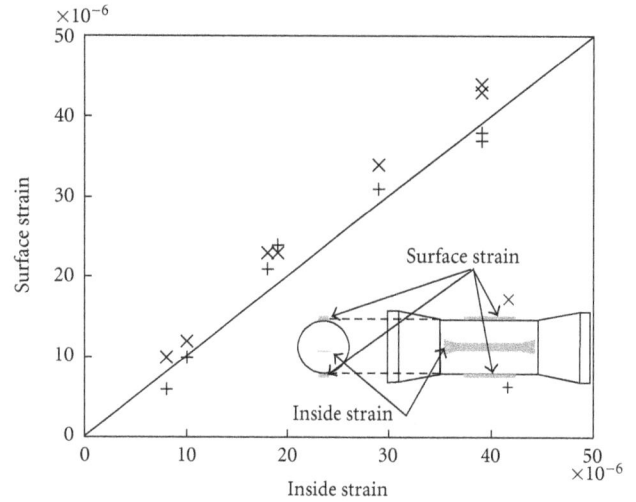

FIGURE 7: Comparison of inside strain and surface strain.

Counto model: $\dfrac{E}{E_m} = 1 + \dfrac{V_g}{\sqrt{V_g - V_g + E_m/(E_g - E_m)}}$, (3)

Hashin model: $\dfrac{E}{E_m} = \dfrac{V_m E_m + (1 + V_g) E_g}{(1 + V_g) E_m + V_m E_g}$, (4)

where E_i and V_i represent the Young's modulus and the volume fraction of concrete constituents (e.g., mortar, coarse aggregate, etc.), n is the number of constituents, and the subscripts m and g refer to mortar and coarse aggregate, respectively. The Counto and Hashin models are based on a 2-phase composite (mortar and aggregate). While these models are generally more accurate than simple parallel and series models, they may estimate the Young's modulus inappropriately for concretes containing high aggregate volumes or high air content.

In the present work a new triphase model is proposed based on random distribution of elements within a 2-dimensional 80×80 grid (Figure 6). Each element in the model is composed of mortar, coarse aggregate, or air. The number elements corresponding to each phase is proportional to the volume fraction of that phase in concrete. Elements are placed randomly in the model using a Monte Carlo procedure. To determine the tensile modulus of the grid, simple micromechanical calculations are performed based on the series and parallel models. First the tensile modulus of each row of elements is determined using the parallel model and then the modulus of the grid is determined by combining all rows using the series model. Alternatively, the modulus of each column can first be determined using the series model and then the columns are combined using the parallel model.

4. Experimental Results and Discussion

4.1. Evaluation of the Reliability of the Embedded Strain Sensor. It is important to ensure proper measurements of the tensile strain using the embedded strain sensor. For this purpose, a dog-bone concrete specimen is tested in tension

and the tensile strain is measured by both the embedded gage as well as 2 wire strain gages mounted on the surface of the concrete specimen. Figure 7 presents the results showing that the internally measured strain is practically equal to the surface measured strains. This implies that the tensile stress is applied uniformly to the specimen and the embedded sensor can be used to monitor concrete's tensile strain. Based on this conclusion, further measurements in this study employ only the embedded sensor.

4.2. Tensile Stress-Strain Responses. Figure 8 presents examples of the tensile stress-strain responses of each concrete. As shown, the slope of each stress-strain regression line (i.e., the tensile modulus) develops with increasing age and reducing the water-cement ratio (W/C) of concrete. Figure 8(a) shows the stress-strain response at 1 day; the results indicate that the residual strain after unloading of specimens is zero (i.e., plastic strain at age of 1 day when specimen are loaded to 60×10^{-6} is negligible). Note that the response of mixture O57 ($W/C = 57\%$) at 1 day could not be obtained because the concrete was too weak to allow performing the direct tension test. Figure 8(b) presents the stress-strain responses at 7 days; the maximum residual strain after loading specimens to 60×10^{-6} is 3×10^{-6} corresponding to the mixture O39. For this mix, had the loading branches of the stress-strain response been used to determine the modulus, the tensile modulus would be estimated as 31.2 GPa, comparing with 33.0 GPa obtained from using the unloading stress-strain branches. The ratio of the modulus obtained by the two methods is approximately 0.95.

4.3. Time Dependent Development of the Mechanical Properties. The results of the compressive and splitting tensile strength measurements of the three concrete mixtures are presented in Figure 9. At 7 days, the concrete mixtures have compressive strengths in the range of 20 to 45 MPa, and splitting tensile strengths in the range of 2 to 3.3 MPa. The time-dependent compressive and tensile Young's moduli are

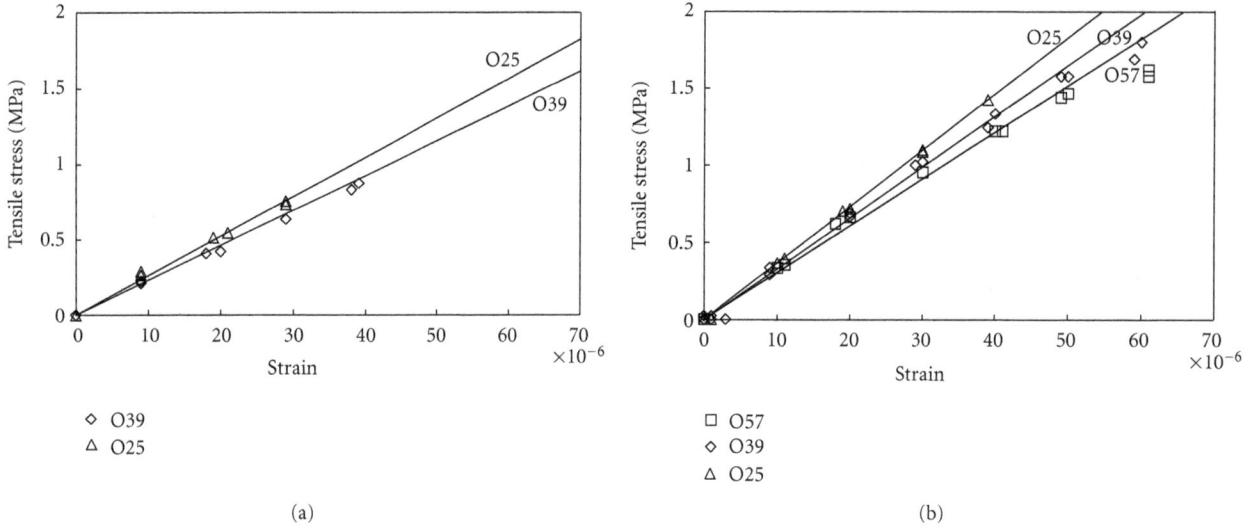

FIGURE 8: Tensile stress-strain responses: (a) age of 1 day and (b) age of 7 days.

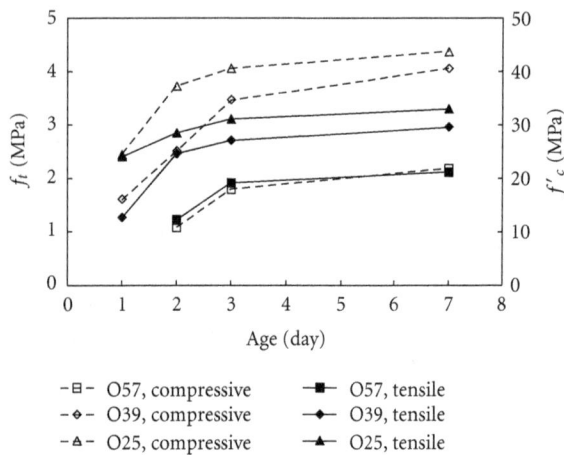

FIGURE 9: Strengths versus age of concrete.

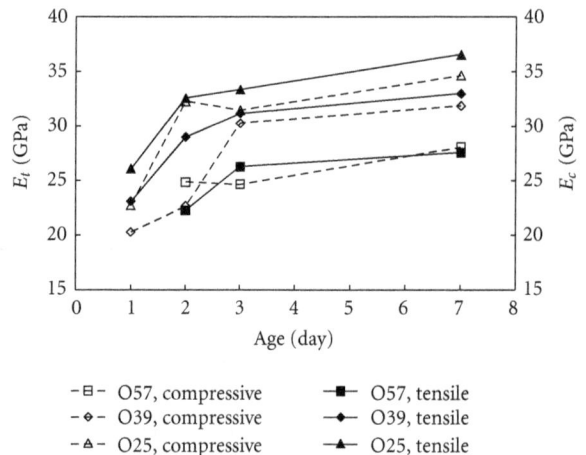

FIGURE 10: Young's moduli versus age of concrete.

presented in Figure 10. The compressive moduli in the graph demonstrate the secant moduli under 33% of the maximum stress. Based on Figure 10, the tensile Young's moduli are approximately 1.0–1.3-times larger than the compressive moduli. The result may be caused by different response to stress applied to each specimen, such that the tensile stresses were less than 10% of the compressive stresses. This difference can be especially significant at early ages when the large macropores dictate the tensile response of concrete.

A comparison between the tensile modulus of concrete and the corresponding mortar specimens are provided in Figure 11. The results indicate that the concrete tensile modulus is always higher than the mortar modulus due to the stiffness provided by the coarse aggregates in concrete. However, the difference is narrowed as concrete ages due to hydration of cement which results in an increased stiffness of the mortar.

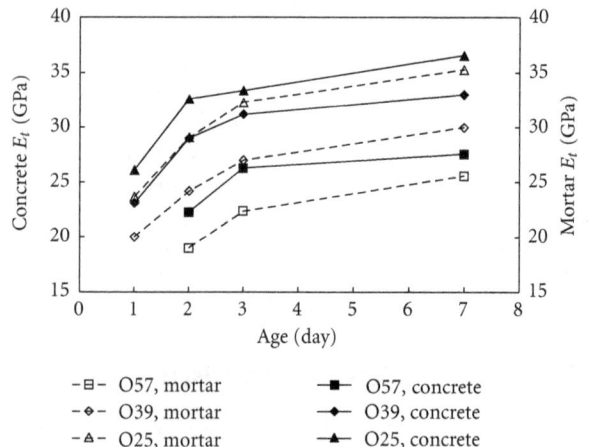

FIGURE 11: Tensile Young's modulus predicted by employing grids of different fineness.

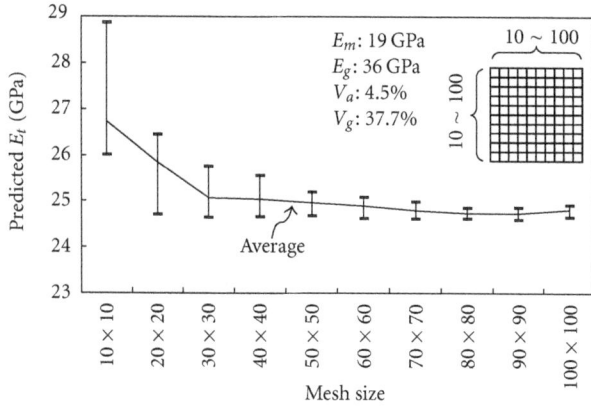

FIGURE 12: Comparison of experiment and predicted tensile Young's modulus.

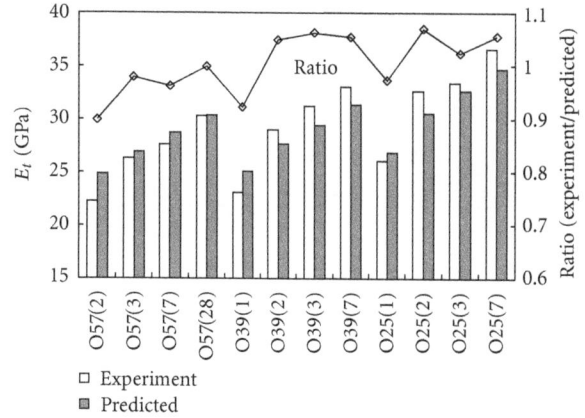

FIGURE 13: Tensile Young's modulus predicted by employing grids of different fineness.

5. Prediction of the Tensile Young's Modulus at Early Age

5.1. Input Data for Development of the Simplified Composite Model. The tensile Young's moduli of coarse aggregates and mortar are needed as input parameters for use in the composite models. While the moduli of mortar is directly measured (Figure 11), it is difficult to directly obtain the tensile modulus of the aggregate due the size and number of specimens required for tensile testing, difficulty in using a proper tensile grip method of the rock specimen, as well as the variability in the material properties due to layering and impurities of the rock. Thus, the tensile Young's modulus of the coarse aggregate is indirectly calculated by employing the Counto and Hashin models in the present study. The study estimates the tensile Young's modulus (E_g) of 36.0 GPa from the experimental result of O25 at age of 7 days, for which the modulus of mortar was almost equal to the modulus of concrete. That is, the estimated value to the modulus is little affected by the models used for obtaining the modulus of the aggregate. Similar values are obtained based on testing O25 at earlier ages or by testing the other two mixtures.

Table 4 provides the input data for the composite model. The volume fractions of each component (mortar, coarse aggregates, and air) are obtained from the mix proportions given in Table 3. To determine the appropriate number of elements in the model, the study estimates the tensile Young's moduli of concrete by employing models of various mesh sizes. Figure 12 presents the resulting tensile moduli predicted as a function of mesh fineness. The graph shows the maximum, minimum, and average moduli obtained from 10 consecutive simulations in each model. Based on these results, a 80 × 80 model was chosen for the remaining simulations in this work.

5.2. Quality of the Model Predictions. Figure 13 shows the tensile Young's moduli of all concrete specimens predicted by using the proposed composite model. Each bar presents the average of 10 simulations for each concrete mixture and at each age. The bar graph indicates that the predicted

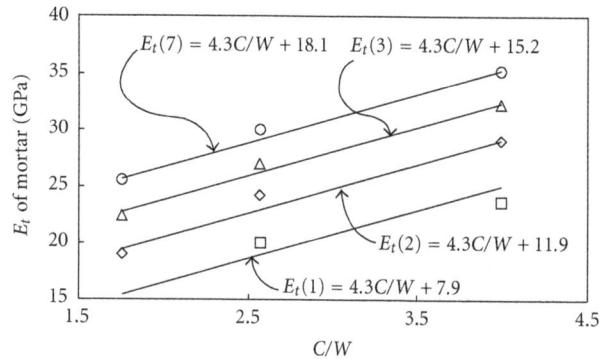

FIGURE 14: Relations between cement-water ratio and tensile Young's moduli of mortar.

Young's moduli are in good agreement with the experimental values; the ratio of the two is in range of 0.90 to 1.07. This implies that the tensile modulus of concrete can be predicted appropriately by employing the composite model when the volume fraction of constituents is known.

5.3. Empirical Formula for the Tensile Modulus of Mortars in This Study. Considering that the Young's modulus of coarse aggregate and air content of concrete are age-independent, the Young's modulus of concrete at early age may be predicted if the modulus of mortar can be estimated appropriately. For the mortar studied in this work, empirical correlations between the experimental measurement modulus, the W/C, and the age of mortars are established as presented in Figure 14. The figure shows a linear correlation between the inverse water to cement ratio (C/W) and the tensile moduli of mortar (E_t) at ages of 1, 2, 3, and 7 days. Interestingly, the slope k_i of all regression lines in the graph is approximately 4.3 GPa:

$$\text{Relation of } E_t - \frac{C}{W}: E_t = k_i \cdot \frac{C}{W} + C_i, \tag{5}$$

where E_t (GPa) is the tensile modulus of mortar, and C_i (GPa) is an age-dependent parameter in each regression line.

TABLE 4: Input data for the composite model.

Mix. ID	V_m	V_g	V_a	E_t of mortar (GPa) shown in Figure 11				E_g (GPa)
				1 day	2 days	3 days	7 days	
O57	57.7%	37.7%	4.5%	N/A	19.0	22.4	25.6	36.0
O39	62.0%	34.2%	3.8%	20.0	24.2	27.0	30.0	36.0
O25	66.4%	30.0%	3.6%	23.7	29.1	32.3	35.3	36.0

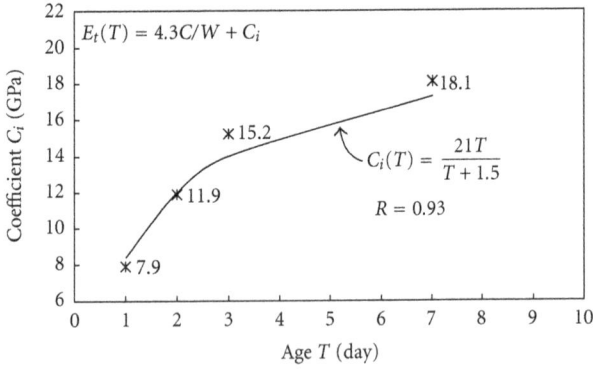

FIGURE 15: Coefficient C_i of the regression line versus age of mortar.

The change in C_i with age of mortar is shown in Figure 15. A Goral curve [16], which is often used for estimating concrete strength development with age, is fit to the data points resulting in the following:

$$\text{Coefficient } C_i \text{ with age: } C_i = \frac{21T}{T + 1.5}, \quad (6)$$

where T is mortar age in days. Combining (5) and (6) results in

$$\text{Tensile Young's modulus of mortar:}$$
$$E_t(T) = 4.3\frac{C}{W} + \frac{21T}{T + 1.5}, \quad (7)$$

where E_t is the estimated tensile Young's modulus of mortar (GPa).

It must be noted that similar to modulus of concrete, the tensile modulus of mortar is a function of the volume fraction and stiffness of sand, volume fraction and modulus of cement paste (itself a function of age and W/C), and the air content of the mortar. By accounting for age and W/C, (7) can provide an estimate for the tensile modulus of mortars with similar volume fraction and stiffness of the fine aggregates. Young's modulus of coarse aggregate can be considered as an influencing factor to the tensile modulus of concrete, so this equation including the effect of fine aggregate may be useful for normal concrete using sea sand when an appropriate value of the modulus for coarse aggregate is provided. To be applicable to mortars other than those used here, the most significant remaining parameter is the volume fraction of sand which must be taken into consideration.

5.4. *Prediction of the Tensile Modulus of Concrete by the Composite Model.* By combining (7) with the composite

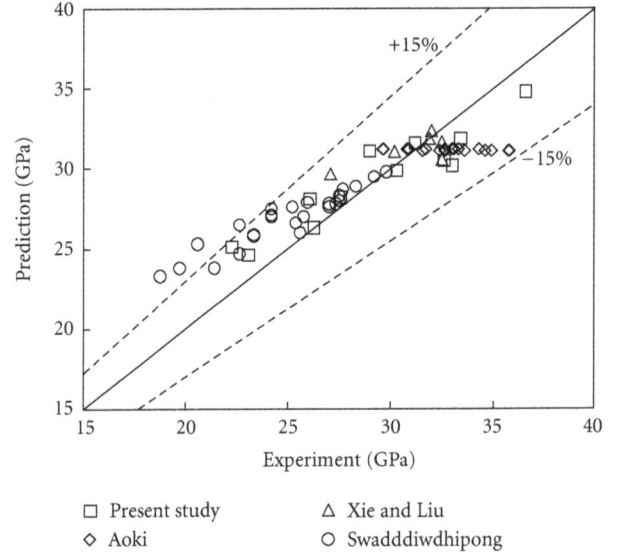

□ Present study   △ Xie and Liu
◇ Aoki   ○ Swadddiwdhipong

FIGURE 16: Comparison of experiments and predictions.

model (Figure 6), the age-dependent tensile moduli of concrete can be estimated from its mixture proportions (i.e., volume fraction of constituents) and the aggregate modulus. Figure 16 presents a comparison between the model predictions and the experimental data from this study as well as those from previous investigations [2, 4, 7]. Herein, these predicted data were obtained with the assumption that the aggregate properties in other studies are equal to the values employed in this study, because the modulus of coarse aggregate used is not reported in the previous investigations. The figure shows that the proposed method can predict the tensile modulus of concrete, with reasonable accuracy, solely based on the mixture proportions.

Figure 17(a) describes characteristics of tensile moduli of concrete with $W/C = 55\%$ and coarse aggregates of different volume fractions. The vertical axis in the graph presents the tensile Young's moduli ratio of concrete to coarse aggregate. Herein, the model results are obtained from the assumptions shown in Table 5. The figure demonstrates that the tensile modulus of concrete increases with increasing the volume of coarse aggregates since the aggregates have a higher Young's modulus than the mortars. It is also noted that tensile moduli of concrete having more coarse aggregates (i.e., larger vol. fractions) than the solid volume content of coarse aggregates [17] are unavailable. Figure 17(b) presents a similar set of curves corresponding to concrete with $W/C = 30\%$. Both figures describe that the tensile modulus of concrete develops rapidly at early age and gradually plateaus at later ages.

TABLE 5: Conditions for simulation of tensile Young's modulus.

Composite model	See Figure 6
Elements	80×80
Air content	4.5%
Tensile Young's modulus of mortar	see (7)
Tensile Young's modulus of coarse aggregate	36 GPa
Solid volume content of coarse aggregate	60%
Tensile Young's modulus of concrete	Average of 10 simulations

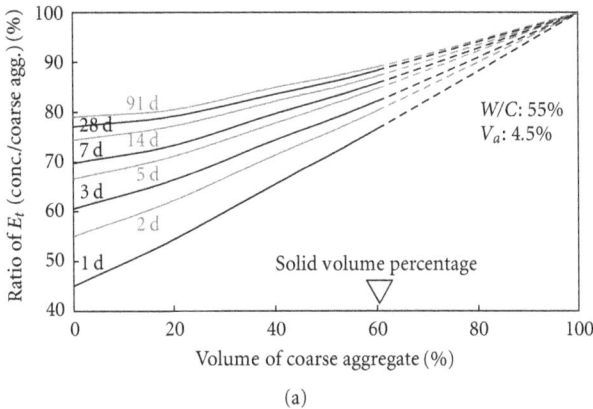

(a)

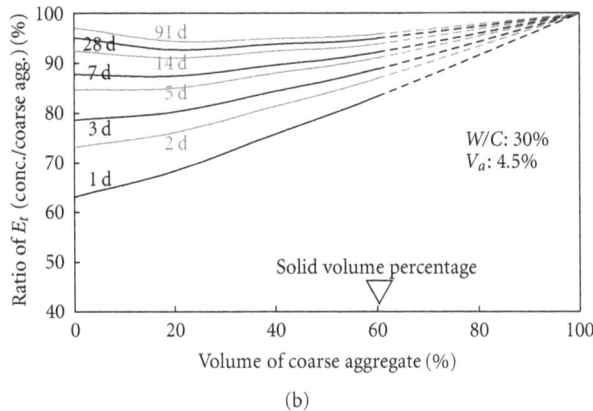

(b)

FIGURE 17: Prediction results for tensile Young's modulus of concrete: (a) $W/C = 55\%$ and (b) $W/C = 30\%$.

6. Conclusions

This paper describes the experimental measurement of tensile Young's modulus of concrete at early age using a direct tension setup. Moreover, a predictive composite model was developed to estimate the age-dependent tensile modulus of concrete using the volume fractions and properties of the constituents. The main conclusions are summarized as follows.

(1) The tensile stress-strain response of concrete was observed to be very linear even at early ages (e.g., 1 day old). The residual strains after repeated loading of specimens up to strains of 60×10^{-6} were negligible. The tensile Young's modulus obtained from

the stress-strain response develops according to the water-cement ratio and the age of concrete.

(2) The tensile modulus of concrete is approximately 1.0–1.3-times larger than its compressive modulus within the material's first week of age. As such, estimation of the tensile modulus based on empirical correlations with the compressive strength of concrete can be inaccurate.

(3) The age-dependent tensile moduli of concrete could be predicted appropriately by the proposed composite model as long as the volume fractions of coarse aggregates, mortar, and air, and the modulus of aggregates are known. In comparison with the experimental results, the model predictions showed accuracy better than ±15%.

References

[1] A. M. Neville, *Properties of Concrete: Fourth and Final Edition*, Pearson/Prentice Hall, 1995.

[2] Y. Aoki, K. Shimano, D. Iijima, and Y. Hirano, "Examination on simple uniaxial tensile test of concrete," *Proceedings of the Japan Concrete Institute*, vol. 29, no. 1, pp. 531–536, 2007 (Japanese).

[3] Architectural Institute of Japan, *Recommendations of Practice of Crack Control in Reinforced Concrete Buildings-Design and Construction*, Architectural Institute of Japan, 2006.

[4] N. Xie and W. Liu, "Determining tensile properties of mass concrete by direct tensile test," *ACI Materials Journal*, vol. 86, no. 3, pp. 214–219, 1989.

[5] F. A. Oluokun, E. G. Burdette, and J. H. Deatherage, "Elastic modulus, Poisson's ratio, and compressive strength relationships at early ages," *ACI Materials Journal*, vol. 88, no. 1, pp. 3–10, 1991.

[6] S. Hagihara, S. Nakamura, Y. Masuda, and M. Kono, "Experimental study on mechanical properties and creep behavior of high-strength concrete in early age," *Concrete Research and Technology*, vol. 11, no. 1, pp. 39–50, 2000 (Japanese).

[7] S. Swaddiwudhipong, H. R. Lu, and T. H. Wee, "Direct tension test and tensile strain capacity of concrete at early age," *Cement and Concrete Research*, vol. 33, no. 12, pp. 2077–2084, 2003.

[8] B. Bissonnette, M. Pigeon, and A. M. Vaysburd, "Tensile creep of concrete: study of its sensitivity to basic parameters," *ACI Materials Journal*, vol. 104, no. 4, pp. 360–368, 2007.

[9] I. Yoshitake, Y. Ishikawa, H. Kawano, and Y. Mimura, "On the Tensile Young's Moduli of early age concrete," *Journal of Materials, Concrete Structures and Pavements*, vol. 63, no. 4, pp. 677–688, 2007 (Japanese).

[10] T. J. Hirsh, "Modulus of elasticity of concrete affected by elastic moduli of cement paste matrix and aggregate," *ACI Journal*, vol. 59, no. 3, pp. 427–452, 1962.

[11] U. J. Counto, "The effect of the elastic modulus of the aggregate on the elastic modulus, creep and creep recovery of concrete," *Magazine of Concrete Research*, vol. 16, pp. 129–138, 1964.

[12] Z. Hashin, "The elastic moduli of heterogeneous materials," *Journal of Applied Mechanics, ASME*, vol. 29, no. 1, pp. 143–150, 1962.

[13] H. Mihashi and J. P. D. B. Leite, "State-of-the-art report on control of cracking in early age concrete," *Journal of Advanced Concrete Technology*, vol. 2, no. 2, pp. 141–154, 2004.

[14] F. P. Zhou, F. D. Lydon, and B. I. G. Barr, "Effect of coarse aggregate on elastic modulus and compressive strength of high performance concrete," *Cement and Concrete Research*, vol. 25, no. 1, pp. 177–186, 1995.

[15] I. B. Topçu, "Alternative estimation of the modulus of elasticity for dam concrete," *Cement and Concrete Research*, vol. 35, no. 11, pp. 2199–2202, 2005.

[16] L. M. Goral, "Empirical time-strength relations of concrete," *ACI Journal*, vol. 53, pp. 215–224, 1956.

[17] JIS A 1104, "Methods of Test for Bulk Density of Aggregates and Solid Content in Aggregates," Japan Industrial Standards Committee, 2006.

6

Dual Mode Sensing with Low-Profile Piezoelectric Thin Wafer Sensors for Steel Bridge Crack Detection and Diagnosis

Lingyu Yu,[1] Sepandarmaz Momeni,[2] Valery Godinez,[2] Victor Giurgiutiu,[1] Paul Ziehl,[3] and Jianguo Yu[3]

[1] Department of Mechanical Engineering, University of South Carolina, Columbia, SC 29208, USA
[2] Mistras Group Inc., 195 Clarksville Road, Princeton Junction, NJ 08550, USA
[3] Department of Civil Engineering, University of South Carolina, Columbia, SC 29208, USA

Correspondence should be addressed to Lingyu Yu, yu3@cec.sc.edu

Academic Editor: Piervincenzo Rizzo

Monitoring of fatigue cracking in steel bridges is of high interest to many bridge owners and agencies. Due to the variety of deterioration sources and locations of bridge defects, there is currently no single method that can detect and address the potential sources globally. In this paper, we presented a dual mode sensing methodology integrating acoustic emission and ultrasonic wave inspection based on the use of low-profile piezoelectric wafer active sensors (PWAS). After introducing the research background and piezoelectric sensing principles, PWAS crack detection in passive acoustic emission mode is first presented. Their acoustic emission detection capability has been validated through both static and compact tension fatigue tests. With the use of coaxial cable wiring, PWAS AE signal quality has been improved. The active ultrasonic inspection is conducted by the damage index and wave imaging approach. The results in the paper show that such an integration of passive acoustic emission detection with active ultrasonic sensing is a technological leap forward from the current practice of periodic and subjective visual inspection and bridge management based primarily on history of past performance.

1. Introduction

According to the Federal Highway Administration (FHWA) National Bridge Inventory (NBI) of 2007, the number of structurally deficient and functionally obsolete bridges is 72,524 and 79,792, respectively [1]. While there are about 10,000 bridges being constructed, replaced, or rehabilitated annually in the United States at a cost of over $5 billion, the total annual costs including maintenance and routine operation are significantly higher [1]. As the inventory continues to age, routine inspection practices will not be sufficient for the timely identification of areas of concern and to provide enough information to bridge owners to make informed decisions for safety and maintenance prioritization. Continuous monitoring is therefore desirable for long-term evaluation; monitoring areas of concern, such as retrofits or previous repairs or monitoring an area with known flaws, before a scheduled inspection. Continuous monitoring can also be used in cases where there is a concern about vandalism,

terrorism, and/or bridge element integrity. Therefore, monitoring of fatigue cracking in steel bridges is of interest to many bridge owners and agencies.

To address this urgent need, the authors are conducting research on novel and promising sensing approaches together with energy harvesting devices to reduce the dramatic uncertainty inherent into steel bridge inspection and maintenance plan [2]. One of the challenges in this research is focused on the development of dual use piezoelectric wafer active sensors for fatigue crack detection. The combined schematic uses acoustic emission to detect the presence of fatigue cracks in their early stage while active sensing allows for the imaging and quantification of cracks using minimum number of sensors.

1.1. Crack Detection on Steel Bridge Structures. The monitoring of fatigue cracking in bridges has been approached with acoustic emission using either resonant or broadband

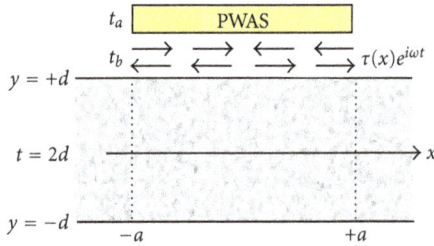

FIGURE 1: PWAS and structure interaction through the interface layer.

sensors. Acoustic emission monitoring is capable of detecting crack growth behavior [3–7] and assessing the integrity of structures such as bridges and aircraft [8, 9]. Acoustic emission data has recently been directly related to crack growth rate in representative steel compact tension (CT) specimens. This development holds promise for the prognosis of in-service bridges [10]. The method has the notable advantage that the precise location of cracking does not need to be known for evaluation purposes. Rather, the sensors together with appropriate algorithms are capable of locating and quantifying active crack activity. It has been reported that acoustic emission techniques are so sensitive that fatigue cracks can be detected successfully even though the crack length may be less than $10\,\mu$m [5, 7]. However, one of the challenges in passive monitoring is that the acoustic emission relies on an active crack growing process. Also, though acoustic emission sensing can detect a crack at its very early stage, it generally cannot provide information about the crack size or crack growth rate unless an initial size is provided.

Historically, AE signals have been captured with specially designed and fabricated AE sensors. Conventional AE sensors are made of piezoelectric crystals as the sensing elements which are encapsulated for protection and coupled together with a wear plate for good acoustic coupling. The frequency content and sensitivity of the sensor are controlled by the geometry and properties of the piezoelectric crystal as well as the housing for the crystal.

1.2. Piezoelectric Wafer Active Sensors.
Piezoelectric wafer active sensors (PWAS) can function as an active sensing or passive device or network using piezoelectric principles and provide a correlation between mechanical and electrical variables. They can be permanently attached to the structure to monitor condition at will and can operate in active guided wave interrogation or passive AE sensing modes. The transmission of actuation and sensing between the PWAS and the host structure is achieved through the bonding adhesive layer. The adhesive layer (Figure 1) acts as a shear layer, in which the mechanical effects are transmitted through shear effects [11].

An important characteristic of PWAS, which distinguishes them from conventional ultrasonic transducers, is their capability of exciting multiple guided Lamb wave modes at a single frequency. There are at least two Lamb modes, A0 and S0, existing simultaneously when the product

of the wave frequency and structure thickness ($f \cdot d$ product) falls in the range of 0~1 MHz-mm. At larger $f \cdot d$ product values, more modes are present. In addition, due to the intrinsic dispersion property, the Lamb wave modes propagate at different speeds and the speeds change with frequency, which complicates the signal interpretation for damage detection. A single mode that is sensitive to the damage is desired for most of the SHM applications. This may be attained through wave tuning [12]. The process of wave tuning attempts to modify the excitation parameters to excite a certain mode for detection of a specified type or instance of damage [12]. By carefully selecting PWAS length at either an odd or even multiple of the half wavelength, a complex pattern of strain maxima and minima emerges (Figure 2). Since several Lamb modes, each with its own different wavelength, coexist at the same time, a selected Lamb mode can be tuned by choosing the appropriate frequency and PWAS dimensions.

An example of PWAS tuning is presented in Figure 2 for a 7-mm square PWAS installed on a 1-mm aluminum alloy 2024-T3 plate. The experimental amplitude plot in Figure 2(a) shows that for the plate being studied, a S0 tuning frequency around 200 kHz can be identified, where the amplitude of the A0 mode is minimized while that of the S0 mode is still strong. Therefore, by choosing the excitation frequency, a single mode can be obtained for damage detection [13]. Theoretical prediction given in Figure 2(b) is consistent with the experimental results.

1.3. PWAS Dual Mode Sensing toward Field Application.
Dupont et al. [14] have demonstrated the possibility of using embedded piezoelectric thin wafers to detect AE signals in composite materials. In the subject project, we are developing a dual mode sensing approach using the low-profile wireless PWAS network with energy harvested from wind and/or ambient vibration energy [15]. To minimize the energy consumption, on one hand, it is envisioned that as few as four PWAS will be employed to monitor the crack growth. On the other hand, the dual mode sensing allows PWAS to operate in passive AE mode throughout the entire monitoring process unless significant AE events are detected, indicating major crack presence in the structure. When an AE event is identified, the PWAS can be switched to active mode to interrogate the bridge structure with propagating guided waves to assess the crack size and location. Eventually, an entire steel bridge may be mapped with the guided wave interrogation with visualized crack damage.

2. PWAS Passive Mode Acoustic Emission Detection

In the subject project, we adapted PWAS as AE sensors to detect stress waves with frequency components concentrated at 150 kHz where the acoustic signals propagating with minimal attenuation and background noise due to the rubbing of structural components.

2.1. PWAS as AE Sensor.
Laboratory tests have been conducted to investigate the PWAS application as an AE sensor.

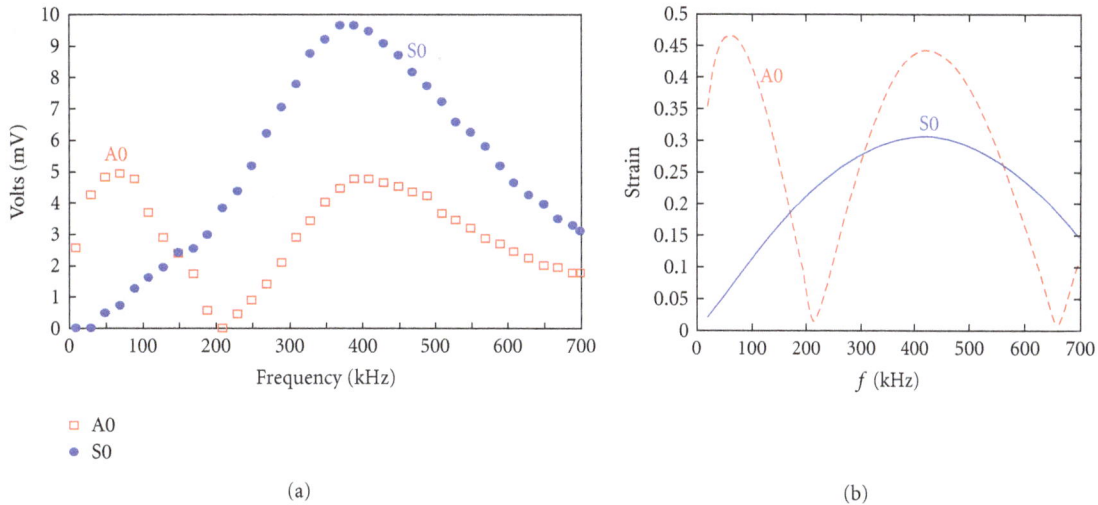

FIGURE 2: Lamb wave mode tuning on a 1-mm thick aluminum alloy 2024-T3 using 7-mm PWAS. (a) Experimental wave amplitude within 0~700 kHz; (b) predicted strain curves [13].

A typical commercial R15I (http://www.pacndt.com/downloads/Sensors/Integral%20Preamp/R15I-AST.pdf) AE sensor was used to calibrate the measurements. Two specimens were used in the tests. One was a 1.6 mm thick 2024 aluminum plate and the other was a 19 mm thick A572 grade 50 structural steel panel. Both specimens were installed with 7-mm diameter 0.2 mm thick round PWAS Material APC 850 by Americanpiezo (http://www.americanpiezo.com/apc-materials/choosing.html) using M-200 bonding adhesive following the standard installation used by strain gauge. The R15I sensor was mounted using hot glue.

The test setup is illustrated in Figure 3. PAC DiSP (http://www.pacndt.com/index.aspx?go=products&focus=/multichannel//disp.htm) system was used to perform data acquisition. PWAS was connected to a preamp then to channel 1. The preamp had a 100–1200 kHz built-in filter and could accommodate a signal amplification of 40 dB. The R15I sensor had a built-in internal preamp with a gain also at 40 dB and was connected to channel 2. AE events were introduced by pencil lead break (PLB). Rubber gloves were used to avoid causing electrical disturbance by touching the plate.

2.1.1. Aluminum Plate Tests. In the first part of the work, a 1.6-mm thick aluminum plate, approximately 300-mm by 300-mm, was used for testing PWAS AE detection. PWAS and R15I were placed adjacently on the plate, about 165 mm away from the plate edge where the PLB was applied. In total, five PLB of various lead sizes were applied. The PWAS transducer detected all of them with comparable amplitudes to those captured by R15I, as summarized in Table 1.

Figure 4 shows the waveforms and the frequency spectra for the test described in the third row in Table 1. In this test, 0.5 mm HB pencil lead was broken at the edge of the plate in the out-of-plane direction (pressing down and springing up). The detected AE signal amplitudes of the two sensors were about the same (78 dB). The PWAS gave a peak signal of

TABLE 1: AE detection on the 1.6-mm aluminum plate.

PLB size	PWAS AE amplitude (dB)	R15I AE amplitude (dB)
0.7 mm	83	82
0.5 mm	89	89
0.5 mm	78	78
0.3 mm	88	86
0.3 mm	84	78

2200 mV against a noise of 200 mV, resulting in a signal-to-noise ratio (SNR) of approximately 11 or 21 in dB. Compared to R15I waveform, the response of PWAS was conspicuously crisper in the earlier part corresponding to the S0 mode arrival which was followed by the arrival of slower A0 mode.

Looking at the AE waveforms by PWAS and R15I present in Figure 4, it is noticed though the two waveforms have comparable signal peak amplitudes, PWAS shows higher floor noise (circled part). For AE detection, it is needed that the sensor shows good signal-to-noise ratio. In this case, floor noise for PWAS is too high and needs to be decreased.

An expanded view of the early PWAS response is shown in Figure 5(a). The fast S0 mode and slow but highly dispersive A0 mode are clearly present. The out-of-plane displacement waveform generated by PlotRLQ (PlotRLQ is a computer program that computes theoretical waveforms by summing the contributions of Lamb wave modes excited by specified moment tensor sources at specified depths within the plate based on classic Lamb wave theory) calculated for this situation is shown in Figure 5(b). The agreement between PWAS waveform and theoretical prediction is discernable, especially for the S0 mode, the earliest part of the response. The small discrepancy comes from the ragged appearance at the beginning of the A0 mode and from the relatively abrupt falloff after the A0 reaches its peak amplitude. This may be credited to the PWAS sensor aperture effect as described in [12].

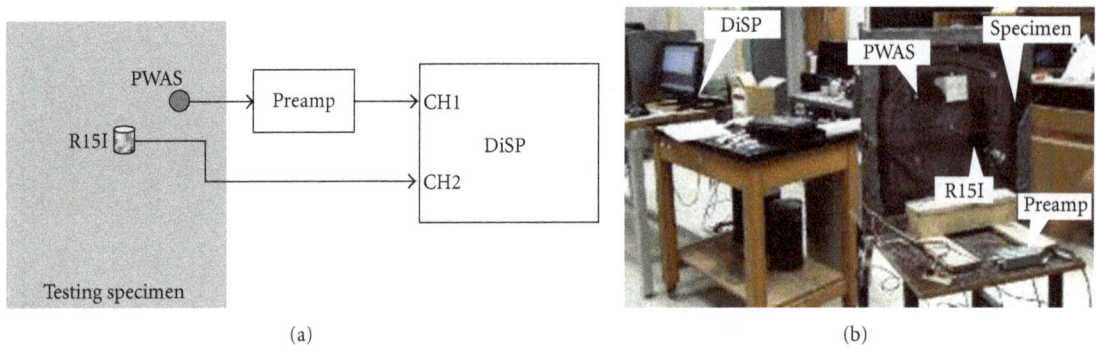

(a)   (b)

FIGURE 3: AE testing setup. (a) Test setup schematic; (b) laboratory test setup.

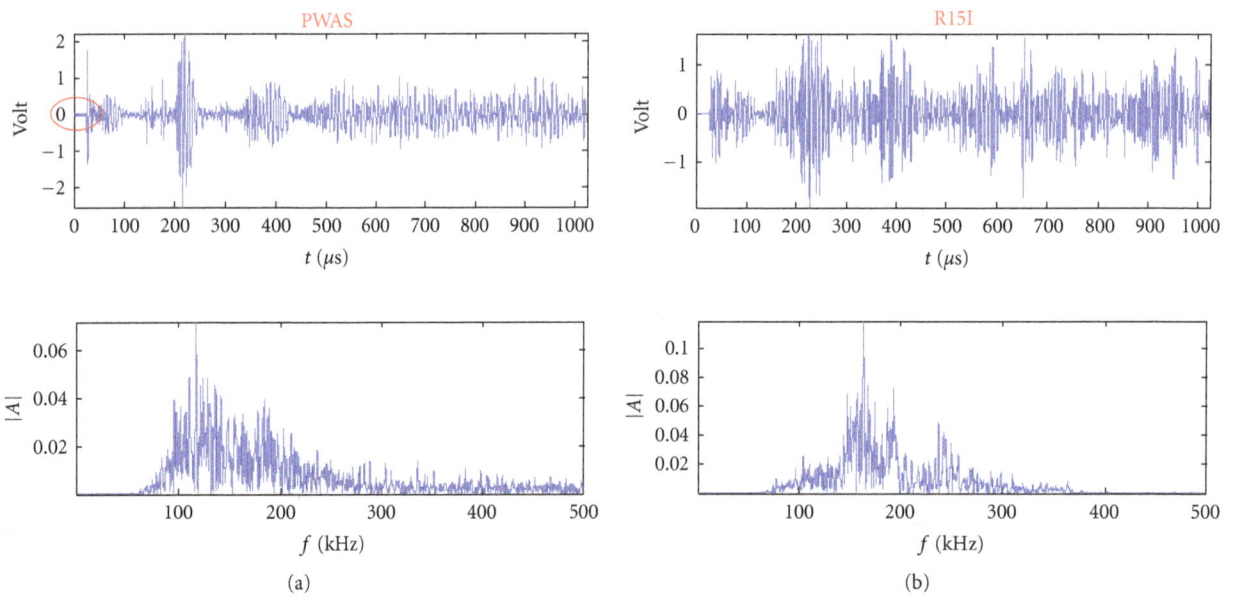

(a)   (b)

FIGURE 4: 0.5 mm PLB detection on 1.6-mm aluminum plate. (a) PWAS AE waveform and its frequency spectrum; (a) R15I AE waveform and its frequency spectrum.

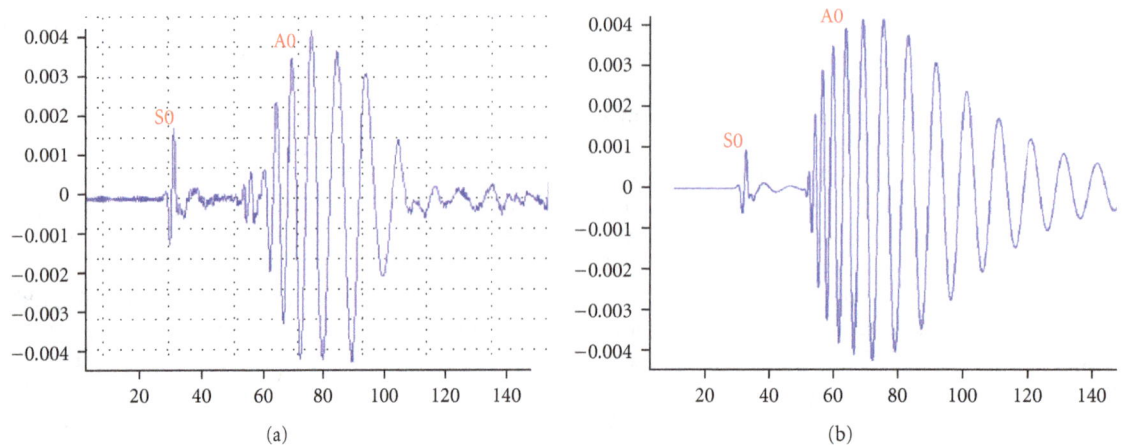

(a)   (b)

FIGURE 5: PWAS response compared to theoretical prediction. (a) PWAS waveform; (b) theoretical out-of-plane displacement by PlotRLQ.

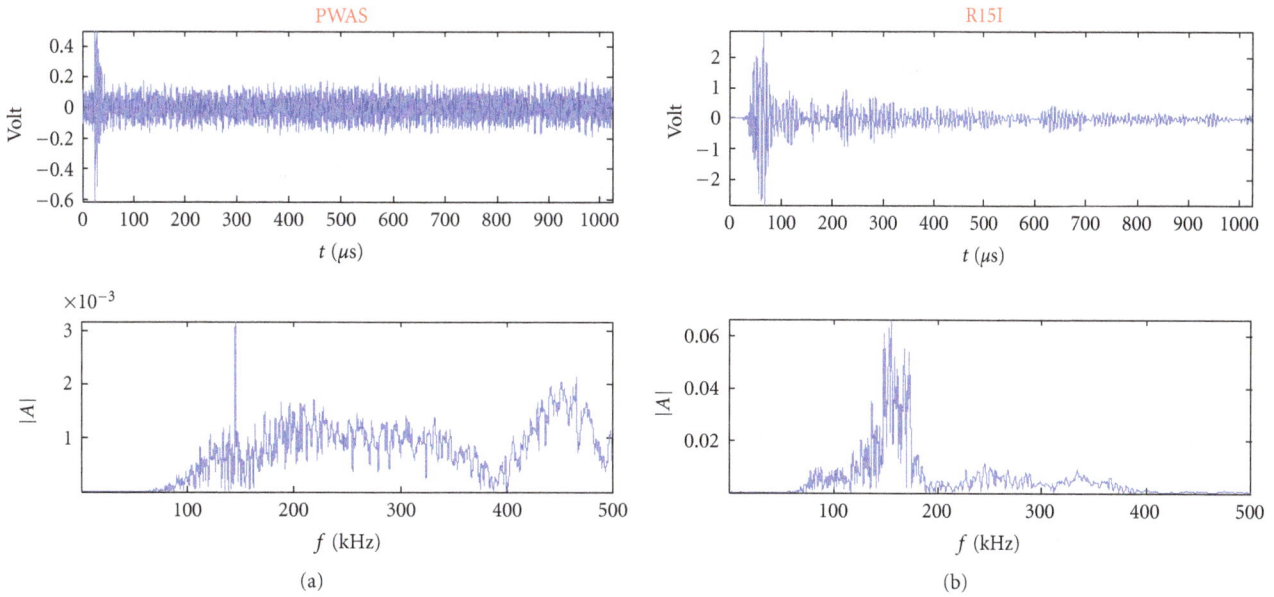

FIGURE 6: 0.5 mm PLB detection on 19-mm steel plate. (a) PWAS AE waveform and its frequency spectrum; (a) R15I AE waveform and its frequency spectrum.

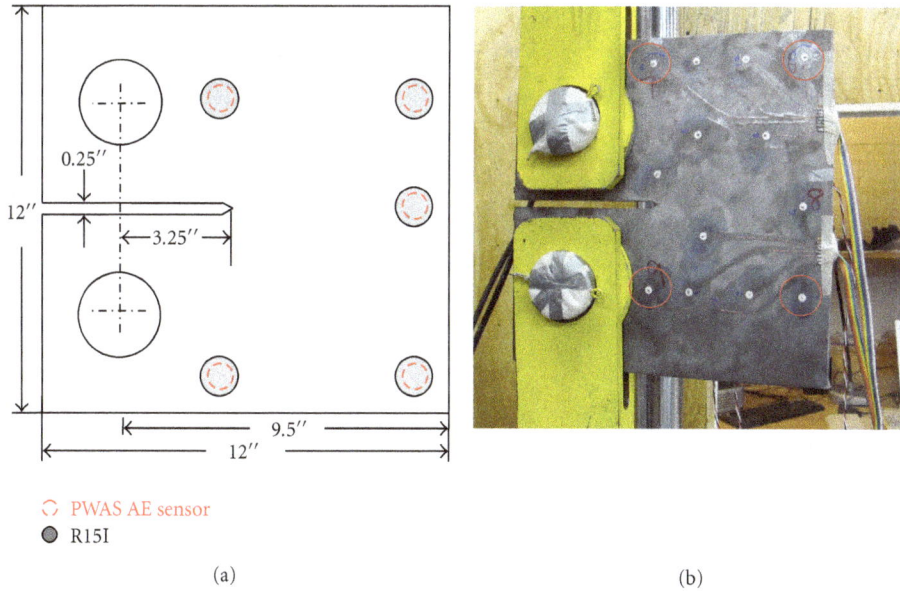

FIGURE 7: AE detection on a 1/2″ CT specimen. (a) Geometry of the specimen and arrangement of transducers; (b) a snapshot of the actual specimen. AE PWAS are circled (rest are for active sensing). The R15I were installed on the other side of the specimen.

2.1.2. Steel Plate Tests. The second part of the work was conducted on a 19-mm thick steel plate. 0.5 mm HB PLB was applied on the surface of the plate about 72 mm away from PWAS. The R15I transducer was glued on the plate with a distance of 98 mm from the PLB.

The PLB was detected by PWAS with an AE amplitude of 73 dB in contrast to the 87 dB detected by R15I. The waveforms and their frequency spectra are provided in Figure 6. The PWAS signal showed a strong negative-going spike accounting for the 73 dB amplitude, fitting in a 600 mV peak. The background noise was about 150 mV, resulting in a

SNR of approximately 4 or about 12 in dB. The SNR of PWAS in steel plate was much lower than that in aluminum plate.

By examining the frequency spectra, it can be noted that PWAS has major frequency components beyond 200 kHz, showing a wider frequency response compared to resonant type R15I AE sensor.

2.2. PWAS AE Detection in Compact Tension Testing. Compact tension (CT) specimens made of the same material as the steel plate used in Section 2.1 were used. The geometry of the specimens is displayed in Figure 7. Custom fixtures

(a)

(b)

(c)

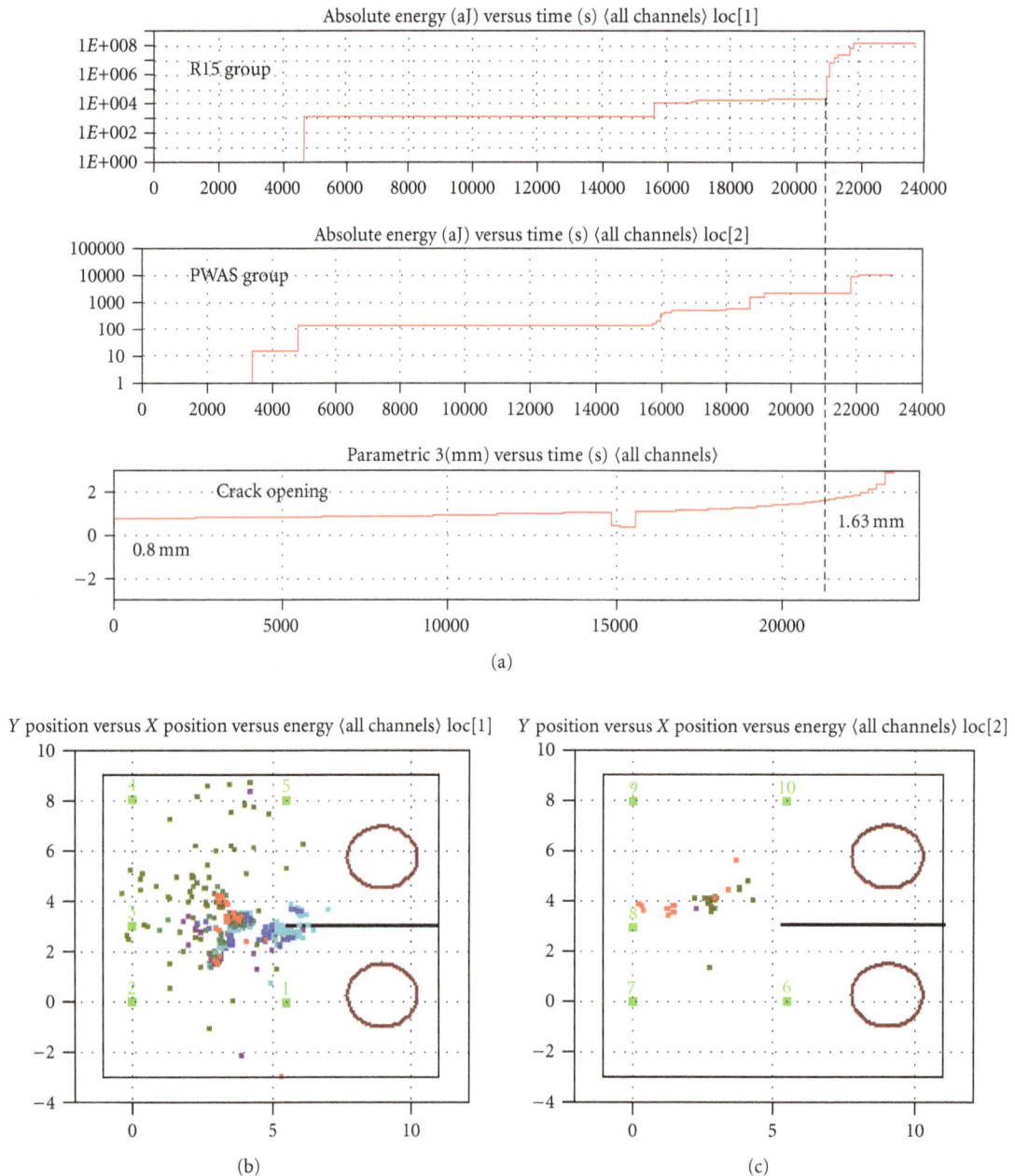

FIGURE 8: Comparison of crack localization in CT test on 1/2″ steel specimen. (a) Cumulative acoustic energy by PWAS and R15I; (b) cracking detection and localization by R15I; (c) cracking detection and localization by PWAS.

were designed and fabricated to mount the CT specimens. The cyclic tension loads of minimum 1 KN and maximum 50 KN were applied to the specimen using an MTS810 servohydraulic mechanical testing machine. Fatigue tests were conducted under load-controlled mode with frequency of 1 HZ. A clip gage was employed to measure the crack mouth opening displacement (CMOD) to clarify crack opening and closure and to determine the magnitude of the CMOD. The surface cracks were also monitored optically with a high-resolution recording microscope. Two separate sets of AE sensors, namely, R15I and PWAS monitored the process, as well. They connected to the Sensor Highway II

(http://www.pacndt.com/products/Remote%20Monitoring/ AE_Sensor_Highway.pdf) data acquisition system through preamps. The data from these two sets of sensors were analyzed separately with AEwin (http://www.pacndt.com/index .aspx?go=products&focus=/software/aewin.htm) software.

The results of crack localization from PWAS sensors and R15I during CT testing are shown in Figure 8. R15I sensors detected 1,171 AE events prior to failure while PWAS detected only 54 events. Figure 8(a) gives the cumulative acoustic energy of R15I and PWAS together with the crack opening displacement. While PWAS detected a fewer number of acoustic activities, they detected the crack growth when

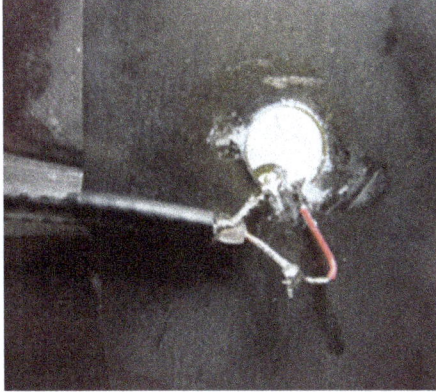

FIGURE 9: PWAS installation using a coaxial cable.

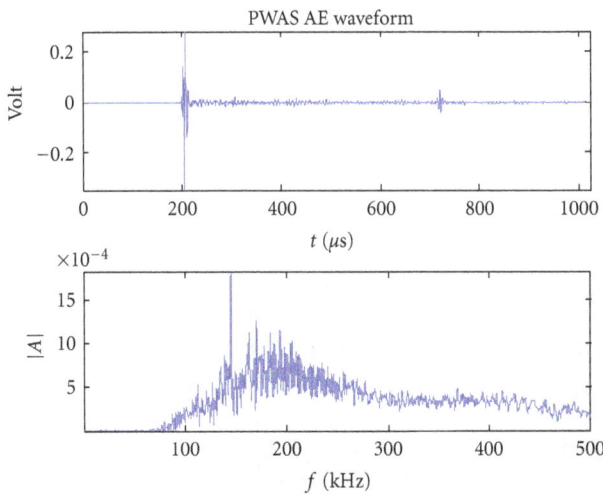

FIGURE 10: PLB detection on steel plate using coaxial cable wired PWAS.

the crack size reaches 0.83 mm. As can be seen in Figure 8(b), PWAS localization was closer and concentrated around the crack tip compared with the R15I detection in Figure 8(c). This was thought to be mainly due to the enhanced sensitivity of the R15I sensors, which can complicate source location in small-scale laboratory specimens due to reflections.

2.3. PWAS Adaptation toward Field Application. In the validation tests, it has been noticed that PWAS on steel specimens exhibited higher floor noise compared to the standard R15I AE sensors, therefore, providing a poor SNR ratio. To enhance the signal quality toward field application and decrease the background floor noise, we improved PWAS installation by using a coaxial cable similar to that used for R15I sensors. The shield of the coax was connected to the steel plate very close to the PWAS while the center conductor was connected to the positive electrode of the PWAS, as shown in Figure 9.

Detection of a 0.3 mm PLB about 20 mm away from a coaxial cable wired PWAS was evaluated (Figure 10). The rest

of the setup remained the same as for the tests presented in Section 2.1.2. The resulting waveform had an amplitude of 71 dB, approximately 400 mV. The background floor noise was discernibly decreased and approximated at 10 mV. Therefore, SNR was measured at 40 or 32 dB, significantly improved compared to the 12 dB presented in Section 2.1.2.

3. PWAS Active Mode Crack Sensing

In the dual mode sensing schematic, after significant cracking has been identified by passive mode AE detection, active mode sensing using pitch-catch interrogation is evoked to quantify crack growth through damage index and array imaging. A PWAS network consisting of several sensors spatially distributed on the plate can be used to interrogate the plate with one sensor generating the guided wave and the others receiving the structural response. When an elastic Lamb wave is transmitted and travels through the structure, wave scattering occurs in all directions where there is a change in the material properties due to damage. The scatter signal is defined as the difference between the measurement during the development of damage and the baseline signal at the initial stage. One advantage of using scatter signals is to minimize the influence caused by boundaries or other structural features which would otherwise complicate the Lamb wave analysis.

3.1. Damage Index Evaluation. We assume that cracking is the sole source of changes in the detected Lamb waves. Also being assumed is that the waves travel in straight paths in the plate structures. Hence, the objective of our Lamb wave signal analysis is to extract damage-related characteristics from the measured sensory data. In this research damage index (DI) is defined as [16]

$$DI = 1 - \frac{C_{XY}}{\sigma_X \sigma_Y}. \tag{1}$$

C_{XY} is the covariance of X and Y given by

$$C_{XY} = \sum_{j=1}^{N} \left(X_j - \mu_X \right) \left(Y_j - \mu_Y \right). \tag{2}$$

where μ is the mean value and N is the length of the dataset. σ_X and σ_Y are the standard deviations of X and Y, respectively, with their product given as

$$\sigma_X \sigma_Y = \sqrt{\sum_{j=1}^{N} \left(X_j - \mu_X \right)^2 \left(Y_j - \mu_Y \right)^2}. \tag{3}$$

For the active sensing implemented during the CT test presented in Section 2.2, four PWAS were used to perform pitch-catch wave propagation interrogation, as marked and numbered in Figure 11(a). Using the definition of damage index defined above, the DI curves were plotted at different crack length as shown in Figure 11(b) for all the pitch-catch paths. The DI increases when the crack grows. The detection along sensors P0 to P1 is most sensitive, followed by the one

(a)   (b)

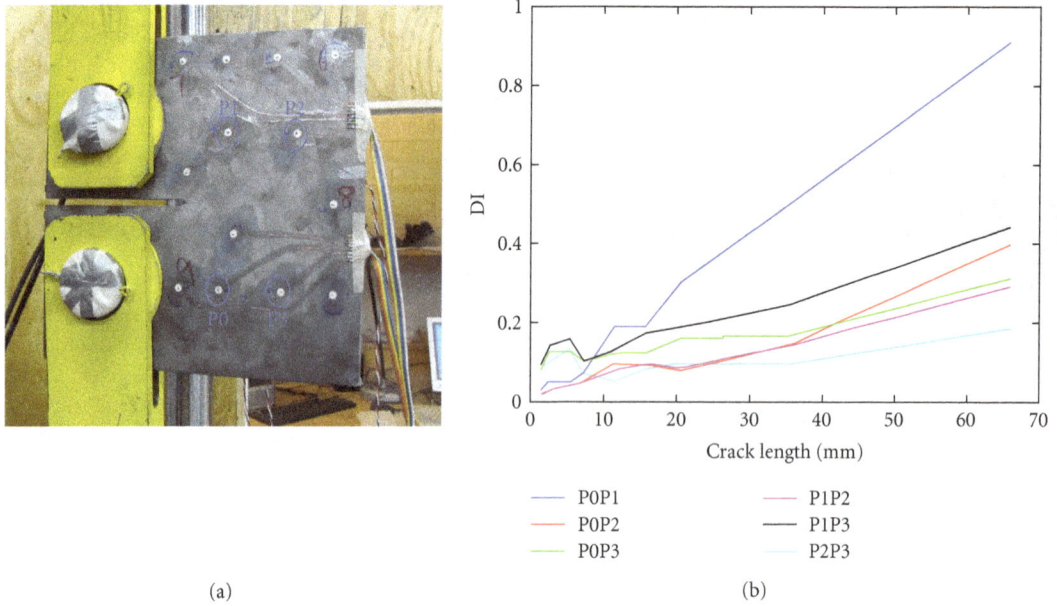

FIGURE 11: AE detection on a 1/2″ CT specimen. (a) Geometry of the specimen and arrangement of transducers; (b) a snapshot of the actual specimen. AE PWAS are circled (rest are for active sensing). The R15I was installed on the other side of the specimen.

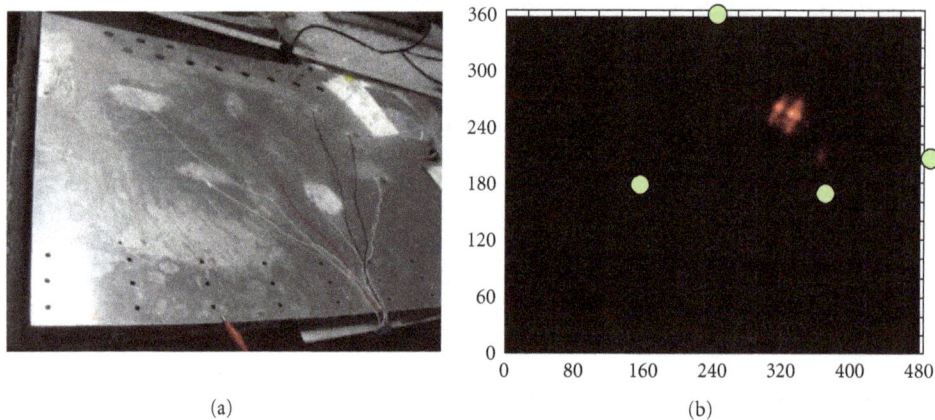

(a)   (b)

FIGURE 12: Array imaging for crack growth detection. (a) Test specimen and sensor network; (b) array imaging of a hairline crack of 18 mm length. Sensor locations are marked as green face circles.

along sensors P1 to P3 since the paths are perpendicular to the crack development path. The increment of the DI curves is also well correlated to the crack growth under fatigue loading.

3.2. Array Imaging. The array imaging methods have the incredible capability to map the structure and existing damage in it, providing a means to qualitatively assess the structural integrity. The sparse array uses scatter signals from a network of sensors to construct a diagnosis image. The image construction is based on triangulation principle and conducted by shifting back the scatter signals at time quantities defined by the transmitter-receiver locations used in the pitch-catch mode. Assuming a single damage scatter is located at point $Z(x, y)$ in the structure, the scatter signal

from transmitter T_i to receiver R_j contains a single wave packet caused by the damage. The total time of traveling τ_Z is determined by the locations of the transmitter T_i at (x_i, y_i), the receiver R_j at (x_j, y_j), and $Z(x, y)$, as

$$\tau_Z = \frac{\sqrt{(x_i - x)^2 + (y_i - y)^2} + \sqrt{(x - x_j)^2 + (y - y_j)^2}}{c_g},$$
(4)

where c_g is the group velocity of the traveling Lamb wave, assuming constant. When a wave packet is shifted back by the quantity defined by the transducers and the exact position of the damage, that is, τ_Z, ideally the peak will be shifted right back to the time origin. If the wave packet is shifted by a quantity defined with otherwise cases (such as τ_i and

τ_O), the peak will not be shifted right at the time origin. For an unknown damage, for a certain scatter signal with τ_Z, the possible locations of the damage constitute an orbit of ellipse with the transmitter and receiver as the foci. To locate the damage, ellipses from other scatters (or transmitter-receiver pairs in the network) are needed (triangulation). For a given network of M transducers, a total of M^2 scatter signals will be used without considering reciprocity. The pixel value at the location $Z(x, y)$ is defined as

$$P_Z(t_0) = \prod_{i=1}^{M} \prod_{j=1}^{M} s_{ij}(\tau_Z), \quad i \neq j, \qquad (5)$$

where s_{ij} is the scatter signal obtained from ith transmitter and jth receiver. Details of principles and applications of ultrasonic array imaging can be found in [17].

The crack detection demonstration was conducted on a 1-mm thick aluminum plate. The imaging was performed by a four-PWAS network to assess a simulated hairline crack centered inside, shown in Figure 12(a). A hairline crack of 18 mm was made on the plate, and then crack length was increased to 23 mm, resulting in a growth of 5 mm. The Lamb wave used to conduct the imaging was the S0 mode at 310 kHz with a wavelength about 17 mm.

The image result of the crack at 18 mm is shown in Figure 12(b), giving a clear detection with two highlighted spots representing the two crack tips; therefore, they could be used to estimate the crack length and monitor the crack growth. The first crack of 18 mm was estimated at 17.09 mm and the second one of 23 mm was estimated at 22.47 mm. A crack growth of 5.38 mm was hence measured with the array imaging method with an error of about 7.6% compared to the actual growth of 5 mm (from 18 mm to 23 mm).

4. Conclusions

Piezoelectric wafer active sensors (PWAS) have made tremendous progress in structural health monitoring during the past decade, but their exposure to civil infrastructure has been little discussed so far. The work presented here intends to explore PWAS applications for in-field monitoring of infrastructure (e.g., civil steel bridges) using both acoustic emission and active wave propagation sensing. Laboratory demonstration on both thin aluminum and thick steel plates, the PWAS have been proved as AE sensors. The use of coaxial wiring cables has greatly improved the PWAS waveform signal-to-noise ratio, making it more suitable for field application. The PWAS AE sensing of fatigue cracking on a CT specimen showed that it can provide concentrated detection around the crack tip with a relatively fewer numbers of acoustic emission events than the R15I AE transducers. The PWAS active mode sensing using propagating guided waves can interrogate the structures at will and provide a clear indication and quantitative estimation of the crack growth through damage index or array imaging. The dual mode sensing of the presented PWAS methodology has shown its promising application to insitu health monitoring and diagnosis of steel bridges.

Acknowledgments

This paper was performed under the support of the US Department of Commerce, National Institute of Standards and Technology, Technology Innovation Program, Cooperative Agreement number 70NANB9H9007. The authors would also like to thank Dr. Adrian Pollock from Mistras Group for his insightful comments on the presented passive sensing.

References

[1] I. M. Friedland, H. Ghasemi, and S. B. Chase, *The FHWA Long-Term Bridge Performance Program*, Federal Highway Administration, Turner-Fairbank Highway Research Center, McLean, Va, USA, 2007.

[2] L. Yu, V. Giurgiutiu, P. Ziehl, P. Pollock, and D. Ozevin, "Steel bridge fatigue crack detection with piezoelectric wafer active sensors," in *Sensors and Smart Structures Technologies for Civil, Mechanical, and Aerospace Systems*, vol. 7647 of *Proceedings of SPIE*, San Diego, Calif, USA, March 2010.

[3] A. C. E. Sinclair, D. C. Connors, and C. L. Formby, "Acoustic emission analysis during fatigue crack growth in steel," *Materials Science and Engineering*, vol. 28, no. 2, pp. 263–273, 1977.

[4] R. I. Stephens, S. G. Lee, and H. W. Lee, "Constant and variable amplitude fatigue behavior and fracture of A572 steel at 25°C(77°F) and -45°C(-50°F)," *International Journal of Fracture*, vol. 19, no. 2, pp. 83–98, 1982.

[5] M. N. Bassim, S. S. Lawrence, and C. D. Liu, "Detection of the onset of fatigue crack growth in rail steels using acoustic emission," *Engineering Fracture Mechanics*, vol. 47, no. 2, pp. 207–214, 1994.

[6] A. Berkovits and D. Fang, "Study of fatigue crack characteristics by acoustic emission," *Engineering Fracture Mechanics*, vol. 51, no. 3, pp. 401–416, 1995.

[7] D. H. Kohn, P. Ducheyne, and J. Awerbuch, "Acoustic emission during fatigue of Ti-6Al-4V: Incipient fatigue crack detection limits and generalized data analysis methodology," *Journal of Materials Science*, vol. 27, no. 12, pp. 3133–3142, 1992.

[8] Z. Gong, E. O. Nyborg, and G. Oommen, "Acoustic emission monitoring of steel railroad bridges," *Materials Evaluation*, vol. 50, no. 7, pp. 883–887, 1992.

[9] H. L. Chen and J. H. Choi, "Acoustic emission study of fatigue cracks in materials used for AVLB," *Journal of Nondestructive Evaluation*, vol. 23, no. 4, pp. 133–151, 2004.

[10] J. Yu, P. Ziehl, B. Zrate, and J. Caicedo, "Prediction of fatigue crack growth in steel bridge components using acoustic emission," *Journal of Constructional Steel Research*, vol. 67, no. 8, pp. 1254–1260, 2011.

[11] E. F. Crawley and J. de Luis, "Use of piezoelectric actuators as elements of intelligent structures," *AIAA Journal*, vol. 25, no. 10, pp. 1373–1385, 1987.

[12] V. Giurgiutiu, *Structural Health Monitoring with Piezoelectric Wafer Active Sensors*, Elsevier Academic Press, 2008.

[13] G. B. Santoni, L. Yu, B. Xu, and V. Giurgiutiu, "Lamb wave-mode tuning of piezoelectric wafer active sensors for structural health monitoring," *Journal of Vibration and Acoustics*, vol. 129, no. 6, pp. 752–762, 2007.

[14] M. Dupont, R. Osmont, R. Gouyon, and D. L. Balageas, "Permanent monitoring of damaging impacts by a piezoelectric sensor based integrated system," in *Proceedings of the 2nd International Workshop on Structural Health Monitoring*, pp.

561–570, Stanford University, Stanford, Calif, USA, September 2000.

[15] L. Yu, V. Giurgiutiu, P. Ziehl, and D. Ozevin, "Piezoelectric based sensing in wireless steel bridge health monitoring," in *Nondestructive Characterization for Composite Materials, Aerospace Engineering, Civil Infrastructure, and Homeland Security*, vol. 7294 of *Proceedings of SPIE*, San Diego, Calif, USA, March 2009.

[16] X. Zhao, H. Gao, G. Zhang et al., "Active health monitoring of an aircraft wing with embedded piezoelectric sensor/actuator network: I. Defect detection, localization and growth monitoring," *Smart Materials and Structures*, vol. 16, no. 4, pp. 1208–1217, 2007.

[17] L. Yu and V. Giurgiutiu, "Piezoelectric wafer active sensor guided wave imaging," in *Smart Sensor Phenomena, Technology, Networks, and Systems*, vol. 7648 of *Proceedings of SPIE*, San Diego, Calif, USA, March 2010.

7

High-Performance Steel Bars and Fibers as Concrete Reinforcement for Seismic-Resistant Frames

Andres Lepage,[1] Hooman Tavallali,[2] Santiago Pujol,[3] and Jeffrey M. Rautenberg[4]

[1] Department of Architectural Engineering, The Pennsylvania State University, 104 Engineering Unit A, University Park, PA 16802, USA
[2] Leslie E. Robertson Associates, 40 Wall Street, 23rd Floor, New York, NY 10005, USA
[3] School of Civil Engineering, Purdue University, 550 Stadium Mall Drive, West Lafayette, IN 47907, USA
[4] Wiss, Janney, Elstner Associates, 2000 Powell Street, Suite 1650, Emeryville, CA 94608, USA

Correspondence should be addressed to Andres Lepage, lepage@psu.edu

Academic Editor: Rajesh Prasad Dhakal

Experimental data are presented for six concrete specimens subjected to displacement reversals. Two specimens were reinforced longitudinally with steel bars Grade 410 (60 ksi), two with Grade 670 (97 ksi), and two with Grade 830 (120 ksi). Other experimental variables included axial load (0 or 0.2 f_c' A_g) and volume fraction of hooked steel fibers (0 or 1.5%). All transverse reinforcement was Grade 410, and the nominal concrete compressive strength was 41 MPa (6 ksi). The loading protocol consisted of repeated cycles of increasing lateral displacement reversals (up to 5% drift) followed by a monotonic lateral push to failure. The test data indicate that replacing conventional Grade-410 longitudinal reinforcement with reduced amounts of Grade-670 or Grade-830 steel bars did not cause a decrease in usable deformation capacity nor a decrease in flexural strength. The evidence presented shows that the use of advanced high-strength steel as longitudinal reinforcement in frame members is a viable option for earthquake-resistant construction.

1. Introduction

For many years, the earthquake-resistant design of reinforced concrete structures in the USA has been dominated by the use of steel reinforcement with specified yield strength, f_y, of 410 MPa (60 ksi). Although higher values of f_y are allowed for non-seismic applications, f_y has been limited to 550 MPa (80 ksi) since the 1971 edition of ACI 318 [1]. Current version of ACI 318 [2] maintains the above limits but allows designs with f_y of 690 MPa (100 ksi) only if used for confining reinforcement.

The terms advanced high-strength steel (AHSS) [3] or ultrahigh strength steel (UHSS) [4] are used to designate high-performance steel bars with a yield strength in excess of 550 MPa (80 ksi) and a fracture strain, ε_{su}, of 6% or more measured in a 203-mm (8-in.) gage length. Figure 1 shows representative stress-strain curves of both conventional Grade 410 and AHSS steels. After the introduction of ASTM A1035 in 2004 [5] and their acceptance as confining

reinforcement in the 2005 version of ACI 318, there has been growing interest in AHSS bars. ASTM A706 [6] introduced its new Grade 550 in 2009 and it is likely that new ASTM designations with higher grades will follow. However, there is a paucity of test data describing the behavior of concrete members reinforced with AHSS bars as longitudinal reinforcement.

A series of experiments was designed to determine the deformation capacity of AHSS-reinforced concrete frame members subjected to displacement reversals [7, 8]. Specimens with high-performance fiber-reinforced concrete (HPFRC) were included. HPFRC is defined here as a class of fiber-reinforced concrete that shows strain hardening after first cracking [9].

The use of high-strength steel bars as reinforcement in concrete elements has the potential to reduce problems associated with congested reinforcement cages and concrete placement, as well as reduce costs associated with the shipment and placement of reinforcing steel. The cost savings

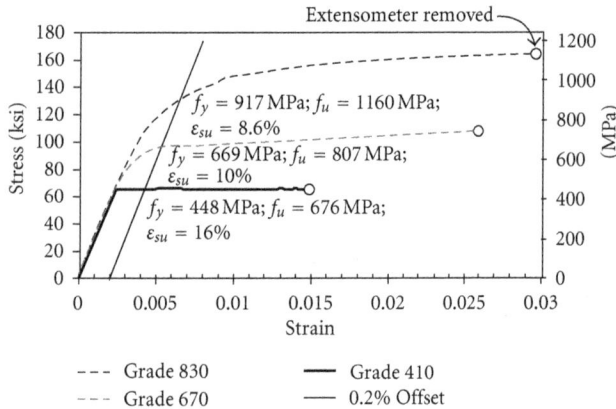

FIGURE 1: Representative tensile properties of reinforcing bars, 1 ksi = 6.9 MPa.

are nearly proportional to the increase in yield strength used in design.

2. Background

There is no evidence in the USA of reinforced concrete structures built or rehabilitated using AHSS longitudinal and transverse reinforcement and designed to take full advantage of f_y in excess of 550 MPa (80 ksi) for stresses induced by combined shear, flexure, and axial forces. This underutilization is predominantly due to the shortage of experimental data and the limitations contained in existing building codes.

The use of high-strength reinforcement in concrete columns was first considered in the early 1930s by Richart and Brown [10] in a series of laboratory tests on columns with circular cross sections and spiral reinforcement. The column tests showed that longitudinal bars with yield strength close to 690 MPa (100 ksi) were fully effective in columns resisting concentric axial loads. The spiral reinforcement allowed the concrete in the core of the column to develop compressive strains large enough for the longitudinal reinforcement to reach its yield point.

Later, in the 1960s, the experimental work at the PCA laboratories [11] led to a series of reports titled "High-strength bars as concrete reinforcement." In Part 5 of the PCA series [12], similar observations to those by Richart and Brown [10] were made for columns reinforced with spirals. But in the case of tied columns with rectangular sections, the PCA report noted: "If the specified yield point of longitudinal reinforcement in tied columns is to be developed at ultimate strength of the columns, then it is necessary that the yield point be reached at or before a strain of 0.003 in./in." [12].

Todeschini et al. [13] confirmed the PCA findings. As a consequence, ACI 318-63 [14] introduced strength design provisions with limits on the specified yield strength of reinforcing bars. ACI 318-63 Section 1505 had a limit of 520 MPa (75 ksi) on the specified yield strength of compression reinforcement and 410 MPa (60 ksi) for tension reinforcement (unless special tests satisfied crack control requirements, in which case 520 MPa was permitted). The limits on the specified yield strength in Section 1505 of the 1963 Code

were justified in the commentary [15] by stating: "High strength steels frequently have a strain at yield strength or yield point in excess of the 0.003 assumed for the concrete at ultimate. The requirements of Section 1505 are to adjust to this condition. The maximum stress in tension of 60,000 psi without test is to control cracking."

In 1971, ACI 318 [1] increased the limit on yield strength of reinforcement to 550 MPa (80 ksi). The limit in current USA design provisions [2] is still 550 MPa (80 ksi). Section 9.4 of ACI 318-11 [2] requires "The values of f_y and f_{yt} used in design calculations shall not exceed 80,000 psi, except for prestressing steel and for transverse reinforcement in 10.9.3 and 21.1.5.4." The exception only applies to transverse reinforcement used for confinement, where f_{yt} up to 690 MPa (100 ksi) is allowed. For earthquake-resistant design, Section 21.1.5 of ACI 318-11 allows only longitudinal reinforcement with f_y of 410 MPa (60 ksi) or lower.

If the use of AHSS bars as primary reinforcement is to be considered for practical use, the issues related to concrete compressive strain and crack control must be addressed. A method of addressing these issues is to add fibers to the concrete mix to control crack widths and to enhance the usable compressive strain of concrete. High tensile and compressive strain capacities are attainable by concrete reinforced with dispersed fibers. Recent developments in high-performance fiber-reinforced cementitious composites include new formulations of concrete matrices and fibers to achieve strain hardening behavior with fiber volume contents in the range of 1% to 2% [16].

Studies on the compressive properties of cementitious composites have shown that the introduction of fibers into the matrix delays spalling of the cover and increases the load capacity and the ductility of columns over that of comparable reinforced concrete (RC) specimens [17]. Several research projects have explored the application of fiber-reinforced composites in earthquake-resistant construction, as summarized by Parra-Montesinos [9]. Results from these studies revealed HPFRC to be effective in increasing shear strength and deformation capacity in members subjected to cycles of large inelastic deformations, but none of these studies incorporated AHSS reinforcement. The experiments presented below were designed to address this gap.

3. Experiments

A collaborative experimental program between The Pennsylvania State University and Purdue University was aimed at studying the behavior of AHSS-reinforced concrete frame members subjected to displacement reversals. Beam tests were conducted at Penn State by Tavallali [7], and column tests were conducted at Purdue by Rautenberg [8]. The scope of this paper is limited to specimens defined by the following ranges of test variables:

(i) nominal yield strength of the longitudinal reinforcement, $f_y = 410$, 670, or 830 MPa (60, 97, or 120 ksi);

(ii) applied axial force, $P = 0$ or $0.2 f_c' A_g$;

(iii) volume fraction of hooked steel fibers, $V_f = 0$ or 1.5%.

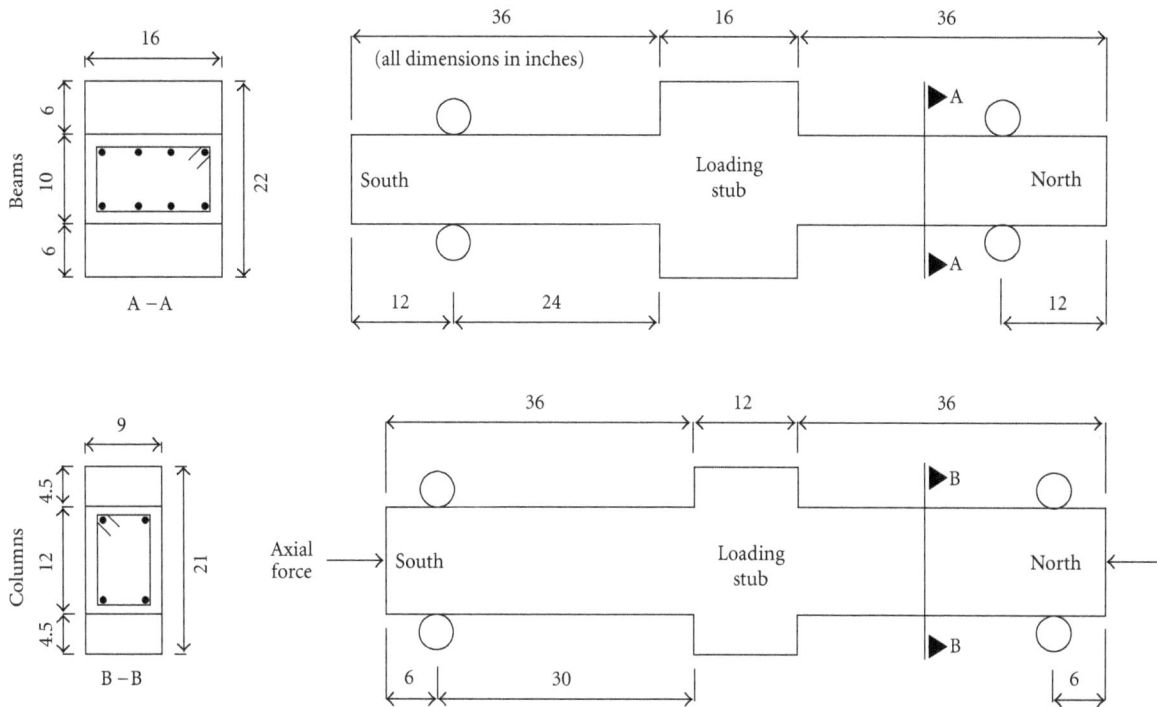

FIGURE 2: Geometry of test specimens, 1 in. = 25.4 mm.

All transverse reinforcement was Grade 410 and the nominal compressive strength of concrete was 41 MPa (6 ksi). The layout of the longitudinal reinforcement was symmetrical with identical top and bottom layers. Table 1 shows a summary of the specimens considered.

The geometry of typical beam and column specimens is shown in Figure 2. The specimens consisted of two beams (or columns) connected to a central stub. The specimens were loaded through the central stub so that they were in single-curvature bending. Each beam (or column) was intended to represent a cantilever with the central stub acting as the base of the cantilever. The shear span-to-effective depth ratio for all specimens was 3. In the column specimens, the axial force was kept constant throughout the loading protocol.

The amount of longitudinal reinforcement in each specimen was chosen so that the nominal flexural strength of all specimens was nearly identical. Thus, specimen CC-3.3-20 with Grade-410 steel used about twice the amount of longitudinal reinforcement as used in specimens UC-1.6-20 and UC-1.6-20F with Grade-830 steel. A similar relationship exists between specimens CC4-X, UC4-X, and UC2-F. The shear stress corresponding to the probable moment, M_{pr}, calculated according to ACI 318-11 [2] was approximately $0.42\sqrt{f_c'}$, MPa ($5\sqrt{f_c'}$, psi) for the beams and $0.58\sqrt{f_c'}$, MPa ($7\sqrt{f_c'}$, psi) for the columns. The longitudinal bars were all continuous through the central stub and had sufficient embedment length to develop $1.25 f_y$ at the faces of the stub. The embedment lengths satisfy the requirements for development lengths in Chapter 12 of ACI 318-11.

HPFRC specimens UC2-F and CC-1.6-20F had 1.5% volume fraction of Dramix RC-80/30-BP hooked steel fibers,

with length-to-diameter ratio of 80 and a length of 30 mm (1.2 in.). The fibers had a nominal tensile strength of 2300 MPa (330 ksi). These fibers are commercially available and manufactured by Bekaert Corporation. The maximum aggregate size in the concrete matrix of specimens CC4-X, UC4-X, UC2-F, and UC-1.6-20F, was limited to 13 mm (0.5 in.), less than half the fiber length. Maximum aggregate size for specimens CC-3.3-20 and UC-1.6-20 was 25 mm (1 in.).

Specimens CC4-X, UC4-X, CC-3.3-20, and UC-1.6-20, without fibers, used Grade-410 transverse reinforcement spaced at $d/4$, in compliance with Chapter 21 of ACI 318-11 [2]. Specimens UC2-F and UC-1.6-20F, cast using HPFRC, used Grade-410 transverse reinforcement spaced at approximately $d/2$. The spacing in the HPFRC specimens was increased to account for the enhanced shear strength and confinement provided by fibers.

The drift ratio history applied to each specimen follows the protocol of FEMA 461 [18] described in Table 2. The drift ratio was defined as the lateral displacement of the specimen divided by the shear span, corrected for the rotation of the loading stub. Two cycles at each drift target were applied to the stub at increasing amplitudes. After step 12, the displacement was increased monotonically until failure (defined here as a reduction in the lateral-load resistance of more than 25% from the peak value).

4. Measured Response

The measured data for shear versus drift ratio of the controlling end of the specimens (north or south) are presented for the 12 steps (24 cycles) of the loading protocol.

TABLE 1: Description of test specimens.

Specimen		Axial load $A_g f_c'$	f_c' MPa (ksi)	Bars per layer[a]	Longitudinal reinforcement[a]				Transverse reinforcement[a]		
					d_b mm (in.)	f_y[b] MPa (ksi)	f_u MPa (ksi)	ε_{su} %	s mm (in.)	d_b mm (in.)	f_{yt}[b] MPa (ksi)
					Beams[c]						
#1	CC4-X	0	41 (6.0)	4	22 (7/8)	448 (65)	676 (98)	15.9	51 (2.0)	9.5 (3/8)	469 (68)
#2	UC4-X	0	43 (6.2)	4	18 (0.71)[d]	669 (97)	807 (117)	10.4	51 (2.0)	9.5 (3/8)	469 (68)
#3	UC2-F	0	44 (6.4)	4	18 (0.71)[d]	669 (97)	807 (117)	10.4	102 (4.0)	9.5 (3/8)	469 (68)
					Columns[e]						
#4	CC-3.3-20	0.20	54 (7.8)	3	22 (7/8)	441 (64)	634 (92)	20.3	64 (2.5)	9.5 (3/8)	428 (62)
#5	UC-1.6-20	0.21	43 (6.3)	2	19 (3/4)[f]	917 (133)	1160 (168)	8.6	64 (2.5)	9.5 (3/8)	434 (63)
#6	UC-1.6-20F	0.19	51 (7.4)	2	19 (3/4)[f]	917 (133)	1160 (168)	8.6	114 (4.5)	9.5 (3/8)	434 (63)

[a] Bar layout is symmetrical with identical top and bottom layers. Transverse reinforcement consists of single rectilinear hoops spaced at $s = d/4$ (specimens #1, #2, #4, and #5) or $s \approx d/2$ (specimens #3 and #6 with fibers).
[b] Defined using the 0.2%-offset method.
[c] Beams tested at The Pennsylvania State University: $b = 406$ mm (16 in.), $h = 254$ mm (10 in.), and $d = 203$ mm (8 in.).
[d] Provided by SAS Stressteel.
[e] Columns tested at Purdue University: $b = 229$ mm (9 in.), $h = 305$ mm (12 in.), $d = 254$ mm (10 in.).
[f] Provided by MMFX Technologies.

TABLE 2: Loading protocol.

Step[a]	1	2	3	4	5	6	7	8	9	10	11	12
Drift ratio, %	0.15	0.20	0.30	0.40	0.60	0.80	1.0	1.5	2.0	3.0	4.0	5.0

[a] Two symmetrical cycles of loading in each step, with equal drifts in the positive and negative direction, as recommended in FEMA 461 [18].

Beams are shown in Figure 3 and columns in Figure 4. The figures do not include the data associated with the monotonic push to failure (after step 12 in Table 2). The maximum measured shears and drift ratios are presented in Table 3.

4.1. Beam Specimens. Specimen CC4-X is the control RC beam specimen compliant with the provisions for special moment frame beams in Chapter 21 of ACI 318-11 [2]. All reinforcing bars were Grade 410, see Table 1. The widths of flexural cracks exceeded 0.4 mm (0.016 in.) at a drift ratio of 0.8%. First yield of the longitudinal reinforcement was measured between 0.76% and 1.17% with a mean of 0.97% (Table 4). The peak shear force of 242 kN (54.4 kip) was reached during the final push at a drift ratio in excess of 10%, an indication of a stable hysteretic behavior throughout the test, see Figure 3(a).

Specimen UC4-X had similar properties to specimen CC4-X with the exception that it was reinforced longitudinally with Grade-670 bars, see Table 1. Compared with specimen CC4-X, specimen UC4-X showed a reduced postcracking stiffness and increased yield deformation, see Figures 3(a) and 3(b). Due to the reduced amount of longitudinal reinforcement in specimen UC4-X, a smaller drift ratio (0.6%) was associated with crack widths exceeding 0.4 mm (0.016 in.), as opposed to 0.8% in specimen CC4-X. First yield of the longitudinal reinforcement occurred between 1.36% and 1.68% with a mean of 1.49% (Table 4). The peak shear force of 234 kN (52.5 kip) was reached at a drift ratio of 2.6%. During the final push, the peak shear was

215 kN (48.3 kip) at a drift ratio of 7.1%. Specimen UC4-X exceeded 10% drift without failure.

Specimen UC2-F had similar properties to specimen UC4-X except that it was cast using HPFRC. Spacing of the transverse reinforcement was increased to $d/2$ to investigate the effects of fibers. Crack widths exceeded 0.4 mm (0.016 in.) at a drift ratio of 1%, a significant improvement compared to the other beam specimens. The drift ratio at first yield of the longitudinal reinforcement varied between 1.19% and 1.37% with a mean of 1.28% (Table 4). The south beam deviated from the loading protocol at the first cycle of step 9 because the north beam was controlling through step 8, see Figure 3(c). During the first cycle of step 9, the specimen reached a peak shear force of 273 kN (61.4 kip). During the final push, the peak shear was 223 kN (50.2 kip) at a drift ratio of 6.3%. Specimen UC2-F exceeded 10% drift without failure.

The reduction of transverse reinforcement by 50% in specimen UC2-F did not have a negative impact on its behavior. HPFRC was effective in enhancing the shear capacity and confinement. However, crack openings concentrated in a single flexural crack near the face of the stub at drift ratios exceeding 3%. The presence of this wide flexural crack resulted in the accumulation of residual strains in longitudinal bars crossing the crack. Figures 5 and 6 show specimens UC4-X and UC2-F at the end of the second cycle at 5% drift (end of step 12). Figures 7 and 8 show the strain measurements in the longitudinal bars of these specimens at face of the central stub during steps 10 and 11. The strain values measured at the same location for the same drift ratios are

TABLE 3: Maximum measured shear force and drift ratio.

Specimen		V_{max}[a] kN (kip)		v_{max}[b] $\sqrt{f_c'}$ MPa (psi)		θ_{max}[c] %	
		+	−	+	−	+	−
		Beams					
#1	CC4-X	242 (54.4)	217 (48.8)	0.46 (5.5)	0.41 (4.9)	>10	5.0
#2	UC4-X	234 (52.5)	226 (50.7)	0.43 (5.2)	0.42 (5.0)	>10	5.1
#3	UC2-F	273 (61.4)	232 (52.1)	0.50 (6.0)	0.42 (5.1)	>10	5.1
		Columns					
#4	CC-3.3-20	252 (56.6)	266 (59.9)	0.59 (7.1)	0.63 (7.5)	5.1	5.0
#5	UC-1.6-20	228 (51.3)	228 (51.3)	0.60 (7.2)	0.60 (7.2)	5.2	4.8
#6	UC-1.6-20F	260 (58.4)	247 (55.5)	0.63 (7.5)	0.60 (7.2)	5.3	5.4

[a] Maximum measured shear force during the loading protocol (Table 2).
[b] Shear stress calculated using $V_{max}/(b\,d)$, expressed as a fraction of $\sqrt{f_c'}$, where b, d, and f_c' are given in Table 1.
[c] Maximum drift ratio reached while maintaining a shear force not less than 75% of V_{max} in the same direction of loading. Maximum reported value includes the drift ratio attained during the final monotonic push in the positive direction.

TABLE 4: Measured yield point.

Specimen				Shear[a] kN (kip)	Mean	Drift ratio %	Mean	Secant stiffness[b] kN/mm (kip/in.)
Beams	#1	CC4-X	South +	206 (46.3)		0.93		
			South −	176 (39.6)	174 (39.2)	1.03	0.97	29.5 (168)
			North +	180 (40.5)		1.17		
			North −	134 (30.2)		0.76		
	#2	UC4-X	South +	202 (45.3)		1.52		
			South −	194 (43.7)	198 (44.6)	1.38	1.49	21.8 (125)
			North +	199 (44.7)		1.68		
			North −	198 (44.6)		1.36		
	#3	UC2-F	South +	202 (45.4)		1.37		
			South −	185 (41.7)	196 (44.1)	1.26	1.28	25.1 (144)
			North +	218 (49.1)		1.29		
			North −	178 (40.1)		1.19		
Columns	#4	CC-3.3-20	South +	238 (53.4)		0.67		
			South −	252 (56.7)	244 (54.8)	0.78	0.78	42.0 (240)
			North +	242 (54.4)		0.83		
			North −	243 (54.7)		0.85		
	#5	UC-1.6-20	South +	—[c]		—[c]		
			South −	204 (45.9)	208 (46.7)	1.39	1.41	20.2 (115)
			North +	210 (47.3)		1.44		
			North −	209 (47.0)		1.39		
	#6	UC-1.6-20F	South +	—[c]		—[c]		
			South −	221 (49.6)	224 (50.4)	1.28	1.42	21.6 (123)
			North +	—[c]		—[c]		
			North −	228 (51.2)		1.55		

[a] Shear force measured at first yield of the longitudinal reinforcement. Top and bottom bars were instrumented with strain gages at the north and south faces of the central stub. The yield point is based on the value of f_y reported in Table 1.
[b] Secant stiffness to first yield of the longitudinal reinforcement. In columns, P-delta effects are removed.
[c] Measurement not available.

(a) Specimen CC4-X north

(b) Specimen UC4-X south

(c) Specimen UC2-F south

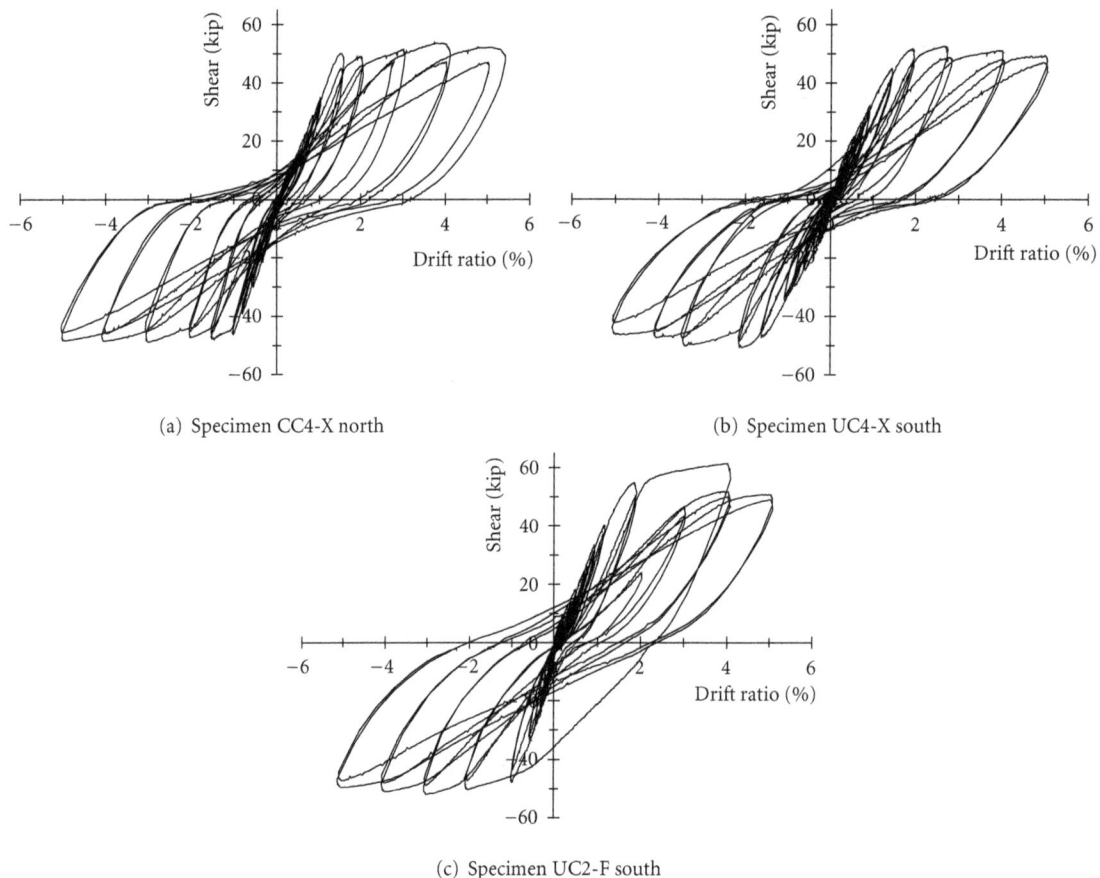

FIGURE 3: Measured shear versus drift ratio, beams (1 kip = 4.45 kN).

approximately 50% larger in specimen UC2-F, an indication of higher strain concentration in the plastic hinge of the HPFRC specimen. As a result of strain accumulation, the longitudinal bars fractured at 11% drift in specimen UC2-F, while similar bars in specimen UC4-X fractured at a drift ratio of 15%. In any case, the beam specimens all reached drift ratios well in excess of what would be expected for a modern building structure subjected to strong ground motion.

4.2. Column Specimens. Specimen CC-3.3-20 is the control RC column specimen compliant with the provisions for special moment frame columns in Chapter 21 of ACI 318-11 [2]. All reinforcing bars were Grade 410, see Table 1. First yield of the longitudinal reinforcement was measured between 0.67% and 0.85% with a mean of 0.78% (Table 4). Figure 4(a) shows the measured shear-drift response. The north column followed the loading protocol (Table 2) through the first cycle of step 12. During the second cycle to 5% drift, the longitudinal bars buckled at a drift ratio of about 1%. A plausible explanation for the failure of the specimen is that cracks on both sides of the specimen were still open at low drift ratios, and the axial load was carried predominantly by the longitudinal reinforcement, leading to bar buckling.

Specimen UC-1.6-20 had similar properties to specimen CC-3.3-20 with the exception that it was reinforced

longitudinally with Grade-830 bars using about half as much longitudinal reinforcement. Figure 4(b) shows the measured shear-drift response. Compared with specimen CC-3.3-20, specimen UC-1.6-20 showed reduced postcracking stiffness and increased yield deformation. First yield of the longitudinal reinforcement occurred between 1.39% and 1.44% with a mean of 1.41% (Table 4). The north column completed the first half-cycle to 5% drift, but the longitudinal bars buckled during the second half cycle at that drift ratio. Testing was continued, and the remaining longitudinal bars buckled at a drift ratio of about 2% during the second cycle to 5% drift. Again, a plausible explanation is that the longitudinal bars carried a larger fraction of the axial load at low drift ratios when cracks were still open.

Specimen UC-1.6-20F had similar properties to specimen UC-1.6-20, with the exception that the concrete matrix consisted of HPFRC. The spacing of the transverse reinforcement was nearly doubled to evaluate the influence of HPFRC in shear strength, confinement, and bar buckling. Figure 4(c) shows the measured shear-drift response. The drift ratio at first yield of the longitudinal reinforcement varied between 1.28% and 1.55% with a mean of 1.42% (Table 4), which is similar to the yield drift of specimen UC-1.6-20, but the yield force in UC-1.2-20F was 8% greater.

The HPFRC column specimen was the only column (with axial load of 0.2 $f'_c A_g$) that successfully completed

(a) Specimen CC-3.3-20 north

(b) Specimen UC-1.6-20 north

(c) Specimen UC-1.6-20F north

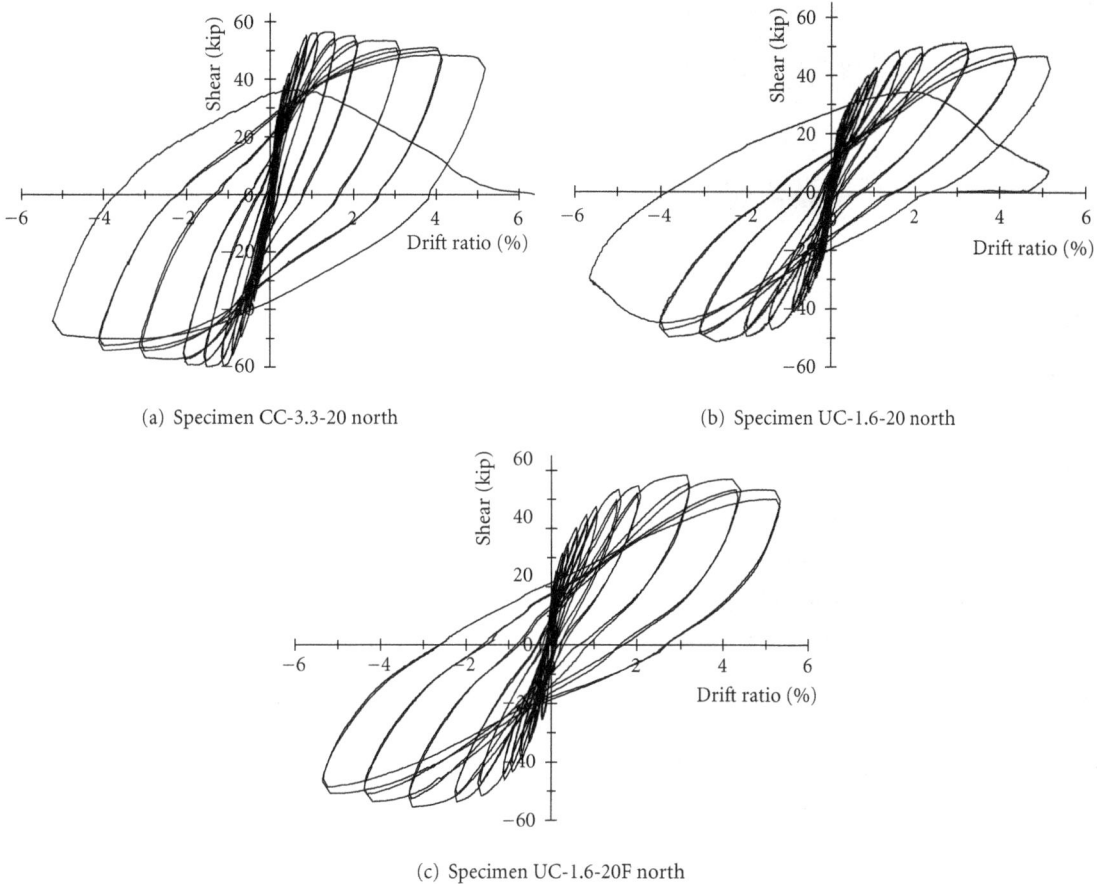

FIGURE 4: Measured shear versus drift ratio, columns (1 kip = 4.45 kN).

the 12-step loading protocol. The fibers were effective in reducing the amount of spalling of the concrete cover and providing lateral support to the longitudinal reinforcement. Unlike specimens CC-3.3-20 and UC-1.6-20, the reinforcement in specimen UC-1.6-20F did not buckle. However, in the final push, one of the longitudinal bars fractured in tension at a drift of about 3%. Adding fibers changed the mode of failure from buckling of the compression bars to fracture of the tension bars. It is important to note that the measured fracture strain, ε_{su}, of the bars in specimen UC-1.6-20F was 8.6% (Table 1), the lowest of the bars tested. To reduce the vulnerability of bar fracture, reinforcing bars with ε_{su} greater than 10% are recommended in HPFRC applications as suggested by the stable response of the beam specimens through drift ratios of 10%. In any case, specimens reached drift ratios of 5% or greater, well in excess of what would be expected for a modern building structure subjected to strong ground motions.

5. Stiffness Comparisons

Stiffness characteristics for the test specimens are inferred from the measured shear versus drift curves (Figures 3 and 4) with special emphasis on secant stiffness to first yield, postyield stiffness, and unloading stiffness. The shear force

and drift ratio associated with first yield of the longitudinal reinforcement are presented in Table 4. The reported secant stiffness to first yield for each specimen corresponds to the average of the measured yield for the north and south ends of each specimen and for the positive and negative direction of loading (see Figure 9). For clarity, the measured response in Figure 9 only includes data through the end of step 8 controlled by the north beam. The yield points identified in Figure 9 were obtained from strain gages placed on longitudinal bars at the locations of maximum moment (at opposite faces of the loading stub). The yield strain was defined using f_y/E_s, where f_y was based on the 0.2%-offset method as reported in Table 1.

It is important to recognize that there is no consensus about the definition of yield displacement. The definition used here was chosen simply because it was convenient. It is clear that the 0.2%-offset rule was not defined having estimation of hysteretic response in mind. It is also known that the displacement associated with strain-gage readings approaching the yield strain tends to be smaller than the displacement associated with yielding of the specimen [19].

5.1. Beam Specimens. The secant stiffness to first yield of specimen UC4-X was about 3/4 of the stiffness of specimen CC4-X and about 7/8 of the stiffness of specimen UC2-F

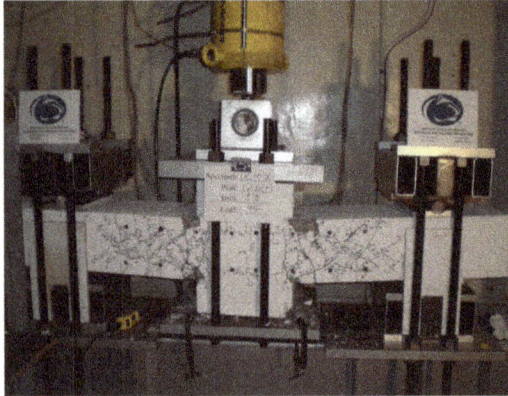

FIGURE 5: Specimen UC4-X at drift ratio of 5%.

FIGURE 6: Specimen UC2-F at drift ratio of 5%.

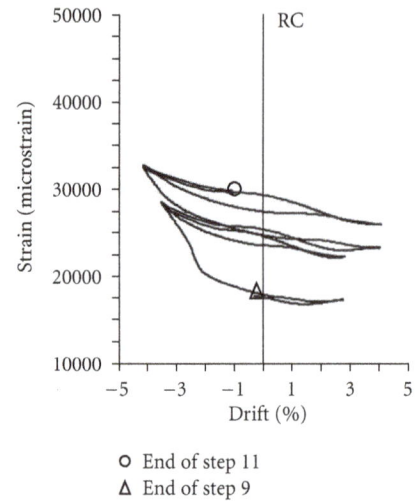

FIGURE 7: Top bar tensile strain at face of loading stub, specimen UC4-X south.

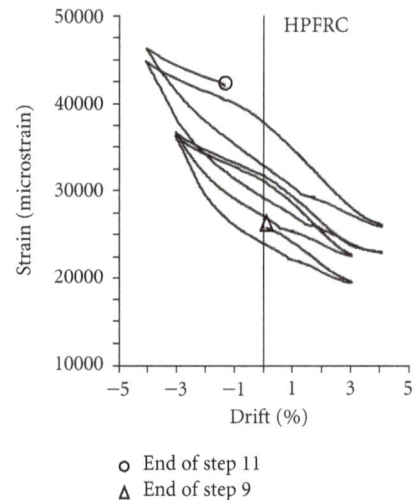

FIGURE 8: Top bar tensile strain at face of loading stub, specimen UC2-F south.

(see Table 4). After yielding, specimen CC4-X showed a small increase in flexural strength while specimen UC4-X had a nearly flat postyield shear-drift curve resembling the stress-strain curve of the rebar (Figure 1). The conventional Grade-410 bars are characterized by a tensile-strength-to-yield-strength ratio (f_u/f_y) of 1.5, while the AHSS Grade-670 bars have a ratio of 1.2. The hysteretic energy dissipated in specimen UC4-X during a given postyield hysteretic loop was smaller than the energy dissipated in the same cycle for specimen CC4-X. With the use of HPFRC in specimen UC2-F, the amount of energy dissipated increased when compared to specimen UC4-X. At each cycle, the secant slope of the shear-drift curve, measured from peak drift to zero shear (unloading stiffness), was about 20% higher in specimen UC2-F than in specimen UC4-X.

5.2. Column Specimens. The secant stiffness to first yield of specimens UC-1.6-20 and UC-1.6-20F was about 1/2 of the stiffness of specimen CC-3.3-20, see Table 4. The effect of HPFRC on the stiffness of the column was small. After yielding, specimens UC-1.6-20 and UC-1.6-20F (with Grade-830 bars) showed a small increase in flexural strength and reached peak shear near 3% drift while specimen CC-3.3-20 reached peak shear near 1.5% drift. The hysteretic

energy dissipated during a given postyield hysteretic loop was smaller in specimens UC-1.6-20 and UC-1.6-20F (with Grade-830 bars) than for specimen CC-3.3-20 (with Grade-410 bars). At each cycle, the secant slope of the shear-drift curve of specimens UC-1.6-20 and UC-1.6-20F, measured from peak drift to zero shear (unloading stiffness), was about 1/2 of that for specimen CC-3.3-20.

6. Calculated Seismic Response

The experimental results in Section 4 show that for the beams reinforced with Grade-670 bars, the secant stiffness to first yield is about 3/4 of that for beams reinforced with Grade-410 bars, see Table 4. For columns reinforced with Grade-830 bars, the secant stiffness to first yield is about 1/2 of that for columns reinforced with Grade-410 bars.

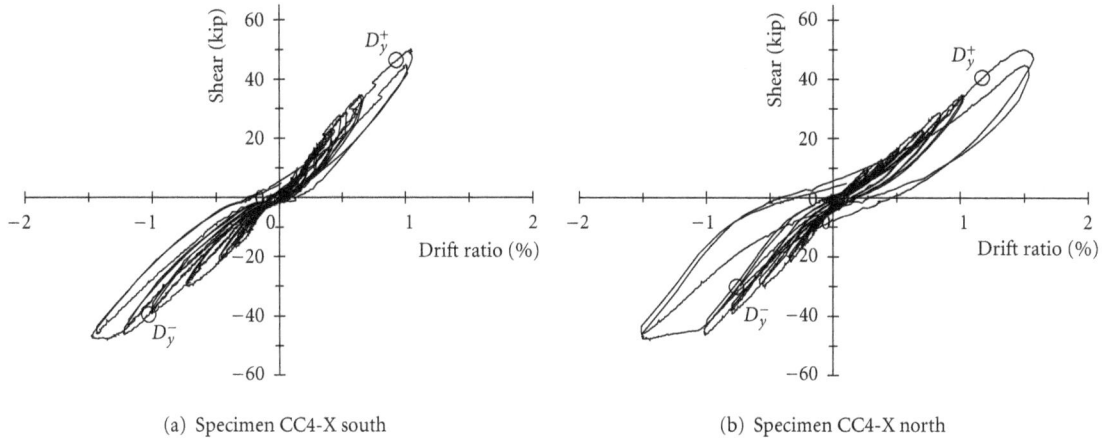

(a) Specimen CC4-X south

(b) Specimen CC4-X north

FIGURE 9: Measured yield points in specimen CC4-X (1 kip = 4.45 kN).

This section investigates how changes in the hysteretic response of RC members, due to the use of Grade-410 reinforcement versus AHSS reinforcement (Grade 670 or Grade 830), affect the displacement demand in single-degree-of-freedom (SDOF) systems subjected to strong ground motions.

6.1. Properties of SDOF Systems. Three groups of six SDOF systems were selected. Each group represented reinforced concrete with a different type of reinforcement: Grade 410, Grade 670, and Grade 830. The Grade-410 systems included three different periods of vibration (T = 0.6, 0.9, and 1.2 s) and two strength ratios (SR = 1/3 and 1/6). The strength ratio measures the ratio of the maximum force induced in a nonlinear SDOF system to the maximum force induced in a linear SDOF system. The target design spectrum, defining the linear-response demand, was based on S_{DS} = 1.0 and S_{D1} = 0.75 as defined in ASCE/SEI 7-10 [20]. The properties of the SDOF systems considered are described in Table 5. Note that the Grade-670 and Grade-830 systems were patterned after the Grade-410 systems using equivalent strength but with reduced stiffness as indicated by the difference in periods of vibration.

The force-displacement relations that characterize the SDOF systems are based on a simplified version of the Takeda hysteresis model [21], as shown in Figure 10. The model is defined by four parameters: the initial stiffness, K_y; the yield strength, V_y; the postyield stiffness, K_{py} and the unloading stiffness coefficient, α. The modified Takeda model adopted here is based on a bilinear primary curve where the initial uncracked stiffness is ignored. This model can produce displacement waveforms very similar to that of more elaborate models [22]. The stiffness of the model is K_y, until the force exceeds the yield force V_y. The postyield stiffness is defined here as 5% of the initial stiffness. The stiffness K_u, during unloading from a point of maximum displacement (see Figure 10) is defined using

$$K_u = K_y \left(\frac{D_y}{D_{\max}} \right)^\alpha. \tag{1}$$

A value of α = 0.4 is assigned to the Grade-410 systems, and α = 0.5 is used to represent the Grade-670 and Grade-830 systems. The selected values are in agreement with the beam data presented in Figure 11, where for CC4-X the value of α approaches 0.4 for $D_{\max}/D_y$ between 4 and 5, while for AHSS-reinforced members the value of α is about 0.5 for $D_{\max}/D_y$ between 3 and 4. The data in Figure 11 correspond to the average value of α (average for the positive and negative direction of loading) derived from the available measured response during the second cycles of steps 9 through 12 (see Table 2). The data associated with steps 9 and 10 were excluded for specimen UC2-F because during these steps the test deviated from the loading protocol. The value of α for specimen UC2-F was similar to that of UC4-X because at step 11 localized fiber pullout had occurred. Figure 11 suggests that the value of α for columns is nearly insensitive to the steel grade.

The viscous damping assigned to the SDOF systems is based on a damping coefficient of 5%, assumed constant (mass-proportional damping) during the calculated nonlinear response. The β-Method by Newmark [23], with β = 1/6, was used to evaluate the dynamic response.

6.2. Ground Motions and Scaling. The SDOF systems were subjected to the suite of 10 strong-motion acceleration records described in Table 6. The selected ground motions are representative of major earthquakes in the United States. These earthquake records were obtained from the Center for Engineering Strong Motion Data (CESMD) [24]. The CESMD is part of the California Strong Motion Instrumentation Program (CSMIP) and may be found at http://www.strongmotioncenter.org. Although the recorded raw data are made available by the CESMD, the data used here correspond to the processed (corrected) ground accelerations.

Each of the earthquake records was scaled linearly to a peak ground velocity of 508 mm/s (20 in./s). The 5%-damped linear-response acceleration spectra for the 10 records after scaling are presented in Figure 12. The figure shows that the average of the spectral accelerations, in

TABLE 5: Properties of SDOF systems considered.

System[a] MPa (ksi)	No.	Period of vibration[b] T, s	Spectral acceleration coefficient[c] S_A, g	Strength ratio[d] SR	Yield strength coefficient[e] C_y
Grade 410 (60)	1	0.60	1.00	1/3	0.33
	2			1/6	0.17
	3	0.90	0.83	1/3	0.28
	4			1/6	0.14
	5	1.20	0.63	1/3	0.21
	6			1/6	0.10
Grade 670 (97)	7	0.69	1.00		0.33
	8				0.17
	9	1.04	0.72		0.28
	10				0.14
	11	1.39	0.54		0.21
	12				0.10
Grade 830 (120)	13	0.85	0.88		0.33
	14				0.17
	15	1.27	0.59		0.28
	16				0.14
	17	1.70	0.44		0.21
	18				0.10

[a]Grade-670 and Grade-830 systems represent equivalent alternatives to Grade-410 systems. All systems target identical strength; however, the stiffness of Grade-670 or Grade-830 systems is 0.75 or 0.50 times the stiffness of Grade-60 systems. In all cases, the postyield stiffness was defined as 5% of the initial stiffness.
[b]Target periods of vibration for Grade-410 systems are set to 0.6, 0.9, and 1.2 s for a unit mass, from which the stiffness is derived. The stiffness of Grade-670 or Grade-830 systems is 0.75 or 0.50 times that of Grade-410 systems.
[c]Linear-response acceleration (divided by g) of a 5%-damped SDOF system of period T. It is defined using $S_A = S_{D1}/T \leq S_{DS}$, where $S_{DS} = 1.0$ and $S_{D1} = 0.75$, refer to ASCE/SEI 7-10 [20].
[d]Yield force, V_y, of nonlinear SDOF system (of initial period T) divided by the force induced in a 5%-damped linear SDOF system of period T.
[e]$C_y = S_A \cdot SR = V_y/W$, where SR is the strength ratio and S_A is the spectral acceleration coefficient obtained from the design spectrum (5% damping coefficient) for a system of period T. The value of C_y corresponds to the yield strength, V_y, divided by the weight, W, of the SDOF system.

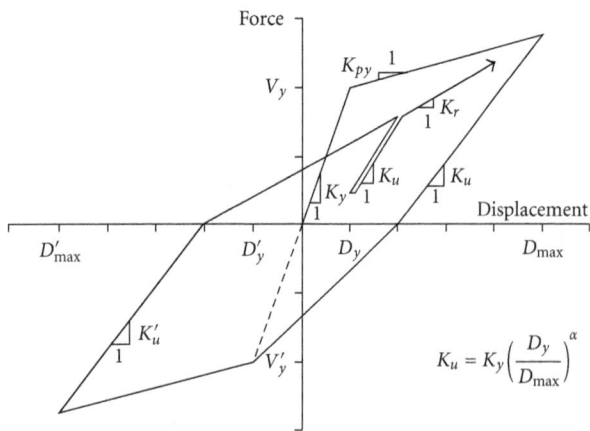

FIGURE 10: Hysteresis model considered.

$$K_u = K_y \left(\frac{D_y}{D_{max}}\right)^{\alpha}$$

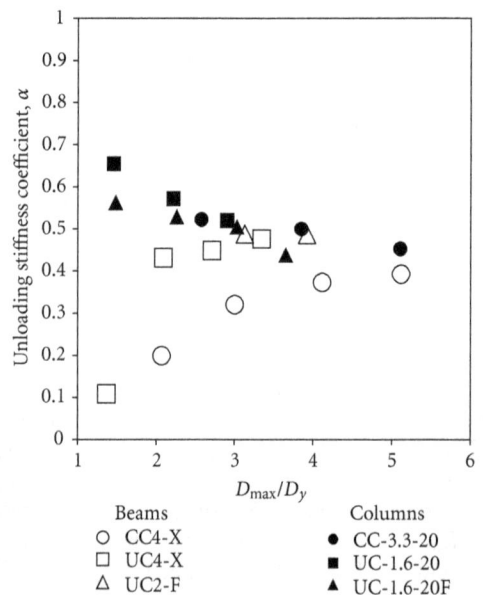

FIGURE 11: Unloading stiffness coefficient for test specimens.

the period range between 0.6 and 1.2 s, is within 10% of the idealized target spectrum based on $S_{DS} = 1.0$ and $S_{D1} = 0.75$ as defined in ASCE/SEI 7-10 [20].

TABLE 6: Ground motions considered.

Station[a]	Earthquake	Magnitude	Epicentral distance km	Site class[b]	PGA[c] g	PGV[d] cm/s
Berkeley, NS (Station 58471) Lawrence Berkeley Lab, Calif., USA	Loma Prieta 10-17-1989	7.1	99.0	C	0.117	22.0
Beverly Hills, NS (Station 00013) 14145 Mulholland Dr., Calif., USA	Northridge 01-17-1994	6.1	12.7	D	0.443	59.3
El Centro, NS (Station 117) Imperial Valley Irrigation District, Calif., USA	Imperial Valley 05-18-1940	6.9	16.9	D	0.348	33.2
El Centro, NS (Station 01335) Imperial Co. Center Grounds, Calif., USA	Superstition Hills 11-24-1987	6.6	36.0	D	0.341	46.6
Lake Hughes, N21E (Station 125, File 1) Fire Station #78, Calif., USA	San Fernando 02-09-1971	7.5	31.3	C	0.148	18.5
Lancaster, NS (Station 24475) Fox Airfield Grounds, Calif., USA	Northridge 01-17-1994	6.1	66.0	D	0.064	5.44
Los Angeles, NS (Station 24303) Hollywood Storage Building Grounds, Calif., USA	Northridge 01-17-1994	6.1	23.0	D	0.231	18.2
Richmond, S10E (Station 58505) City Hall Parking Lot, Calif., USA	Loma Prieta 10-17-1989	7.1	108.0	D	0.106	14.7
Santa Barbara, S48E (Station 283) Courthouse, Calif., USA	Kern County 07-21-1952	7.5	87.8	C	0.131	19.3
Wrightwood, NS (Station 23590) Jackson Flat, Calif., USA	Northridge 01-17-1994	6.1	76.0	C	0.056	5.06

[a] Center for Engineering Strong Motion Data, CESMD [24]. Sensors are part of the California Strong Motion Instrumentation Program, CSMIP.
[b] Based on the classifications of sites in ASCE/SEI 7-10 [20].
[c] PGA: Peak ground acceleration.
[d] PGV: Peak ground velocity.

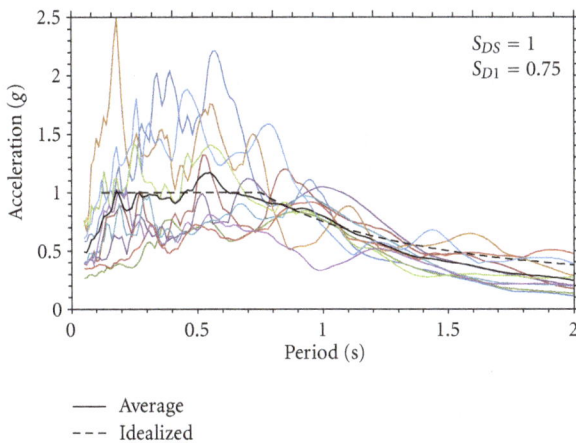

FIGURE 12: Scaled acceleration response spectra.

6.3. Displacement Response Comparison. The calculated maximum displacement will be generally larger for the SDOF systems representing Grade-830 and Grade-670 reinforced structures because of their lower initial and unloading stiffness. Figure 13 compares the calculated maximum displacement of the Grade-670 systems with those of the Grade-410 systems when subjected to the suite of 10 earthquake records. The mean for the ratios of the maximum displacement of the Grade-670 systems to the maximum

displacement of the Grade-410 systems was 1.13 with a coefficient of variation of 0.13.

Similarly to Figure 13, the data in Figure 14 show the comparison of calculated displacement maxima of Grade-830 systems with those of Grade-410 systems for the same suite of ground motions. The mean for the ratios of the maximum displacement of the Grade-830 systems to the maximum displacement of the Grade-410 systems was 1.28 with a coefficient of variation of 0.23. The ratios reported in Figures 13 and 14 should be considered as upper-bound estimates of the increase in displacement demands for moment frames where the reinforcement is replaced with reinforcement of higher grade while maintaining the member cross sections. These estimates were obtained ignoring the effects of initial uncracked stiffness on hysteretic response. Otani [22] has shown that these effects are not always negligible.

Numerical simulations by Rautenberg [8] of multistory concrete frames under strong ground motions have indicated that models of frame buildings with columns reinforced with Grade-830 longitudinal steel and beams reinforced with Grade-410 steel produced roof drifts 1.03 times larger (on average) than the roof drifts computed for models of buildings with columns reinforced with twice as much Grade-410 longitudinal steel. This is because in seismic-resistant frames, a relatively larger number of plastic hinges form in the beams than in the columns. Additional studies are underway to evaluate the nonlinear seismic response

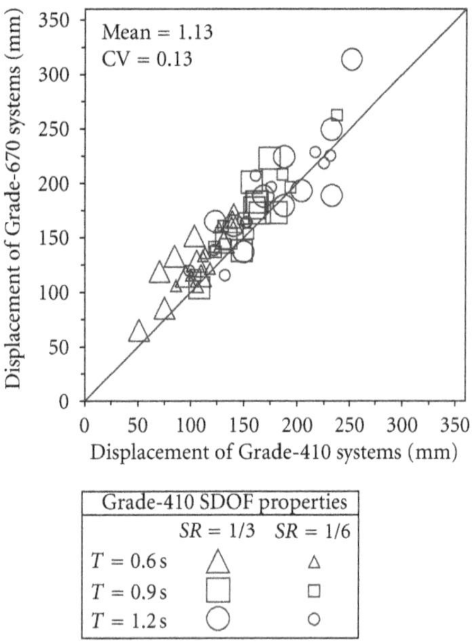

FIGURE 13: Calculated maximum displacement response, Grade-670 versus Grade-410 systems.

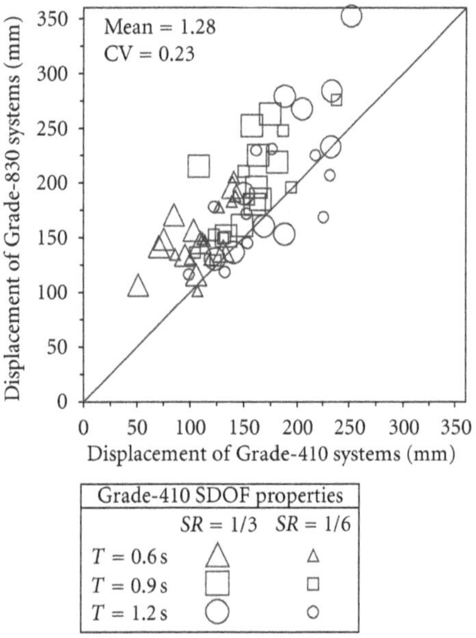

FIGURE 14: Calculated maximum displacement response, Grade-830 versus Grade-410 systems.

of multistory frames where both beams and columns are reinforced with AHSS.

7. Summary and Conclusions

Observations on the nonlinear cyclic response of concrete frame members reinforced with advanced high-strength steel (AHSS) are summarized as follows.

(1) Replacing conventional Grade-410 longitudinal reinforcement with reduced amounts of AHSS reinforcement maintained flexural strength and did not decrease the usable member deformation capacity. The tested beams tolerated drift ratios in excess of 10% without failure while the column specimens tolerated drift ratios of 5% before failure. Column failures in RC specimens were due to buckling of the longitudinal reinforcement while failures in HPFRC specimens were due to fracture of the longitudinal reinforcement.

(2) Increasing the spacing of Grade-410 transverse reinforcement from $d/4$ in RC specimens to $d/2$ in HPFRC specimens, did not reduce the member deformation capacity.

(3) Reducing the amount of longitudinal reinforcement while increasing the yield strength of the reinforcement decreased the postcracking stiffness and increased the yield deformation of the member, leading to a reduction of the area inside the load-deformation hysteresis loops. Reductions in the amount of longitudinal reinforcement achieved by reducing bar diameter may lead to increased vulnerability to bar buckling.

(4) Nonlinear seismic analyses of SDOF systems with identical strength indicated that the mean ratio of calculated maximum displacements for systems representing RC with AHSS reinforcement to those calculated for RC with conventional Grade-410 reinforcement were about 1.1 for Grade-670 systems and 1.3 for Grade-830 systems.

These observations suggest that AHSS reinforcement is a viable option for frame members in earthquake-resistant construction. Additional studies are needed to investigate the nonlinear seismic response of multistory concrete frames reinforced with AHSS bars.

Abbreviation

The following symbols are used in the paper:
A_g: Gross area of concrete section
b: Width of concrete section
C_y: Base shear strength coefficient, ratio of base shear strength to total weight
d: Distance from extreme compression fiber to centroid of longitudinal tension reinforcement
d_b: Diameter of reinforcing bar
D_{max}:   Maximum displacement
D_y: Yield displacement
E_s: Modulus of elasticity of steel, taken as 200,000 MPa (29,000 ksi)
f'_c: Compressive strength of concrete
f_u: Tensile strength of longitudinal reinforcement
f_y: Yield strength of longitudinal reinforcement

f_{yt}: Yield strength of transverse reinforcement
g: Acceleration due to gravity
h: Total depth of concrete section
k_{py}: Post-yield stiffness
k_r: Reloading stiffness
k_u: Unloading stiffness
k_y: Secant stiffness to yield point
M_{pr}: Probable flexural strength of members, determined using a stress of 1.25 f_y in the longitudinal bars
P: Applied axial load
s: Center-to-center spacing of transverse reinforcement
S_A: Design spectral acceleration, 5%-damped linear response
S_{DS}: Design spectral acceleration parameter at short periods, 5%-damped linear response
S_{D1}: Design spectral acceleration parameter at a period of 1 s, 5%-damped linear response
SR: Strength ratio, ratio of base shear strength to 5%-damped linear-response base shear
T: Period of vibration
V_{max}: Maximum shear
V_y: Yield strength
α: Unloading stiffness coefficient
ε_{su}: Fracture strain of reinforcing steel measured in a 203 mm (8-inch) gage length
v_{max}: Maximum shear stress, $V_{max}/(b\,d)$.

Acknowledgments

The support provided by The Pennsylvania State University, Concrete Reinforcing Steel Institute, SAS Stressteel Inc., MMFX Technologies Corporation, Bekaert Corporation, Neturen Corporation, and Kenny Construction Company is greatly appreciated.

References

[1] ACI 318-71, "Building code requirements for reinforced concrete (ACI 318-71)," ACI Standard, American Concrete Institute, Detroit, Mich, USA, 1971.

[2] ACI 318-11, "Building code requirements for structural concrete (ACI 318-11) and commentary," ACI Standard, American Concrete Institute, Farmington Hills, Mich, USA, 2011.

[3] International Iron and Steel Institute, "Advanced High Strength Steel (AHSS) Application Guidelines," 2007, http://www.worldautosteel.org.

[4] ASTM A1011/A1011M-10, *Standard specification for steel, sheet and strip, hot-rolled, carbon, structural, high-strength low-alloy, high-strength low-alloy with improved formability, and ultra-high strength*, ASTM International, West Conshohocken, Pa, USA, 2010.

[5] ASTM A1035/A1035M-04, *Standard specification for deformed and plain, low-carbon, chromium, steel bars for concrete reinforcement*, ASTM International, West Conshohocken, Pa, USA, 2004.

[6] ASTM A706/A706M-09a, *Standard specification for low-alloy steel deformed and plain bars for concrete reinforcement*, ASTM International, West Conshohocken, Pa, USA, 2009.

[7] H. Tavallali, *Cyclic response of concrete beams reinforced with ultrahigh strength steel*, Ph.D. thesis dissertation, The Pennsylvania State University, University Park, Pa, USA, 2011.

[8] J. Rautenberg, *Drift capacity of concrete columns reinforced with high-strength steel*, Ph.D. thesis dissertation, Purdue University, West Lafayette, Ind, USA, 2011.

[9] G. J. Parra-Montesinos, "High-performance fiber-reinforced cement composites: an alternative for seismic design of structures," *ACI Structural Journal*, vol. 102, no. 5, pp. 668–675, 2005.

[10] F. E. Richart and R. L. Brown, "An investigation of reinforced concrete columns," Bulletin 267, Engineering Experiment Station, University of Illinois, Urbana, Ill, USA, 1934.

[11] E. Hognestad, "High strength bars as concrete reinforcement, part 1—introduction to a series of experimental reports," *Journal of the PCA Research and Development Laboratories*, vol. 3, no. 3, pp. 23–29, 1961.

[12] J. F. Pfister and A. H. Mattock, "High strength bars as concrete reinforcement, part 5—lapped splices in concentrically loaded columns," *Journal of the PCA Research and Development Laboratories*, vol. 5, no. 2, pp. 27–40, 1963.

[13] C. E. Todeschini, A. C. Bianchini, and C. E. Kesler, "Behavior of concrete columns reinforced with high strength steels," *ACI Journal Proceedings*, vol. 61, no. 6, pp. 701–715, 1964.

[14] ACI 318-63, "Building code requirements for reinforced concrete (ACI 318-63)," ACI Standard, American Concrete Institute, Detroit, Mich, USA, 1963.

[15] ACI SP-10, "Commentary on building code requirements for reinforced concrete (ACI 318-63)," Standard Building Code, American Concrete Institute, Detroit, Mich, USA, 1965.

[16] W. C. Liao, S. H. Chao, S. Y. Park, and A. E. Naaman, "Self-consolidating high-performance fiber reinforced concrete (SCHPFRC)—preliminary investigation," UMCEE 06-02, University of Michigan, Ann Arbor, Mich, USA, 2006.

[17] S. J. Foster, "On behavior of high-strength concrete columns: cover spalling, steel fibers, and ductility," *ACI Structural Journal*, vol. 98, no. 4, pp. 583–589, 2001.

[18] FEMA 461, *Interim Testing Protocols for Determining the Seismic Performance Characteristics of Structural and Nonstructural Components (FEMA 461)*, Applied Technology Council for the Federal Emergency Management Agency, Washington, DC, USA, 2007.

[19] J. K. Wight and M. A. Sozen, "Strength decay of RC columns under shear reversals," *Journal of the Structural Division*, vol. 101, no. 5, pp. 1053–1065, 1975.

[20] ASCE/SEI 7-10, "Minimum design loads for buildings and other structures," ASCE Standard, American Society of Civil Engineers, Reston, Va, USA, 2010.

[21] T. Takeda, M. A. Sozen, and M. N. Nielsen, "Reinforced concrete response to simulated earthquakes," *Journal of the Structural Division*, vol. 96, no. 12, pp. 2557–2573, 1970.

[22] S. Otani, "Hysteresis models of reinforced concrete for earthquake response analysis," *Journal of the Faculty of Engineering*, vol. 36, no. 2, pp. 125–159, 1981.

[23] N. M. Newmark, "A method of computation for structural dynamics," *Journal of the Engineering Mechanics Division*, vol. 85, no. 3, pp. 67–94, 1959.

[24] CESMD, "Center for Engineering Strong Motion Data," 2011, http://www.strongmotioncenter.org.

8

Quantitative Acoustic Emission Fatigue Crack Characterization in Structural Steel and Weld

Adutwum Marfo, Ying Luo, and Chen Zhong-an

Faculty of Civil Engineering and Mechanics, Jiangsu University, Zhenjiang 212013, China

Correspondence should be addressed to Chen Zhong-an; chenzhan@ujs.edu.cn

Academic Editor: Ghassan Chehab

The fatigue crack growth characteristics of structural steel and weld connections are analyzed using quantitative acoustic emission (AE) technique. This was experimentally investigated by three-point bending testing of specimens under low cycle constant amplitude loading using the wavelet packet analysis. The crack growth sequence, that is, initiation, crack propagation, and fracture, is extracted from their corresponding frequency feature bands, respectively. The results obtained proved to be superior to qualitative AE analysis and the traditional linear elastic fracture mechanics for fatigue crack characterization in structural steel and welds.

1. Introduction

Paris and Erdogan [1] demonstrated that linear elastic fracture mechanics (LEFM) is a useful tool for characterizing crack growth by fatigue. Since that time, application of fracture mechanics to fatigue problems has become a fair routine. Acoustic emission technology is the most appropriate nondestructive testing (NDT) method for studying fatigue crack growth in civil engineering structure because it can monitor its health in real time [2]. Effective crack detection may lead to an early warning. The AE technique can be used to continuously detect slight deformation and damage in the interior of materials. In other words, sampling AE signals and analyzing their characteristics may contribute to the understanding of the real-time failure behavior of materials [3].

The AE parametric analyses have been commonly employed during fatigue crack growth characterization. Ohtsu and Tomoda [4] reported that the AE waveform shape depends on the cracking mode, enabling the classification of cracks in different materials. Shear cracks generally follow tensile as the material approaches to final failure. Yoneda and Ye [5] report that failure phenomena in metals can be interpreted by evaluating the amplitude distribution, AE event count, and total AE energy. Aggelis et al. [6] discuss the application of other AE parameters, such as rise angle (RA) value, rise time (RT), AE hit rate, and duration damage

characterization of metal. They realized that as the duration and RT increase, there is a shift of cracking mode from tensile to shear. Boinet et al. [7] correlated the AE parameters like rise time and duration with corrosive processes in aluminium.

A good correlation between AE parameters and fracture mechanics principles during fatigue has been reported by [8, 9]. Grosse et al. [10] reported the pros and cons of the parametric AE analysis. They postulated that in practical applications it can be difficult to discriminate an AE signal from noise after the signal has been reduced to a few parameters.

Quantitative techniques that deal with the study of AE signal waveform have been applied in various engineering fields for damage evaluation. The fast Fourier transform (FFT) has been used to decompose a time-domain sequence in terms of a set of basic functions. A major problem in using the FFT results from the fact that the transform is the result of integration in the continuous time domain over the entire signal length [11–13]. This problem led to the evolution of the time-frequency data processing methods, such as the short time Fourier transform (STFT). Neild et al. [14] provided a thorough review of various time-frequency techniques for structural vibration analysis. The wavelet transform which is the main interest of this paper has been successfully combined with AE signal parameter for analysis of real-time failure process, such as differentiation of crack types, quantification of damage, and identification of AE source locations [15–17]. This paper discusses the fatigue crack growth characterization in structural steel and weld using quantitative methods. The frequency feature bands corresponding to crack growth sequence, that is, initiation, crack propagation, and fracture, are extracted and compared with qualitative AE analysis and LEFM.

2. Wavelet Packet Theory

Wavelet transform, a contemporary technique for data and signal analysis, has already found its distinguished place in nondestructive testing. Wavelets are mathematical functions that cut up data into different frequency components and then study each component with a resolution matched to its scale. Several types of wavelets which could be successfully applied to AE analysis are Shannon wavelets, Haar wavelets, Daubechies wavelets, Meyer wavelets, Gaussian wavelets, Mexican hat wavelets, Morlet wavelets, and complex frequency B-spline wavelets.

The continuous wavelet transform of an arbitrary function $f(t)$ is defined as

$$WT_x(a,b) = \frac{1}{\sqrt{|a|}} \int_{-\infty}^{\infty} x(t)\psi^*\left(\frac{t-b}{a}\right)dt$$
$$= \langle x(t), \psi_{a,b}(t)\rangle, \tag{1}$$

where

$$\psi_{a,b} = \frac{1}{\sqrt{|a|}}\psi\left(\frac{t-b}{a}\right) \quad a,b \in R, \ a \neq 0. \tag{2}$$

The parameters "a" and "b" stand for scale and shift of the basic wavelet. For each scale "a" and position "b" the time-domain signal is multiplied by shifted and scaled versions of the wavelet function.

The discrete wavelet transform (DWT) of a discrete time sequence $x(n)$ is given as

$$C_{j,k} = 2^{(-j/2)}\sum_n x(n)\psi\left(2^{-j}n-k\right), \tag{3}$$

where $\psi(n)$ is the wavelet function and $2^{(-j/2)}\psi(2^{-j}n-k)$ are scaled and shifted versions of $\psi(n)$ based on the values of scaling coefficient "j," and shifting coefficient "k" wavelet packets transform (WPT) is a generalization of wavelet analysis offering a richer decomposition procedure. WPT can be used to decompose signals into different components in different time windows and frequency bands which can be considered as features of the original signals. Wavelet packets decompose the low frequency component as well as the high frequency component in every subband. The Gabor function is used as the analyzing wavelet because it provides a small window in the time as well as in the frequency domain.

Consider

$$\Psi_{\text{GABOR}}(t) = \frac{1}{\sqrt[4]{\pi}}\sqrt{\frac{\omega_o}{\gamma}}\exp\left[-\frac{(\omega_o/\gamma)^2 t^2}{2} + i\omega_o t\right]. \tag{4}$$

TABLE 1: Mechanical property of 16Mn steel and weld.

	σ_b (MPa)	σ_s (MPa)	δ (%)	E (MPa)
Base metal	360	560	37	206000
Weld	294	441	31	206000

σ_b: yielding stress, σ_s: ultimate tensile stress, δ: percentage of elongation, E: Young's modulus.

TABLE 2: Chemical properties of 16Mn steels.

C	Mn	Si	S	P	Ca
0.16	1.42	0.31	0.033	0.022	0.10

In the frequency domain, the Gabor function is represented as

$$\Psi_{\text{GABOR}} = \frac{2\pi}{\sqrt[4]{\pi}}\sqrt{\frac{\gamma}{\omega_o}}\exp\left[-\frac{(\gamma/\omega_o)^2}{2}(\omega-\omega_o)^2\right]. \tag{5}$$

3. Experimental Details

3.1. Specimens Design. Received hot-rolled 16Mn steel was supplied in the standard thermomechanical heat treatment condition in the form of 16 mm thickness plates. The standard three-point bending (SENB) specimens were designed from the steel and the butt welded connection, as a representative part of a steel bridge in accordance with ASTM E647 standards. The mechanical properties and chemical composition of the specimens used are shown in Tables 1 and 2, respectively. The specimens were notched using electrical discharge machining (EDM) to an initial crack length of 18 mm. The specimen was fatigue precracked using the MTS machine at a frequency of 10 Hz of length 0.5 mm. The specimens' surfaces were mechanically polished by grinding and buffing to permit observations of the crack path. The detailed geometry of the specimen is illustrated in Figure 1.

3.2. Test Instrument and Procedure. A servohydraulic testing machine with a maximum load capacity of 250 kN was used for the fatigue tests at an ambient temperature of 300 K (Figure 2). The specimens were tested under sinusoidal cyclic loading at a frequency of 7.5 Hz. The specimens were tested with different peak loads (16 kN and 10 kN) at a load R-ratios of 0.1. At least three specimens were tested under each condition in other to ensure regularity in the experiment. A SWAES full-waveform acoustic emission detector was used to record the AE signals generated during the fatigue tests.

The AE signals were detected by using 2 broadband piezo-electric sensors with frequency range of 10 kHz to 2 MHz. Vaseline was used at the interface between the sensors and the specimen surface to obtain proper signals. A preamplifier of

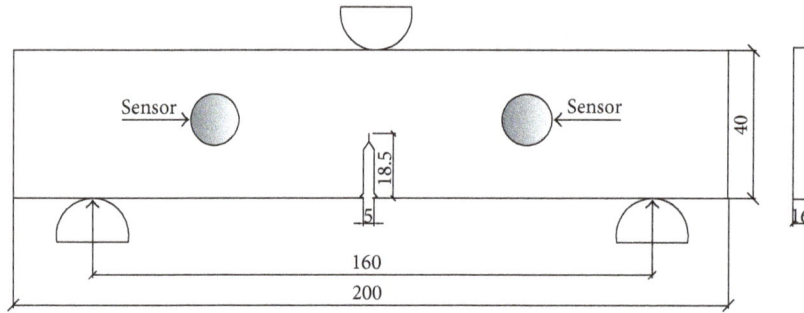

FIGURE 1: Specimen geometry.

FIGURE 2: Experimental setup.

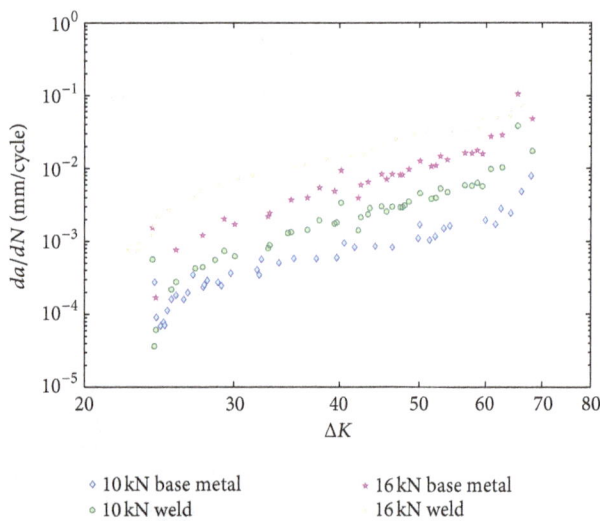

FIGURE 4: Relationships between crack length and number of cycles.

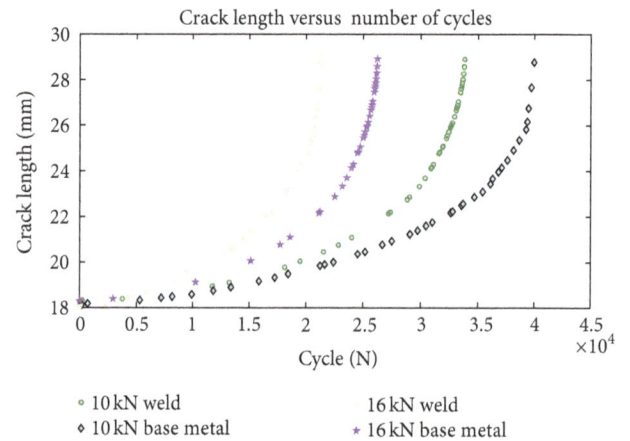

FIGURE 3: Relationships between crack growth rates and stress intensity factor range.

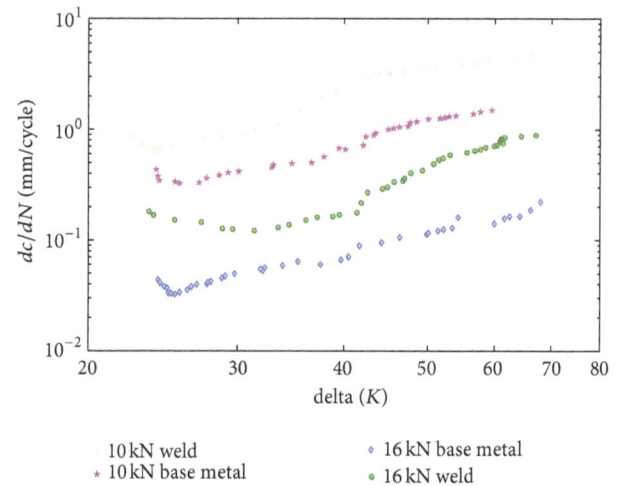

FIGURE 5: Relationship between cumulative count rate and stress intensity factor range.

40 db gain was used to capture the AE signals. The crack-tip opening-displacement gauge (CTOD) was used to monitor the fatigue crack growth in the structure. The crack length at any given cycle was arithmetically computed using

$$\frac{a}{w} = 0.999748 - 3.9504U_v + 2.981U_V{}^2 - 3.21408U_V{}^3$$
$$+ 51.51564U_V{}^4 - 113.031U_V{}^5,$$

(6)

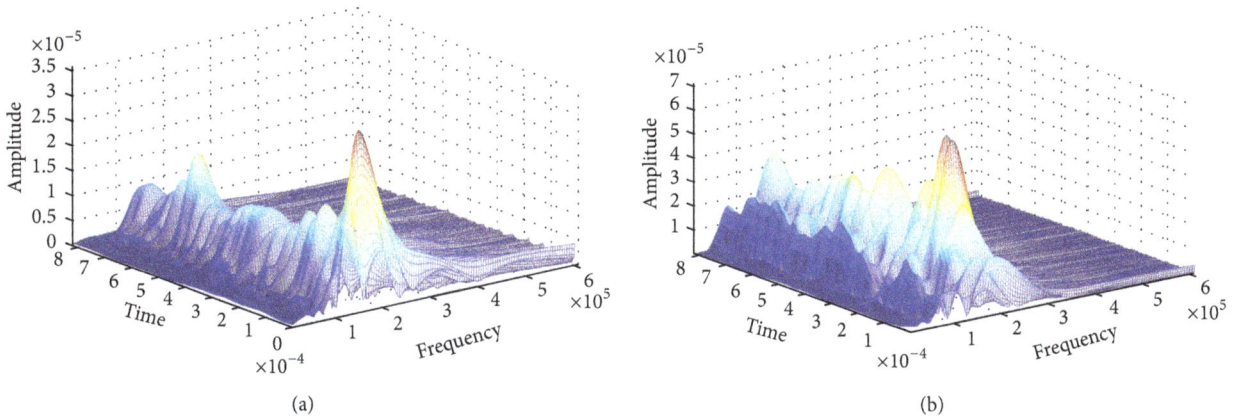

FIGURE 6: Fatigue crack initiation (a) base metal and (b) welded specimen.

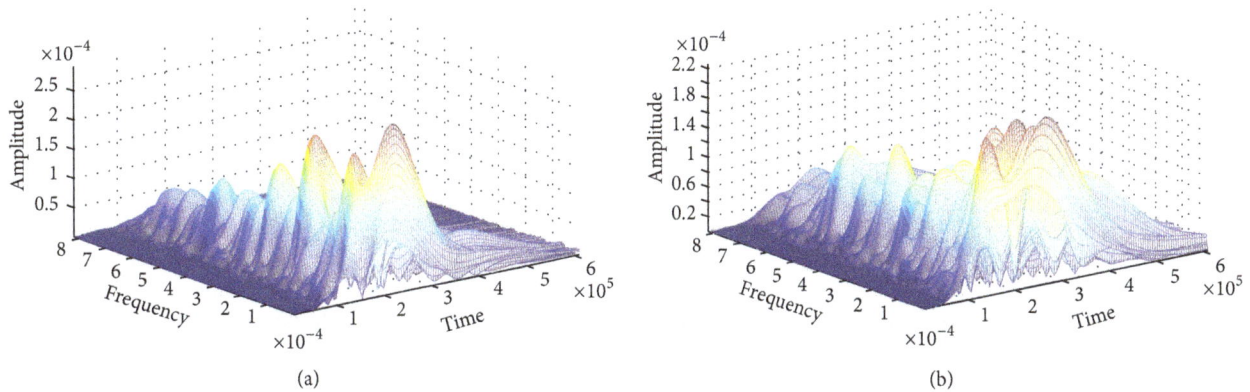

FIGURE 7: Fatigue crack propagation (a) base metal and (b) welded specimen.

where $U_v = 1/\sqrt{(4z_v w/s) + 1}$, $Z_v = BEV/P$, $V =$ crack-mouth-opening displacement, $a =$ crack length, $W =$ specimen width, $B =$ specimen thickness, $S =$ span, and $E =$ modulus of elasticity.

4. Results and Discussions

4.1. Linear Elastic Fracture Mechanics. Paris and Erdogan [1] discovered the power-law relationship for the fatigue crack growth and proposed an exponent of 4 for the constant m after series of experiments. Figures 3 and 4 show the relationship between d the fatigue crack growth rates (da/dN) and stress intensity factor ranges (ΔK) and crack length and number of cycles under different load ratios, respectively.

In nearly the whole fatigue lives, they obeyed the Paris law (7) for the steel and welds

$$\frac{da}{dN} = C\Delta K^m, \qquad \log\left(\frac{da}{dN} = \log C + m\log\Delta K\right), \quad (7)$$

where a is a representative crack length, n is the number of fatigue cycles, ΔK is the applied stress intensity factor range, and C and m are assumed to be constants for a particular material. However, the crack growth rates were increased in the weld relative to in the base metal, suggesting that the

cracks propagated more rapidly in the weld due to the changes in the microstructure. We realized from the diagram that peak load was less significant on the crack propagation rate.

4.2. Acoustic Emission during Fatigue Crack Propagation. The fatigue crack growth characteristics can be analyzed by studying the parameters of the AE signals so generated. Figure 5 shows the relationships between the cumulative AE counts rates (dC/dN) and stress intensity factor ranges ΔK for the welded specimens under different peak loads. The AE counts rates increased in a linear relationship with the increase in ΔK on the log-log axes, which is well consistent with (7). Compared to the results from LEFM, the peak load had influence on dC/dN on the welded specimen. Moreover, higher AE counts rates were also observed in the welded specimens than in the base metal specimens. In addition, the slopes of the lines for the welded specimens were somehow higher than the base metal specimens, also suggesting that the weld generated more AE signals during fatigue crack propagation.

4.3. Wavelet Packet Analysis. The characteristics of fatigue crack propagation during the three-point bending testing of the steel beam and weld connection are classified under

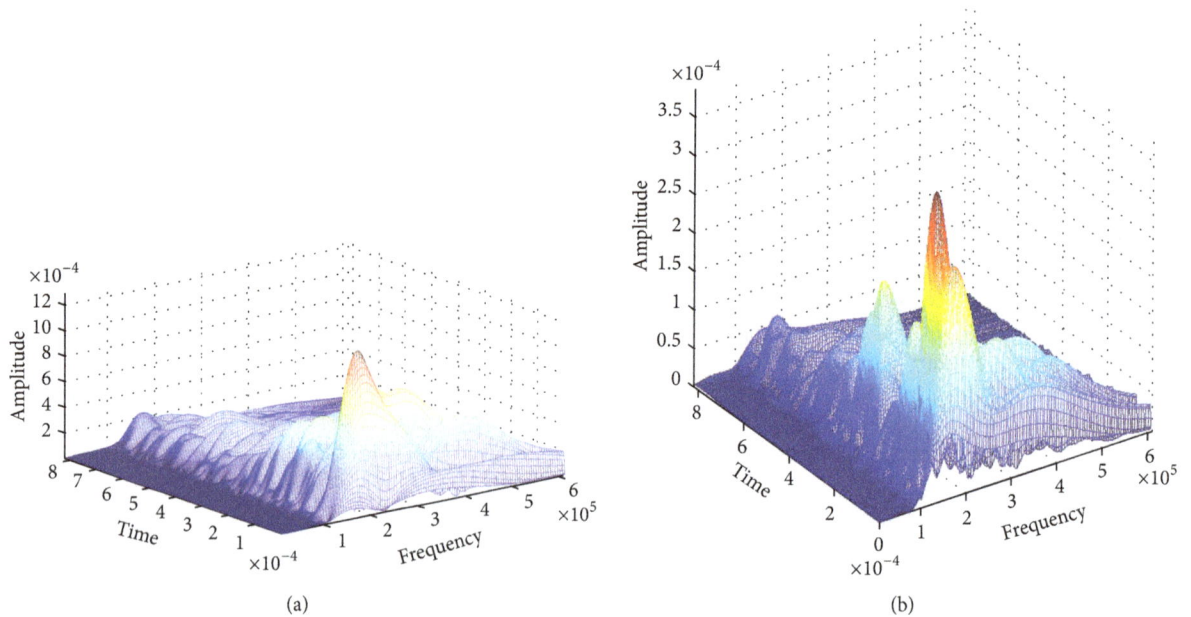

FIGURE 8: Fracture (a) base metal and (b) welded specimen.

3 stages: crack initiation, crack propagation, and failure at various peak loads corresponding to regions I, II, and III, respectively.

Figures 6(a) and 6(b) show the AE wave in region 1 which corresponds to fatigue source initiation for both weld metal and base metal, respectively. The waveform is low amplitude, wide pulse with a narrow frequency scale, mostly located at 80 kHz to 180 kHz. At this stage, the AE signals are generated by the formation of crack source and plastic deformation on the tip of the notch which generate intense AE events. Higher amplitudes are recorded for the weld than the base metal due to the release of residual stresses. Figures 7(a) and 7(b) show the AE in region 2 corresponding to fatigue crack propagation. The waveform at this stage as compared to the fatigue source has high amplitude, narrow pulse, with a wide energy scale, and a main frequency scale of 250 kHz to 350 kHz for the base metal and 250 kHz to 450 kHz for the welded specimen.

Figures 8(a) and 8(b) show the AE waveform at the rapid crack propagation stage; the energy of the AE signal increased until the specimen completely failed, and the AE waveform amplitude is higher than those of the earlier stages. The waveform characteristic shows that this type of AE signal is a burst signal. The fracture waveform is a high amplitude narrow pulse, with a wide energy scale, and a main frequency scale of 300 kHz to 400 kHz for the base metal and between 300 kHz and 600 kHz for the welded specimen.

5. Conclusion

From the above discussions, it is evident that fatigue crack growth rates for the welded specimen are higher than the base metal; this was enhanced by the presence of inclusions and heterogeneous microstructure of the welds. The effect

of peak load on fatigue characterization was found to be insignificant. Furthermore, the quantitative technique of the wavelet transform provided clear results for crack propagation characterization in both the steel and welded specimen.

References

[1] P. C. Paris and F. Erdogan, "Critical analysis of crack propagation Laws," *Journal of Basic Engineering*, vol. 85, pp. 528–534, 1960.

[2] T. M. Roberts and M. Talebzadeh, "Fatigue life prediction based on crack propagation and acoustic emission count rates," *Journal of Constructional Steel Research*, vol. 59, no. 6, pp. 679–694, 2003.

[3] L. Yang, Y. C. Zhou, W. G. Mao, and C. Lu, "Real-time acoustic emission testing based on wavelet transform for the failure process of thermal barrier coatings," *Applied Physics Letters*, vol. 93, no. 23, Article ID 231906, 2008.

[4] M. Ohtsu and Y. Tomoda, "Phenomenological model of corrosion process in reinforced concrete identified by acoustic emission," *ACI Materials Journal*, vol. 105, no. 2, pp. 194–199, 2008.

[5] K. Yoneda and J. Ye, "Crack propagation and acoustic emission behavior of silver-added Dy123 bulk superconductor," *Physica C*, vol. 445–448, no. 1-2, pp. 371–374, 2006.

[6] D. G. Aggelis, E. Z. Kordatos, and T. E. Matikas, "Acoustic emission for fatigue damage characterization in metal plates," *Mechanics Research Communications*, vol. 38, no. 2, pp. 106–110, 2011.

[7] M. Boinet, J. Bernard, M. Chatenet, F. Dalard, and S. Maximovitch, "Understanding aluminum behaviour in aqueous alkaline solution using coupled techniques. Part II: acoustic emission study," *Electrochimica Acta*, vol. 55, no. 10, pp. 3454–3463, 2010.

[8] P. Johan Singh, C. K. Mukhopadhyay, T. Jayakumar, S. L. Mannan, and B. Raj, "Understanding fatigue crack propagation in AISI 316 (N) weld using Elber's crack closure concept:

experimental results from GCMOD and acoustic emission techniques," *International Journal of Fatigue*, vol. 29, no. 12, pp. 2170–2179, 2007.

[9] K. Bruzelius and D. Mba, "An initial investigation on the potential applicability of Acoustic Emission to rail track fault detection," *NDT and E International*, vol. 37, no. 7, pp. 507–516, 2004.

[10] C. U. Grosse, H. Reinhardt, and T. Dahm, "Localization and classification of fracture types in concrete with quantitative acoustic emission measurement techniques," *NDT and E International*, vol. 30, no. 4, pp. 223–230, 1997.

[11] S. Park, N. Stubbs, and R. W. Bolton, "Damage detection on a steel frame using simulated modal data," in *Proceedings of the 16th International Modal Analysis Conference (IMAC '98)*, pp. 616–622, Santa Barbara, Calif, USA, February 1998.

[12] W. X. Ren and G. de Roeck, "Structural damage identification using modal data. I: simulation verification," *Journal of Structural Engineering*, vol. 128, no. 1, pp. 87–95, 2002.

[13] S. W. Doebling, C. R. Farrar, and M. B. Prime, "A summary review of vibration-based damage identification methods," *Shock and Vibration Digest*, vol. 30, no. 2, pp. 91–105, 1998.

[14] S. A. Neild, P. D. McFadden, and M. S. Williams, "A review of time-frequency methods for structural vibration analysis," *Engineering Structures*, vol. 25, no. 6, pp. 713–728, 2003.

[15] H. Kim and H. Melhem, "Damage detection of structures by wavelet analysis," *Engineering Structures*, vol. 26, no. 3, pp. 347–362, 2004.

[16] M. A. Hamstad, A. O'gallagher, and J. Gary, "A wavelet transform applied to acoustic emission signals: part 1: source identification," *Journal of Acoustic Emission*, vol. 20, pp. 39–61, 2002.

[17] R. Khamedi, A. Fallahi, and A. R. Oskouei, "Effect of martensite phase volume fraction on acoustic emission signals using wavelet packet analysis during tensile loading of dual phase steels," *Materials and Design*, vol. 31, no. 6, pp. 2752–2759, 2010.

9

Bridge Deck Load Testing Using Sensors and Optical Survey Equipment

Hubo Cai,[1] Osama Abudayyeh,[2] Ikhlas Abdel-Qader,[3] Upul Attanayake,[2] Joseph Barbera,[2] and Eyad Almaita[3]

[1] Division of Construction Engineering and Management, School of Civil Engineering, Purdue University, 550 Stadium Mall Drive, West Lafayette, IN 47907, USA
[2] Department of Civil and Construction Engineering, Western Michigan University, Kalamazoo, MI 49008, USA
[3] Department of Electrical and Computer Engineering, Western Michigan University, Kalamazoo, MI 49009, USA

Correspondence should be addressed to Hubo Cai, hubocai@purdue.edu

Academic Editor: Polat Gülkan

Bridges are under various loads and environmental impacts that cause them to lose their structural integrity. A significant number of bridges in US are either structurally deficient or functionally obsolete, requiring immediate attention. Nondestructive load testing is an effective approach to measure the structural response of a bridge under various loading conditions and to determine its structural integrity. This paper presents a load-test study that evaluated the response of a prefabricated bridge with full-depth precast deck panels in Michigan. This load-test program integrates optical surveying systems, a sensor network embedded in bridge decks, and surface deflection analysis. Its major contribution lies in the exploration of an embedded sensor network that was installed initially for long-term bridge monitoring in bridge load testing. Among a number of lessons learned, it is concluded that embedded sensor network has a great potential of providing an efficient and accurate approach for obtaining real-time equivalent static stresses under varying loading scenarios.

1. Introduction

Bridges are a critical component of a nation's ground transportation infrastructure system that allows the movement of people and goods from one place to another. Bridges are exposed to load effects and weather impacts that lead to their deterioration. Federal Highway Administration (FHWA) statistics show that about 20% of the National Highway System (NHS) bridges and 27% of non-NHS bridges in US are either structurally deficient or functionally obsolete [1], resulting in a total of about 152,000 bridges requiring immediate attention in the form of repair or replacement. Therefore, routine inspection and load testing are needed to obtain good estimates on bridge response and to support making decisions of appropriate maintenance and rehabilitation activities to ensure their structural integrity and safety [2, 3].

A recent trend in bridge management is the utilization of sensors embedded in the bridge structure to monitor the long-term performance of bridges, under various loads and environmental impacts. Given the monitored data, meaningful information regarding the bridge reliability can be extracted [4]. This is particularly true with the growing adoption of the prefabrication technology in bridge construction. This technology allows the prefabrication of structural components to be conducted offsite under a well-controlled environment with the benefit of higher-quality precast structural components that are expected to perform better with lower maintenance needs, compared to conventional cast-in-place concrete bridges [5]. By reducing the onsite construction time, this prefabrication technology can also save construction time and minimize traffic disruption [6]. Due to its relatively short history in bridge construction, long-term performance data of prefabricated bridges are rarely existent, and sensor technology is many times utilized to monitor their long-term performance. This scenario of embedding sensors in bridge structure is also the case with the Parkview Bridge in Kalamazoo, Michigan, which was constructed in 2006 as a totally prefabricated system. Vibrating-wire strain gages (sensors) were embedded in the precast deck panels to form a sensor network for continuous health monitoring and condition assessment.

While the original intention of embedding sensors into bridges is to continuously monitor their long-term performance, these sensors may serve as an alternative method for measuring the structural response of a bridge under nondestructive load testing. This paper presents the results of a load testing study to evaluate the response of the Parkview Bridge under varying load configurations via nondestructive load testing and the utilization of an embedded sensor network. In this study, a load testing method that incorporates optical surveying systems, the embedded sensor network, and deflection analysis was designed and implemented. The major contribution of this study is its demonstration of how an embedded sensor network might be utilized to acquire real-time equivalent static stresses under various loading scenarios.

2. Bridge Load Testing

The actual response of a bridge to loads is usually better than what the theory dictates [7]. Factors that contribute to the load capacity difference include unintended composite action, load distribution effects, participation of parapets, railings, curbs, and utilities, material property differences, unintended continuity, participation of secondary members, effects of skew, portion of load carried by deck, and unintended arching action due to frozen bearings [7].

Load testing is recommended by AASHTO [8] as an "effective means of evaluating the structural response of a bridge." The purpose of conducting load testing on existing bridges is to evaluate their structural response without causing damages. Therefore, load testing is usually conducted in a nondestructive manner and is sometimes referred to as nondestructive load testing. The goal of this type of testing is to compare field response of the bridge under test loads with its theoretical response [7]. Nondestructive load testing can be further categorized into diagnostic testing and proof testing. Diagnostic testing methods provide the measurements necessary to analyze differential loading effects (i.e., moment, shear, axial force, deflection, etc.) present in various structural members due to applied loads [9]. Proof-load testing aims at determining the magnitude and configuration of loads that cause critical structural components to approach their elastic limit.

Tasks involved in a program of load testing typically include the determination of testing objectives and load configuration, the selection and placement of instrumentation, the adoption of appropriate analysis techniques, and the evaluation and comparison of test results and analytical results [10]. Load testing is being carried out widely to evaluate and rate bridge response on a case-by-case manner to evaluate new construction materials and technologies [10, 11]. While a number of bridge load testing studies can be found in the literature, such studies on bridges using embedded sensors, which is the focus of this paper, are almost nonexistent.

3. Bridge under Test: Parkview Bridge

The Parkview Bridge is located in Kalamazoo, MI. The 249-foot long bridge over US-131 is particularly relevant concerning the advancement of both bridge construction technology and structural monitoring. Being the first of its kind in Michigan, the concrete superstructure of the Parkview Bridge is almost entirely precast, including girders and deck panels. The substructure also has precast concrete piers. This configuration allowed for the implementation of rapid bridge construction (RBC) technology. A second feature of the structure is a network of sensors that was embedded in the deck panels during casting. This sensor network will continuously monitor the response of the bridge to various loads and environmental effects (temperature changes).

The Parkview Bridge consists of four spans, which from east to west are approximately 38, 83, 83, and 45 feet, respectively. The three lanes of traffic are supported by seven precast prestressed AASHTO type III girders spaced at 8'7". Figure 1(a) provides a cross-sectional view of the bridge to illustrate the girder spacing and the deck-panel dimensions. The bridge deck comprises 48 precast nonprestressed 9-inch thick deck panels, skewed at 23 degrees and posttensioned longitudinally; see Figure 1(b). The sensors within the deck panels are placed at varying locations and orientations with the objective of measuring all relevant stresses that are predicted to develop in the bridge deck. Figure 2 illustrates the locations and orientations of the sensors embedded in the bridge deck panels.

4. Load Testing Design and Implementation Methodology

This section describes the objectives, approaches, testing scenarios, load configurations, and testing procedures related to the design and implementation of the load testing program.

4.1. Load Testing Objectives and Approaches. The overall goal of conducting load tests on the Parkview Bridge is to quantify its response to various loading conditions. Specific objectives include measuring roadway-surface deflections due to various live loads, deriving stresses from measured deflections via analytical models, and comparing analytical results with results derived from sensor readings.

Figure 3 presents a flow chart that describes the approaches and tasks used to carry out the testing. The first task is to determine load configurations that should produce the maximum moments in the bridge girders and the maximum deflections of the bridge deck surface. The next step is to compute stresses using calculated girder deck composite-section properties and measured deflections. Roadway-surface deflections are measured by optical surveying using a Trimble Dini level with a precision of 0.0012 inch (0.003048 cm). Stresses derived from measured deflections are compared to stresses derived from sensor strain readings for verification and further analysis.

4.2. Testing Scenarios. A total of ten load scenarios, including four single directional (one truck) and six bidirectional (two

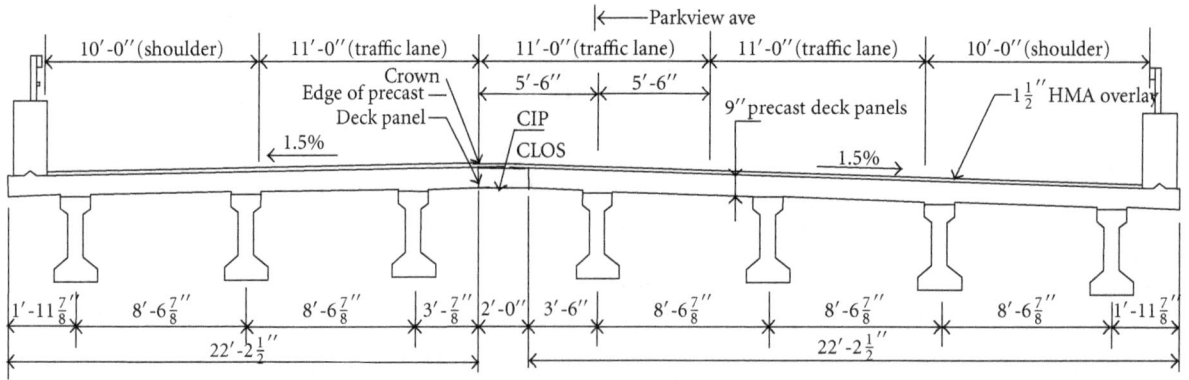

(a) Cross-sectional view of the Parkview Bridge

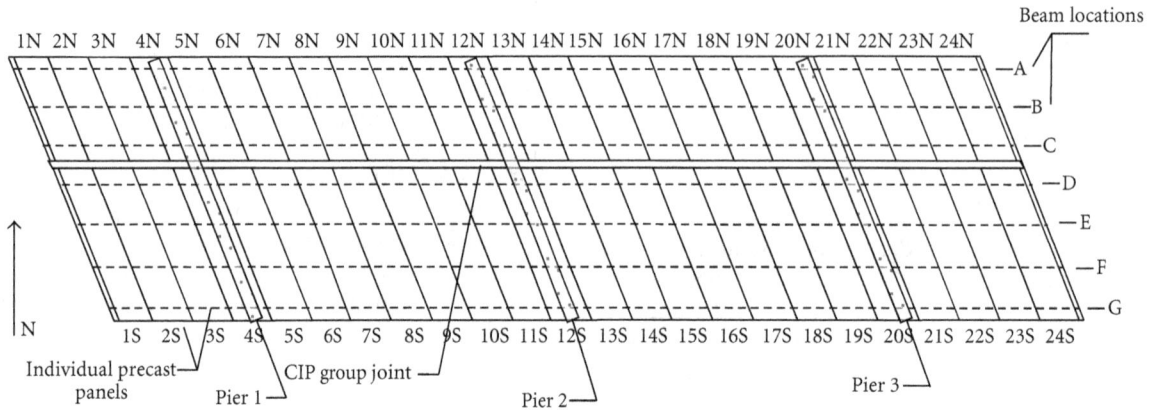

(b) Deck-panel plan view

FIGURE 1: Parkview bridge schematic.

TABLE 1: Testing scenarios.

Scenarios	Truck location (single direction—1 truck)	Truck location (bidirectional—2 trucks)
1	47	—
2	42	—
3	49	—
4	40	—
5	—	45, 44
6	—	47, 42
7	—	49, 40
8	—	51, 38
9	—	47, 40
10	—	45, 38

trucks with opposite heading directions), were designed and implemented in this study. Figure 4 illustrates the midspan locations, where the test loads were placed. Table 1 lists the testing scenarios with their load locations. Before the load testing, roadway surface elevations at these midspan locations, and also at locations above the bridge piers (represented by triangles), were measured before the loads were in place to provide a surface baseline. During the load tests, when the loads were in place, surface elevations at

these locations were measured again to determine surface deflections due to specific test loads.

4.3. Load Configuration. Two types of trucks were used to provide test loads. Figure 5 illustrates the configuration of the trucks used for the single-directional and the bidirectional testing. In this study, single-directional testing means using only one truck, while bidirectional testing means using two trucks simultaneous and when these two trucks were driven to their intended locations facing each other. These configurations were chosen to closely match two trucks that are included in Michigan's set of legal truck-load configurations. The type I truck illustrated in Figure 5 was chosen to approximately match the HS20 design truck used in the design of the bridge. Table 2 provides the measured axle weights for all three trucks (one type I truck and two type II trucks) used in this study.

4.4. Load Testing Procedure. The steps below were followed.

Step 1. Load trucks to approximate AASHTO HS20 and legal Michigan truck configuration and utilize truck weigh stations to determine actual axle weights.

Step 2. Based upon actual axle weight, mark a point on each truck in such a way that when this point is aligned with

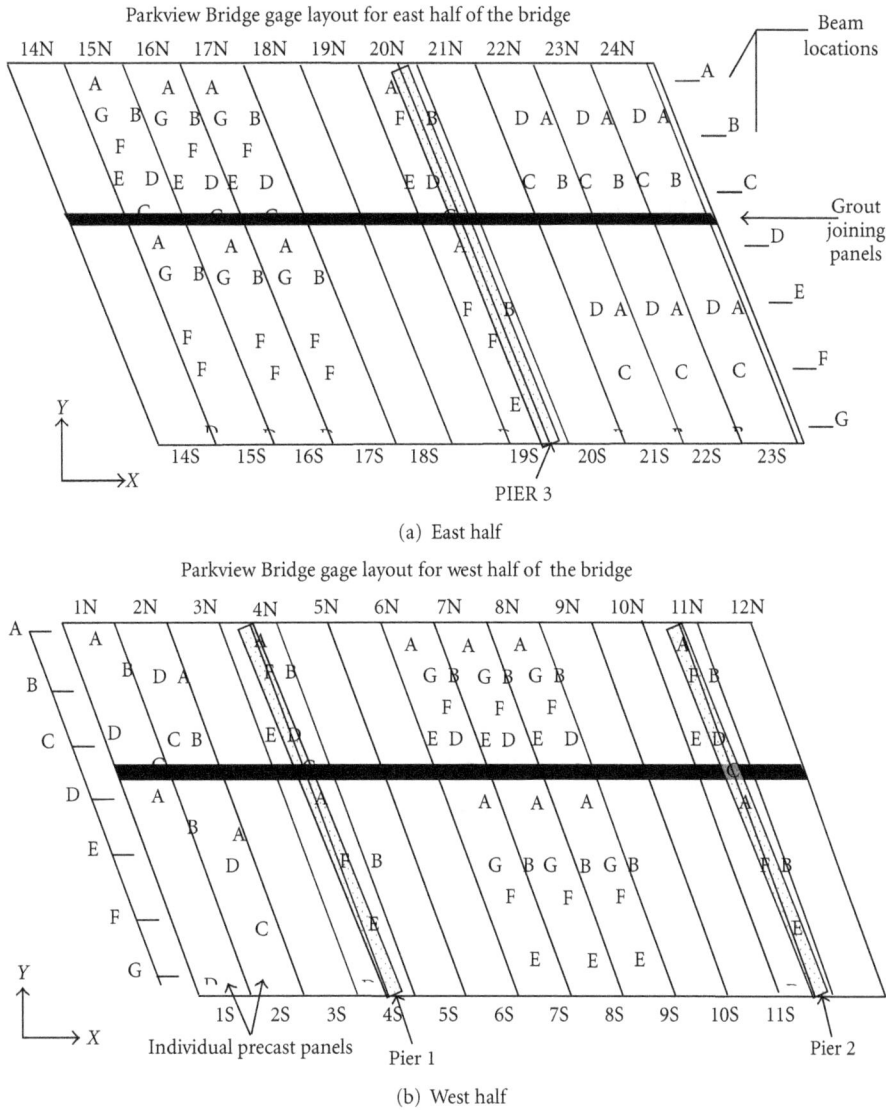

Parkview Bridge gage layout for east half of the bridge

(a) East half

Parkview Bridge gage layout for west half of the bridge

(b) West half

FIGURE 2: Gage layout and orientation for the Parkview Bridge.

TABLE 2: Actual loaded truck weights.

	Loaded truck weight configurations		
Axle no.	Single directional truck type 1 weights (pounds*)	Bidirectional truck type 2 weights (pounds*)	
Front axle	9,640	17,850	18,350
No. 2 axle	35,540	18,050	18,600
No. 3 axle		17,800	18,250
No. 4 axle	34,580	—	—
No. 5 axle		—	—
Gross weight	79,760	53,700	55,200

* 1 pound of mass = 0.4536 kg; 1 pound of force = 4.4482 N.

midspan locations of deck panels, the truck load causes the maximum moment in deck panels.

Step 3. While the trucks are being loaded prior to testing, the survey crew sets up and determines girder midspan locations and pier centerline locations as shown in Figure 4. These locations are marked for future alignment of truck marks and midspan locations in Step 5.

Step 4. Measure surface elevations of points illustrated in Figure 4 to establish a baseline for testing scenario 1.

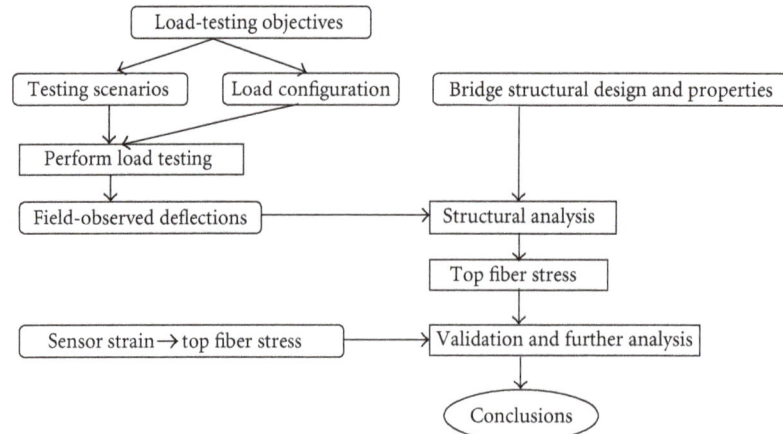

Figure 3: Flow chart of the bridge load testing.

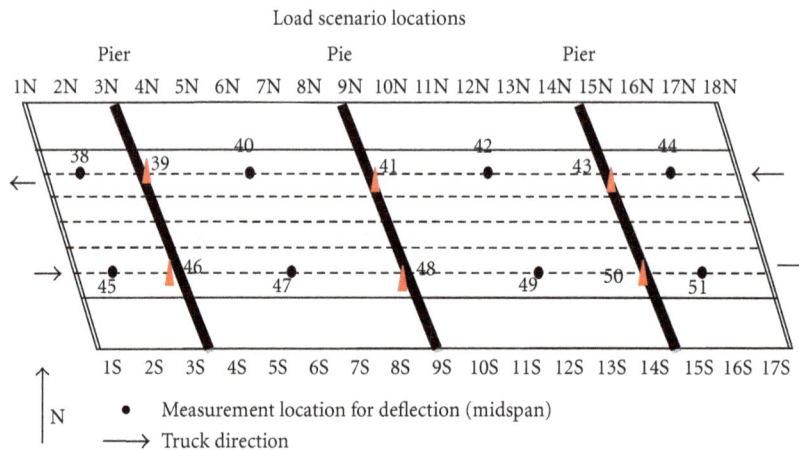

Figure 4: Load locations.

Step 5. Move type I truck to point 47 in Figure 4 (scenario 1). After the truck is in place, measure surface elevations to determine bridge deflection due to the truck load.

Step 6. Record the time the truck is in position and ensure that the truck stays in position for a minimum of 10 minutes to allow the sensors to register their readings (sensors collect data at 5-second increments during the 10-minute time period, and these readings were averaged).

Step 7. Move the truck off the bridge.

Step 8. Repeat Steps 4 to 7 for the remaining testing scenarios.

4.5. Control of Temperature Variations. It is well recognized that environmental factors, particularly temperature variations, can lead to large strains, and when the structure is restrained as in bridges, they can cause large stresses, even though the wire strain gages are self-temperature compensating. Considering the extreme weather conditions that are common in Michigan during winter, load testing

studies shall at least plan for temperature variations. In this load testing study, the original plan of controlling the effects of temperature variation was to start the testing in the morning and monitor the ambient temperature, and when it falls out of the 5° range, the testing pauses and resumes either later that day when the temperature cools down or next morning. In reality, it rained on the day of testing, and the temperature variation was a way below the 5° range to allow all the 10 scenarios to be tested on a single day.

5. Load Testing Results

The load-test program was performed twice: first in September 2008 before the bridge was opened to traffic and again in June 2009. During the first load test, the sensor network was not yet operational. During the second load test, the sensor network was functional and was used in the load test of the bridge. This section presents the results from the second load test only.

Table 3 summarizes surface deflections measured in the field during the second load testing, under the ten loading scenarios.

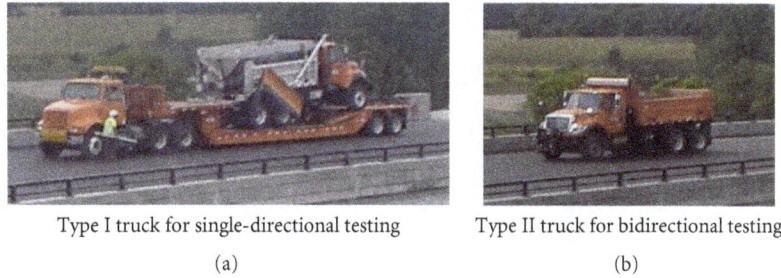

Type I truck for single-directional testing
(a)

Type II truck for bidirectional testing
(b)

FIGURE 5: General configurations of trucks.

TABLE 3: Measured surface deflections.

Location	Surface deflections in 10 scenarios (inches*)									
	1	2	3	4	5	6	7	8	9	10
38	0.01	0.05	−0.05	0.04	0.01	0.00	0.01	−0.06	0.00	−0.05
39	0.01	−0.04	−0.02	0.02	0.02	−0.01	−0.02	0.00	0.00	−0.01
40	0.00	−0.05	−0.02	−0.04	0.00	−0.01	−0.06	0.01	−0.07	0.02
41	0.02	−0.02	−0.04	−0.03	0.00	0.00	−0.02	0.00	0.00	−0.01
42	0.00	−0.09	−0.02	0.04	0.01	−0.07	−0.01	0.02	0.02	0.00
43	0.00	−0.02	−0.03	0.01	0.04	−0.01	0.00	0.01	0.01	−0.01
44	−0.02	−0.01	−0.03	0.00	−0.03	0.00	0.01	0.01	0.00	0.01
45	0.01	−0.05	−0.04	0.02	0.02	0.00	0.00	−0.02	0.01	−0.02
46	0.10	−0.02	−0.04	0.04	0.01	0.02	−0.01	−0.01	0.02	0.21
47	−0.04	−0.04	−0.02	0.01	−0.01	−0.07	0.00	0.02	−0.06	0.01
48	0.02	−0.01	−0.05	0.00	−0.02	0.00	0.02	0.01	−0.02	0.00
49	0.01	−0.01	−0.07	0.01	−0.01	0.01	−0.04	0.03	−0.01	0.00
50	0.00	−0.01	−0.03	−0.01	−0.02	−0.02	0.01	0.01	0.00	0.00
51	−0.02	−0.01	0.00	−0.02	0.00	0.01	0.00	−0.01	−0.01	0.01

*1 inch = 2.54 cm.

(a) Simulated deflected shape

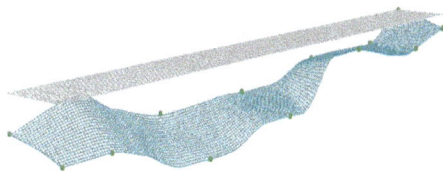

(b) Actual deflected shape

FIGURE 6: Validating scenario 1 results (not drawn to scale).

5.1. Field Measurements Validation. Three-dimensional plots of surface-deflection measurements were developed and compared with the finite element (FE) simulation results. Figure 6 illustrates the comparison for testing scenario 1. The deflected shape of the surface obtained from field observations matches the deflected shape from analytical results, providing confidence in the measurements from testing scenario 1 observations. Similar comparisons were conducted to validate results in all the 10 testing scenarios.

5.2. Top Fiber Live Load Stresses from Deflection Measurements. After validating field observations, moments were derived from surface deflections, and top fiber stresses were calculated utilizing the two PCI equations below and using the simply supported moments at midspans (i.e., assuming zero moments at the piers for conservative results at midspans) [12]. The moments at the piers were then computed from the mid span moments using distribution factors obtained from simulated, unit-force loadings that mimic the truck loads from the 10 scenarios,

$$M_{LL} = D\left(\frac{48EI_c}{5L^2}\right),$$
$$\sigma_{LL} = \frac{M_{LL}y}{I_c},$$

(1)

where M_{LL}: live load moment; D: deflection (inches); E: section modulus of elasticity (4,600,000 psi); I_c: moment of inertia of composite section (438,913 in^4); M_{LL}: moment (lb-in); L: span length (inches); y: distance from the top fiber to the neutral axis (18 inches); σ_{LL}: stress (psi).

The live-load stress results derived from deflection measurements are presented in Tables 4 and 5 (negative values indicate compression).

TABLE 4: Live load stresses from deflections (analytical)—south side panels (psi*).

Scenarios	Midspan (45)	Pier 1 (46)	Midspan (47)	Pier 2 (48)	Midspan (49)	Pier 3 (50)	Midspan (51)
Scenario 1	47	40	−34	12	10	−9	−50
Scenario 2	−186	−159	−29	−10	−11	−10	−17
Scenario 3	−140	−119	−19	−7	−57	−55	−7
Scenario 4	93	80	6	2	6	−6	−63
Scenario 5	70	60	−7	−2	−6	−6	−10
Scenario 6	0	0	−58	21	5	−5	33
Scenario 7	0	0	0	0	−35	34	−3
Scenario 8	−93	−80	14	5	20	19	−30
Scenario 9	47	40	−51	−19	−8	−7	−23
Scenario 10	−96	82	9	−3	−2	2	36

* 1 psi = 6,895 kPa.

TABLE 5: Live load stresses from deflections (analytical)—north side panels (psi*).

Scenarios	Midspan (38)	Pier 1 (39)	Midspan (40)	Pier 2 (41)	Midspan (42)	Pier 3 (43)	Midspan (44)
Scenario 1	47	0	0	0	0	−57	−66
Scenario 2	186	−37	−38	−25	−68	−23	−27
Scenario 3	−186	−18	−19	−5	−12	−60	−70
Scenario 4	140	27	−28	−13	35	9	10
Scenario 5	47	4	−4	−2	5	81	−95
Scenario 6	0	−9	−10	21	−58	−9	10
Scenario 7	47	−5	−5	−4	−12	31	36
Scenario 8	−233	6	7	−7	19	23	27
Scenario 9	0	57	−59	4	10	9	−10
Scenario 10	−143	−15	15	−1	−3	−20	23

* 1 psi = 6,895 kPa.

5.3. Top Fiber Live Load Stresses from Sensor Readings. During the load testing, top fiber strains were recorded by the embedded sensors in the bridge deck panels and downloaded to the laboratory computer for analysis. Even though the sensors are installed throughout the entire bridge deck, only those sensors located along the longitudinal load path were used to derive live load stresses and to compare them to the stresses derived from deflections. Tables 6 and 7 present the stresses derived from the sensors for the south-side and the north-side panels, respectively (negative values indicate compression).

5.4. Total Stresses. Live load stresses are added to dead load stresses calculated based upon structural design and material properties of this bridge to obtain the total top fiber stresses, which are then compared to the allowable compression and tension stresses to ensure that the structure is within design limits under varying loading scenarios. In calculating stresses from load test deflections, we conservatively assumed simply supported span moments as follows:

$$\sigma_{\mathrm{LL}} = \frac{M_{\mathrm{LL}} y}{I_c},$$

$$M_{\mathrm{LL}} = D\left(\frac{48EI_c}{5L^2}\right),$$

(2)

where M_{LL}: live load moment (lb-in); D: deflection (inches); E: section modulus of elasticity (4,600,000 psi); I: moment of inertia of composite section (438,913 in^4); L: span length (inches); y: distance from the top fiber to the neutral axis (18 in); σ_{LL}: stress (psi).

The average 28-day compression strength ($f'c$) was recorded as approximately 8,000 psi from concrete samples taken during the casting. Therefore, maximum allowable stresses in the concrete are as follows:

$$\text{compression } (fc)\text{:} \quad fc \leq 0.45\, f'c \implies 3,600\,\text{psi},$$

$$\text{tension } (ft)\text{:} \quad ft \leq 6\sqrt{f'c} \implies 537\,\text{psi}.$$

(3)

Table 8 provides the deck dead-load stresses that are combined with the live-load stresses given in Tables 4 through 7 to compute total stresses presented in Tables 9, 10, 11, and 12 for all scenarios using the deflection measurements and sensor-readings methods. It is clear from these tables that the total stresses are within the allowable limits for all testing scenarios for both the south and north panels.

6. Discussions and Concluding Remarks

Overall, the load tests were effective in providing information about the bridge's structural response. Stresses in all of the

TABLE 6: Live load stresses from sensors—south side panels (psi*).

Scenarios	Midspan (45)	Pier 1 (46)	Midspan (47)	Pier 2 (48)	Midspan (49)	Pier 3 (50)	Midspan (51)
Scenario 1	1.48	10.60	−11.68	6.52	5.38	−6.82	−0.33
Scenario 2	0.80	−1.50	2.52	1.00	9.57	−14.05	0.45
Scenario 3	1.55	0.65	2.48	11.77	−24.14	8.97	2.80
Scenario 4	3.00	11.70	−2.43	11.95	2.75	−0.65	−0.20
Scenario 5	−2.80	−2.05	0.28	−0.20	−0.27	−0.55	−0.15
Scenario 6	−0.40	6.30	−13.50	22.10	−3.65	11.30	−0.15
Scenario 7	0.93	−0.28	−0.69	11.30	−30.68	4.72	1.18
Scenario 8	−5.95	0.00	0.85	−0.55	0.93	3.35	2.65
Scenario 9	3.50	−2.00	15.63	−21.35	−5.12	3.70	−1.60
Scenario 10	−8.55	−2.40	0.20	0.25	1.38	−0.55	0.05

* 1 psi = 6,895 kPa.

TABLE 7: Live load stresses from sensors—north side panels (psi*).

Scenarios	Midspan (38)	Pier 1 (39)	Midspan (40)	Pier 2 (41)	Midspan (42)	Pier 3 (43)	Midspan (44)
Scenario 1	1.60	8.85	−8.88	4.55	5.47	−0.87	−0.10
Scenario 2	1.95	−1.95	2.32	1.45	14.63	−19.70	0.40
Scenario 3	1.95	0.70	3.82	10.63	−11.06	4.93	4.02
Scenario 4	3.95	18.75	−26.33	17.75	3.53	0.25	0.50
Scenario 5	−2.75	−1.30	−0.10	−0.65	0.65	0.10	0.15
Scenario 6	−0.15	4.45	−6.53	21.90	−25.93	16.60	0.20
Scenario 7	1.55	5.45	−10.87	11.82	−3.51	0.02	1.70
Scenario 8	−6.05	0.35	−0.60	−0.95	−0.92	−2.65	2.55
Scenario 9	4.45	8.65	10.67	−21.30	−10.75	3.45	−0.95
Scenario 10	−6.90	3.10	0.17	−1.25	−1.22	−1.95	0.20

* 1 psi = 6,895 kPa.

TABLE 8: Deck dead load stresses (psi*).

	Top fiber dead load stresses—south							
West abut	Midspan (45)	Pier 1 (46)	Midspan (47)	Pier 2 (48)	Midspan (49)	Pier 3 (50)	Midspan (51)	East abut
0	−436	−466	−773	−430	−760	−379	−428	0
	Top fiber dead load stresses—north							
East abut	Midspan (44)	Pier 3 (43)	Midspan (42)	Pier 2 (41)	Midspan (40)	Pier 1 (39)	Midspan (38)	West abut
0	−428	−379	−760	−430	−773	−466	−436	0

* 1 psi = 6,895 kPa.

TABLE 9: Total stresses based on deflection measurements—south (psi*).

Scenarios	Midspan (45)	Pier 1 (46)	Midspan (47)	Pier 2 (48)	Midspan (49)	Pier 3 (50)	Midspan (51)
Scenario 1	−390	−426	−806	−418	−751	−388	−478
Scenario 2	−623	−626	−801	−440	−771	−389	−444
Scenario 3	−576	−586	−792	−437	−817	−433	−434
Scenario 4	−343	−387	−767	−428	−755	−384	−491
Scenario 5	−366	−407	−779	−432	−766	−384	−438
Scenario 6	−436	−466	−830	−409	−755	−383	−395
Scenario 7	−436	−466	−773	−430	−796	−344	−431
Scenario 8	−529	−546	−758	−425	−740	−359	−458
Scenario 9	−390	−426	−824	−448	−768	−386	−451
Scenario 10	−532	−385	−764	−433	−762	−377	−391

* 1 psi = 6,895 kPa.

TABLE 10: Total stresses based on deflection measurements—north (psi*).

Scenarios	Midspan (38)	Pier 1 (39)	Midspan (40)	Pier 2 (41)	Midspan (42)	Pier 3 (43)	Midspan (44)
Scenario 1	−390	−466	−773	−430	−760	−435	−494
Scenario 2	−250	−503	−811	−455	−828	−401	−454
Scenario 3	−623	−485	−792	−434	−773	−438	−498
Scenario 4	−296	−439	−801	−442	−726	−370	−418
Scenario 5	−390	−463	−776	−432	−755	−298	−522
Scenario 6	−436	−475	−782	−409	−819	−387	−418
Scenario 7	−390	−471	−777	−434	−772	−347	−391
Scenario 8	−669	−460	−766	−437	−741	−356	−401
Scenario 9	−436	−409	−832	−426	−750	−370	−438
Scenario 10	−579	−481	−757	−431	−763	−398	−405

* 1 psi = 6,895 kPa.

TABLE 11: Total stresses based on sensor readings—south (psi*).

Scenarios	Midspan (45)	Pier 1 (46)	Midspan (47)	Pier 2 (48)	Midspan (49)	Pier 3 (50)	Midspan (51)
Scenario 1	−435	−455	−785	−423	−755	−386	−428
Scenario 2	−435	−468	−770	−429	−750	−393	−428
Scenario 3	−434	−465	−771	−418	−784	−370	−425
Scenario 4	−433	−454	−775	−418	−757	−380	−428
Scenario 5	−439	−468	−773	−430	−760	−380	−428
Scenario 6	−436	−460	−787	−408	−764	−368	−428
Scenario 7	−435	−466	−774	−419	−791	−374	−427
Scenario 8	−442	−466	−772	−431	−759	−376	−425
Scenario 9	−433	−468	−757	−451	−765	−375	−430
Scenario 10	−445	−468	−773	−430	−759	−380	−428

* 1 psi = 6,895 kPa.

scenarios were well under the maximum allowable limits. The stress values derived from the strains acquired from the sensor network were compared to those derived from the measured deflections at every location and every scenario. An example comparison of stresses due to live loads for the south deck panels under scenario 1 is illustrated in Figure 7. Such a comparison leads to the following observations:

(i) live load stresses at the deck top fiber are consistently small (both compression and tension stresses are less than 50 psi for scenario 1). In general, stresses due to live load are relatively smaller than stresses due to dead load;

(ii) for a given location in testing scenario 1, the deflection-derived results and sensor-derived results are consistent from the perspective of whether the top-fiber stress is in compression or tension;

(iii) at a given location under a specific testing scenario, the deck-panel top-fiber stresses derived from the sensor readings are consistently smaller than the stresses derived from deflection measurements;

(iv) the stress distribution reveals compression on top fiber (positive moment) at the midspan location where the truck load was located in testing scenario 1 and tension on top fiber (negative moment) at the neighboring pier location. The stresses of the midpoints of nearby spans and pier locations are consistent with the loading configuration.

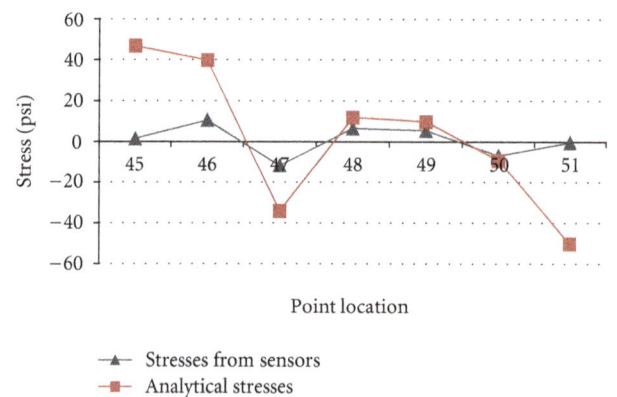

FIGURE 7: Comparison of analytical stresses and sensor stresses—south panels in testing scenario 1.

These comparisons were conducted for all 10 scenarios. It was observed from both sets of results that top fiber stresses due to live load are relatively small when compared to the stresses caused by dead loads and temperature variation. However, the difference between the two sets of calculated stresses is significant, considering the small stress values. Stress values derived from deflection measurements are consistently larger than those derived from sensor readings. Also, the stress types (tension/compression) derived from the deflection measurements do not always match those derived

TABLE 12: Total stresses based on sensor readings—north (psi*).

Scenarios	Midspan (38)	Pier 1 (39)	Midspan (40)	Pier 2 (41)	Midspan (42)	Pier 3 (43)	Midspan (44)
Scenario 1	−426	−370	−769	−425	−768	−467	−436
Scenario 2	−426	−381	−758	−429	−758	−486	−436
Scenario 3	−426	−378	−756	−419	−784	−461	−432
Scenario 4	−424	−360	−786	−412	−769	−466	−436
Scenario 5	−431	−380	−760	−431	−772	−466	−436
Scenario 6	−428	−375	−767	−408	−799	−449	−436
Scenario 7	−426	−374	−771	−418	−777	−466	−434
Scenario 8	−434	−379	−761	−431	−774	−469	−433
Scenario 9	−424	−370	−749	−451	−784	−463	−437
Scenario 10	−435	−376	−760	−431	−774	−468	−436

* 1 psi = 6,895 kPa.

from sensor readings. In other words, some locations may have compression strains based on deflection measurements when the sensors are reading tension strains. These differences may be explained as follows:

(i) the current practice of the analytical methods in calculating stresses is conservative, which confirms the reality that bridges perform better than what theoretical methods predict;

(ii) the conservative assumptions used for computing stresses from deflections, namely using simply supported span moments when the bridge is continuously supported over four spans, result in large computed stresses;

(iii) the surface deflection measures include effects of compression in bearings, slack in component fit, shear deformation, and pier/pile/cap deformations, leading to a larger surface deflection than that of the concrete deck itself;

(iv) the bridge has a 3-inch flexible asphalt overlay on top of the concrete deck panels, possibly resulting in larger surface deflections than what the concrete deck is actually experiencing;

(v) even though the surveying equipment can reach an accuracy of 0.0012 inch (0.003048 cm), the human error cannot be eliminated with methods used. That is, any measured deflections less than 0.01 inch (0.0254 cm) may be considered inaccurate due to the difficulty of eliminating human errors in reading the targets or holding the rod vertically. Therefore, it is believed that sensors provide much more accurate results than the optical survey instrument and whenever possible, sensors, instead of optical survey instrument, shall be used in bridge load testing;

(vi) sensors are not located at the top fiber (18 inches (45.72 cm) from the neutral access of the composite section). Rather, they are 2.5 inches (6.35 cm) below the surface (due to protective-cover requirements) or 15.5 inches (39.37) from the neutral access, resulting in a slightly smaller stress than would exist at the top fiber. Sensor readings and resulting stresses could be adjusted to accommodate this situation and perhaps reduce the differences, but they will remain different.

Advanced deflection-measuring instruments and techniques may eliminate some of the human error, but at a premium cost. In our case, however, since the sensors already existed for the purpose of monitoring the health of the bridge deck, they provided a low-cost, accurate, and quick alternative to the deflection measurement load testing method. Additionally, with sensors installed throughout the bridge, a stress surface can be constructed to provide a complete view of the bridge structure's response to varying loading scenarios, a feature that is difficult to achieve using conventional surveying methods. Therefore, this study concludes that embedded sensors provide a more accurate and precise measuring alternative to measuring the structural response of bridges under various loads.

Acknowledgments

The authors are grateful to the Michigan Department of Transportation (MDOT) for funding this study through MDOT Contract no. 04-0090/Z3. Any opinions, findings, conclusions, or recommendations expressed in this paper are those of the authors and do not necessarily reflect the views of MDOT or Western Michigan University.

References

[1] Federal Highway Administration (FHWA), "FHWA Bridge Programs NBI Data," 2008, http://www.fhwa.dot.gov/bridge/defbr07.cfm.

[2] H. A. Capers and M. M. Valeo, "FHWA's international scan on ensuring bridge safety and serviceability," *Transportation Research Record*, no. 2202, pp. 117–123, 2010.

[3] Federal Highway Administration (FHWA), *Bridge Preservation Guide, Maintaining a State of Good Repair Using Cost Effective Investment Strategies*, FHWA, 2011.

[4] D. M. Frangopol, A. Strauss, and S. Kim, "Bridge reliability assessment based on monitoring," *Journal of Bridge Engineering*, vol. 13, no. 3, pp. 258–270, 2008.

[5] W.-W. Wang, J.-G. Dai, G. Li, and C.-K. Huang, "Long-term behavior of prestressed old-new concrete composite beams," *Journal of Bridge Engineering*, vol. 16, no. 2, pp. 275–285, 2011.

[6] O. Abudayyeh, H. Cai, B. Mellema, and S. Yehia, "Quantifying time and user cost savings for rapid bridge construction technique," *Transportation Research Record*, no. 2151, pp. 11–20, 2010.

[7] NCHRP-234, "Manual for Bridge Rating Through Load-testing," National Cooperative Highway Research Program, Research Results Digest, Number 234, Transportation Research Board, Washington, DC, USA, 1998.

[8] AASHTO, *Manual for Condition Evaluation of Bridges*, Washington, DC, USA, AASHTO, 2nd edition, 2000.

[9] B. Phares, T. Wipf, L. Greimann, and Y. Lee, "Health Monitoring of Bridge Structures and Components Using Smart-Structure Technology," Tech. Rep. 0092-04-14, Wisconsin Highway Research Program, 2005.

[10] D. D. Kleinhans, J. J. Myers, and A. Nanni, "Assessment of load transfer and load distribution in bridges utilizing FRP panels," *Journal of Composites for Construction*, vol. 11, no. 5, pp. 545–552, 2007.

[11] T. Hou and P. J. Lynch, "Rapid-to-deploy wireless monitoring systems for static and dynamic load-testing of bridges: validation on the grove street bridge," in *13th Annual International Symposium on Smart Structures and Materials*, Proceedings of SPIE, San Diego, Calif, USA, February 2006.

[12] PCI, *Bridge Design Manual*, Precast/Prestressed Concrete Institute, Chicago, Ill, USA, 2003.

10

Simple Program to Investigate Hysteresis Damping Effect of Cross-Ties on Cables Vibration of Cable-Stayed Bridges

Panagis G. Papadopoulos, Andreas Diamantopoulos, Haris Xenidis, and Panos Lazaridis

Department of Civil Engineering, Aristotle University of Thessaloniki, 54124 Thessaloniki, Greece

Correspondence should be addressed to Panagis G. Papadopoulos, panaggpapad@yahoo.gr

Academic Editor: Husam Najm

A short computer program, fully documented, is presented, for the step-by-step dynamic analysis of isolated cables or couples of parallel cables of a cable-stayed bridge, connected to each other and possibly with the deck of the bridge, by very thin pretensioned wires (cross-ties) and subjected to variation of their axial forces due to traffic or to successive pulses of a wind drag force. A simplified SDOF model, approximating the fundamental vibration mode, is adopted for every individual cable. The geometric nonlinearity of the cables is taken into account by their geometric stiffness, whereas the material nonlinearities of the cross-ties include compressive loosening, tensile yielding, and hysteresis stress-strain loops. Seven numerical experiments are performed. Based on them, it is observed that if two interconnected parallel cables have different dynamic characteristics, for example different lengths, thus different masses, weights, and geometric stiffnesses, too, or if one of them has a small additional mass, then a single pretensioned very thin wire, connecting them to each other and possibly with the deck of the bridge, proves effective in suppressing, by its hysteresis damping, the vibrations of the cables.

1. Introduction

The pretensioned cables in a typical cable-stayed bridge of medium size [1], as they are very long with a length of magnitude order 100 m, and a pretension axial force of magnitude order 1000 kN, exhibit, perpendicularly to their axis, a very small geometric stiffness, corresponding to their fundamental vibration mode, of a magnitude order only 50 kN/m. Also perpendicularly to their axis, they exhibit a very small intrinsic damping, due to their material internal friction. For the previous reasons, they are often subjected to large amplitude vibrations. And, if the external excitation is approximately periodic, with a period close to a natural period of the cable, for example, the fundamental one, then resonance may happen, and vibration amplitudes increase excessively and are maintained, with no significant reduction for a long time, unless special measures are taken.

Two usual reasons for the previous cable vibrations of cable-stayed bridges are the following.

(1) A pretensioned cable exhibits a sag under its self-weight. Because of traffic, the ends of the cable, on pylon and deck, are subject to a variation of their displacements; thus the elongation and axial force of the cable vary, which implies variation of its geometric stiffness, too, as well as variation of the sag of the cable. This vibration, due to variation of geometric stiffness, is called parametric excitation.

(2) The successive pulses of a wind pressure exert a drag force, perpendicularly to the vertical plane of cables, at one side of the bridge. The variation of this drag force causes vibration of the cables.

In [2], a complete description of the problem of cable vibrations of cable-stayed bridges is presented, as well as a state of the art of various types of dampers for cable vibrations (viscous dampers, cross-ties, and others), along with case studies of dampers on real bridges.

The viscous dampers, although widely mentioned in literature, present some problems: usually, they are installed at the ends of a cable, where they are not very helpful; it seems that their main role is a slight reduction of cable's length, thus a slight increase of its geometric stiffness. Rarely, they

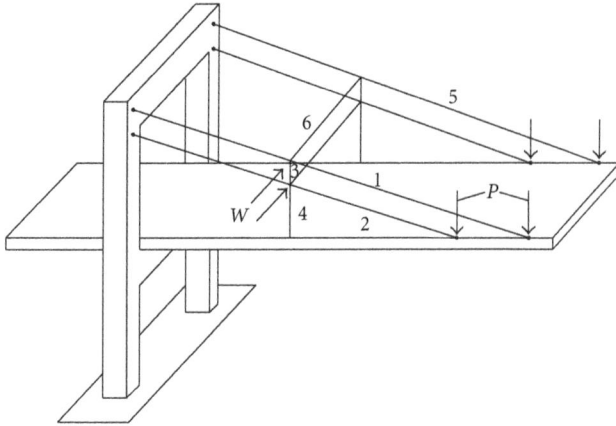

FIGURE 1: Part of a typical cable-stayed bridge [1].

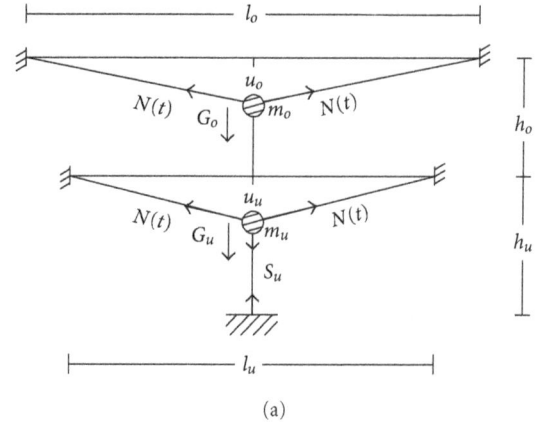

(a)

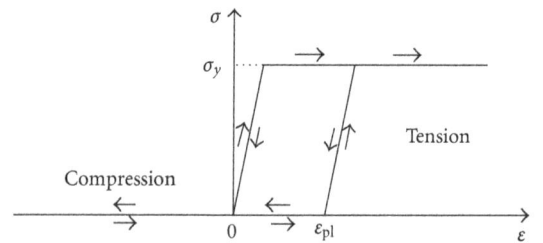

(b)

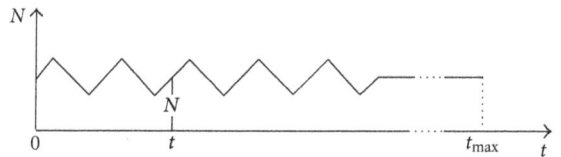

(c)

FIGURE 2: (a) Geometric, static, and dynamic parameters of a cable structure. (b) Primary σ-ε diagram describing the nonlinear axial stress-strain law of a cross-tie. (c) Given time-history of axial forces of cables.

are installed at intermediate points of a cable, where they are more helpful; however, this installation is difficult.

On the other hand, the cross-ties are preferable, for the following reasons: they are light and cheap, they are easily installed and pretensioned, and they easily replaced when damaged. And a great advantage of them is that although they are very thin, with a ratio of cross-section area of the cable to that of the cross-tie of a magnitude order 1000, however, the axial elastic stiffness, of a single pretensioned cross-tie, is comparable in magnitude with the geometric stiffness of a cable, that is of magnitude order 50 kN/m, along the same direction, perpendicularly to cable axis. Also, as the cross-ties are very thin, they are almost invisible, so they do not harm the aesthetics of the bridge.

For the previous reasons, recently many researchers recommend the use of cross-ties to suppress large amplitude cable variations of cable-stayed bridges. In [3–6], analytical studies are performed on cross-ties or hybrid systems consisting of viscous dampers and cross-ties.

Here, a simplified analytical method is proposed [7], in order to investigate the hysteresis damping effect of cross-ties, where, for every individual cable, an SDOF model is adopted, approximating its fundamental vibration mode. The geometric nonlinearity of the cables is taken into account by their geometric stiffness. At same time, the proposed method is accurate, as it includes the material non-linearities of the cross-ties, by their compressive loosening, tensile yielding, as well as hysteresis stress-strain loops.

A short computer program (only about 120 Fortran instructions), fully documented, is presented for the step-by-step dynamic analysis [9] of isolated cables or couples of parallel cables, connected to each other and possibly with the deck of the bridge, by very thin pretensioned wires (cross-ties) and subjected to variation of the axial forces of cables due to traffic [8] or to successive pulses of wind drag force.

Seven numerical experiments are performed. And, based on them, observations are made on the effectiveness of a single pretensioned very thin wire, connecting a couple of cables of a cable-stayed bridge, in suppressing, by its hysteresis damping, their large amplitude vibrations.

2. Equations of the Problem

Figure 1 shows a part of a typical cable-stayed bridge [1], with a pylon consisting of two vertical legs, which are connected by two transverse beams, a part of the deck with a slender rectangular plate section, two couples of pretensioned parallel inclined cables at each side of the bridge, connected by their ends to the pylon and the deck and two pretensioned very thin vertical in-plane cross-ties, at two sides of bridge, which intend to suppress parametric vibration of cables due to traffic loads P, as well as two out-of-plane horizontal cross-ties, which intend to suppress cable vibrations due to wind forces W.

In the following, the equations of nonlinear dynamic analysis will be written, for a specific cable structure consisting of two parallel pretensioned cables (1 and 2 in Figure 1), connected by two very thin pretensioned cross-ties (3 and 4 in Figure 1) to each other and with the deck of the bridge and subjected to parametric vibration, due to traffic. In subsequent applications, by simple and obvious modifications of these equations, other cable structures

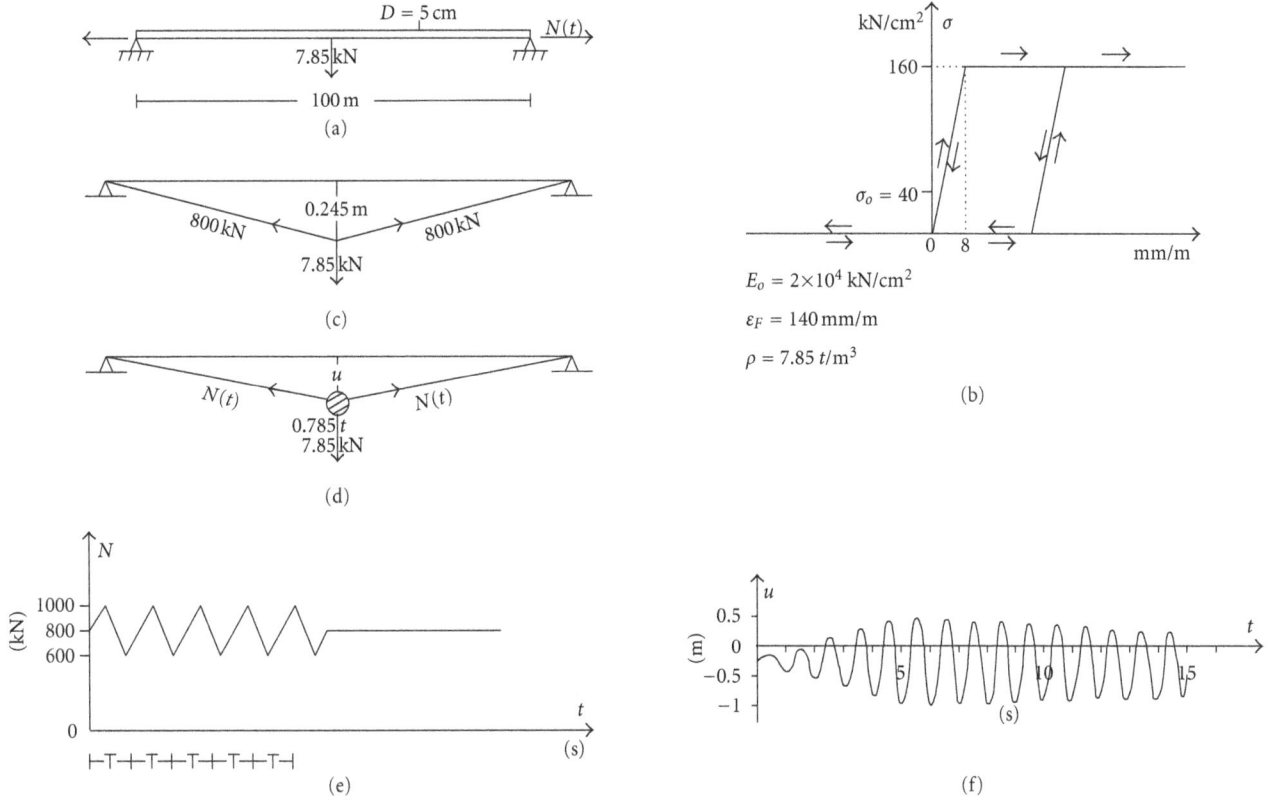

FIGURE 3: First application: isolated cable subject to traffic. (a) Given geometry and loading. (b) Axial stress-strain diagram of high-strength steel. (c) Initial static analysis. (d) Parameters of dynamic analysis. (e) Given time-history $N(t)$ of axial force of cable. (f) Resulting time-history $u(t)$ of vertical displacement at center of cable.

(combinations of main cables and cross-ties) subjected to traffic or wind excitation will be treated, too.

In the following analysis, the inclination of the cables will be ignored for the sake of simplicity. So, in Figure 2(a), the couple of horizontal parallel pretensioned main cables represents the inclined cables 1 and 2 of Figure 1, whereas the very thin pretensioned vertical wires represent the in-plane cross-ties 3 and 4 of Figure 1.

2.1. Geometric Equations.
For every individual cable, a simplified SDOF model is adopted, which approximates its fundamental vibration mode. This unique DOF, for every cable, is the displacement of its center perpendicularly to its axis, that is, the vertical displacements downwards u_o and u_u in Figure 2(a), for the upper and lower cables, respectively. The geometric equations, relating the displacements of cables with the elongations and strains of cross-ties, are the following, according to Figure 2(a):

$$\Delta h_o = h_o - u_o + u_u - h_{o\varphi},$$
$$\Delta h_u = h_u - u_u - h_{u\varphi}, \tag{1}$$

where Δh_o and Δh_u are elongations of upper and lower cross-ties, respectively, h_o and h_u their design (nominal) lengths, and $h_{o\varphi}$ and $h_{u\varphi}$ are their initial undeformed lengths. And the

strains ε_o and ε_u of upper and lower cross-ties, respectively, are

$$\varepsilon_o = \frac{\Delta h_o}{h_{o\varphi}},$$
$$\varepsilon_u = \frac{\Delta h_u}{h_{u\varphi}}. \tag{2}$$

2.2. Constitutive Equations.
Figure 2(b) is the primary stress-strain diagram, which describes the axial nonlinear stress-strain law of a cross-tie, made of the same high-strength steel as the main cables. This σ-ε law includes compressive loosening, tensile yielding, as well as hysteresis stress-strain loops, resulting from the obvious in loading-unloading rule Figure 2(b). There is only one constitutive variable, the plastic strain ε_{pl} of the cross-tie. The present stress σ is an obvious function of present strain ε and present value of plastic strain ε_{pl}:

$$\sigma = \sigma\left(\varepsilon, \varepsilon_{pl}\right), \tag{3a}$$

whereas the variation of plastic strain $\Delta\varepsilon_{pl}$ can be expressed as a function of present strain ε, variation of present strain $\Delta\varepsilon$, as well as present value of plastic strain ε_{pl}, in an

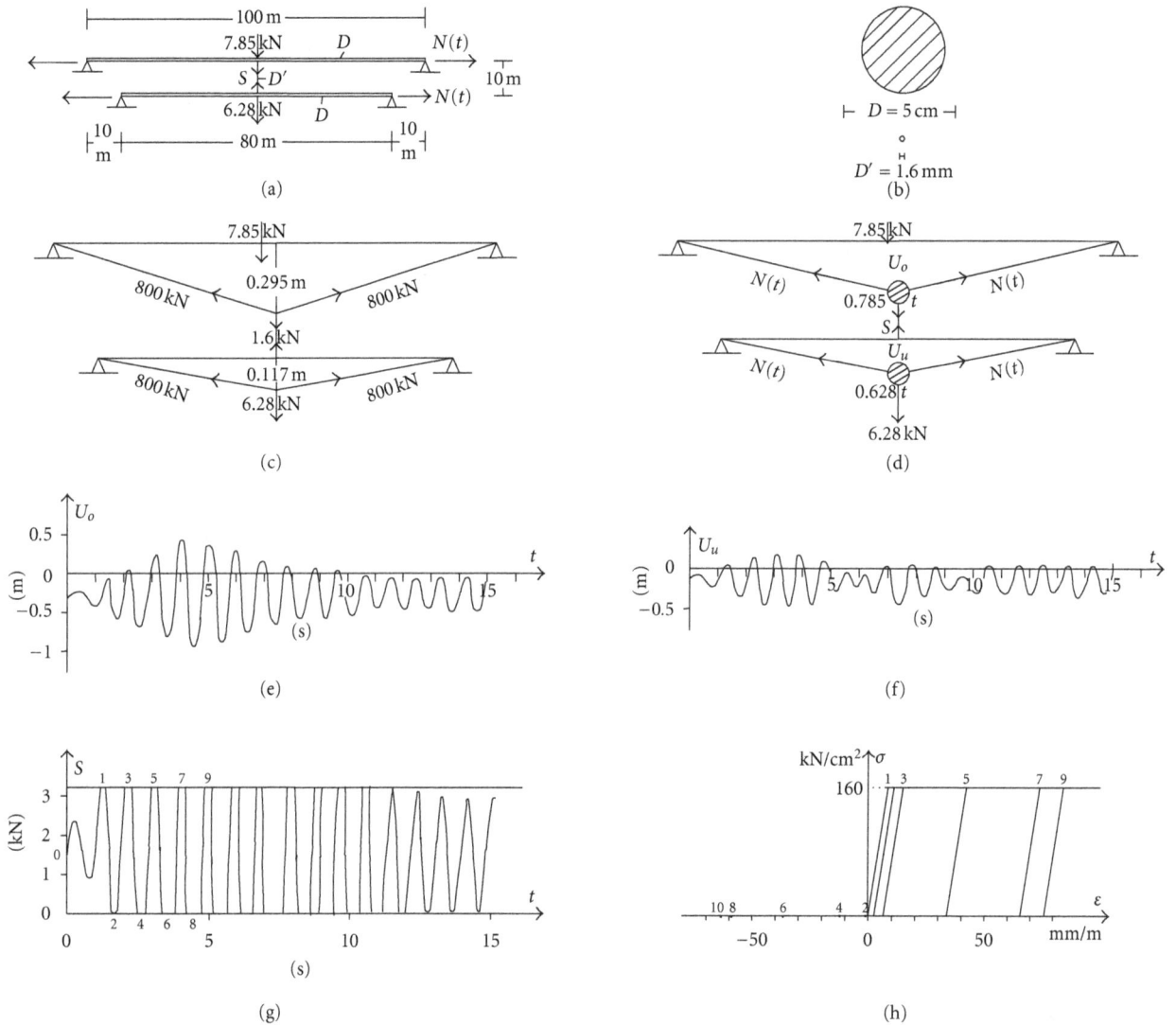

FIGURE 4: Second application: couple of interconnected cables, subject to traffic. (a) Given geometry and loading. (b) Crosssections of main cables and cross-tie. (c) Initial static analysis. (d) Parameters of dynamic analysis. (e), (f) Resulting time-histories of vertical displacements of centers of upper and lower cables. (g) Resulting time-history of axial force of cross-tie. (h) Resulting hysteresis stress-strain loops of the cross-tie.

obvious manner by the loading-unloading-reloading rule of Figure 2(b):

$$\Delta\varepsilon_{pl} = \Delta\varepsilon_{pl}\left(\varepsilon, \Delta\varepsilon, \varepsilon_{pl}\right). \tag{3b}$$

2.3. Static Equations.

The axial forces of the cross-ties are

$$S_o = \sigma_o A_w, \\ S_u = \sigma_u A_w, \tag{4}$$

for the upper and lower ties, respectively, where A_w is the cross-section area of the very thin wires (cross-ties), whereas the vertical nodal forces applied at the centers of the

cables, upper and lower one, respectively, are, according to Figure 2(a)

$$F_o = G_o - K_{Go}u_o + S_o, \\ F_u = G_u - K_{Gu}u_u - S_o + S_u, \tag{5}$$

where the downwards direction has been taken as positive, G_o and G_u are weights at centers of upper and lower cables, respectively, and

$$K_{Go} = 2\frac{N(t)}{l_o/2}, \\ K_{Gu} = 2\frac{N(t)}{l_u/2} \tag{6}$$

are their geometric stiffnesses, where $N(t)$ is the given time-history of the axial forces of the cables.

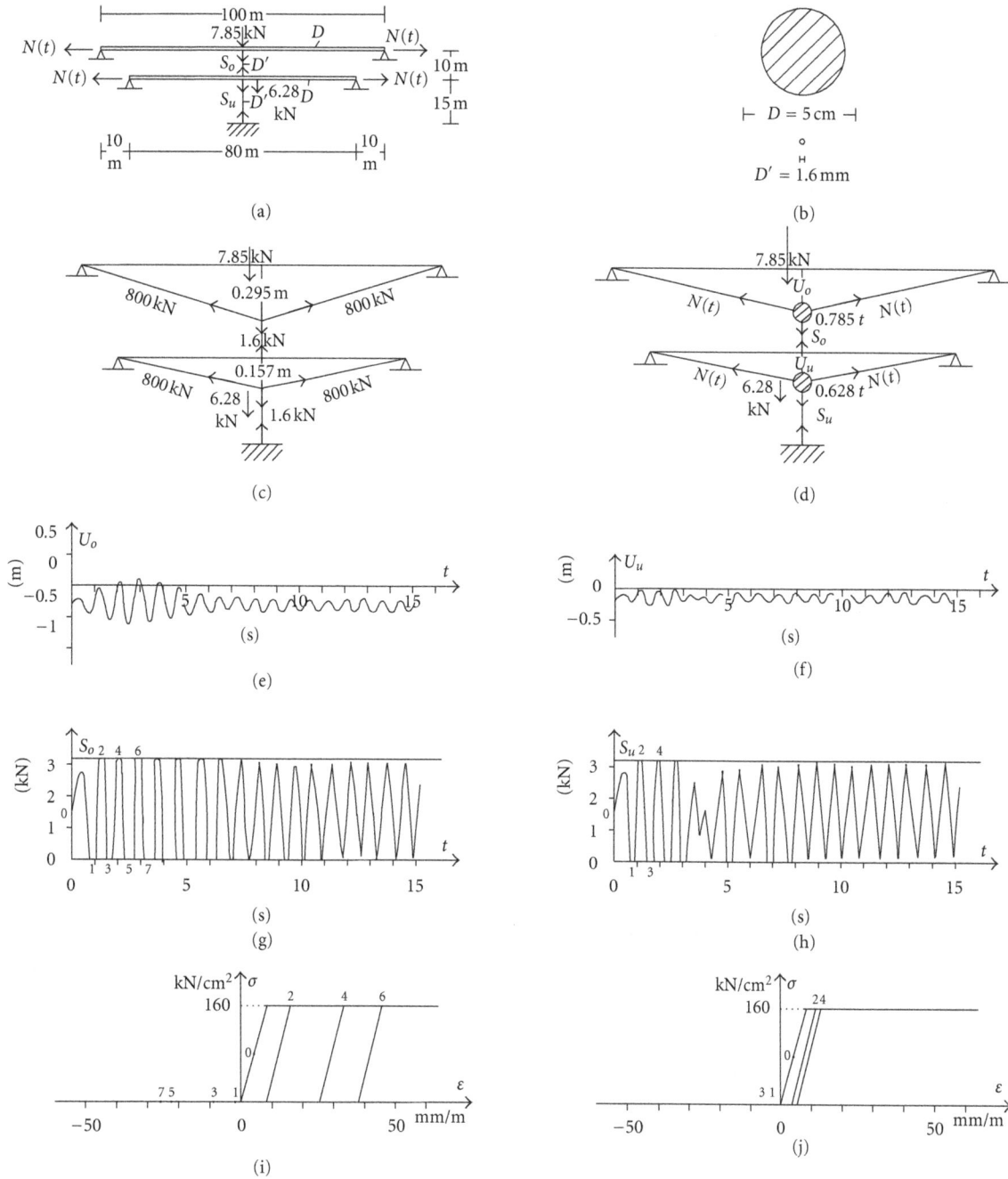

FIGURE 5: Third application: couple of cables, connected to each other and to deck, subject to traffic. (a) Given geometry and loading. (b) Cross sections of main cables and cross-ties. c. Initial static analysis. (d) Parameters of dynamic analysis. (e), (f) Resulting time-histories of displacements of upper and lower cables. (g), (h) Resulting time-histories of axial forces of upper and lower cross-tie. (i), (j) Resulting hysteresis stress-strain loops of upper and lower cross-ties.

2.4. Dynamic Equations. Damping is ignored, as the material internal friction of the cables is meaningless. The vertical accelerations at the centers of upper and lower cable are

$$\ddot{u}_o = \frac{F_o}{m_o},$$
$$\ddot{u}_u = \frac{F_u}{m_u},$$

(7)

where m_o and m_u are lumped masses at centers of cables (Figure 2(a)), whereas the upper dots mean derivation with respect to time.

2.5. External Excitation. Within the input data of the problem, the time-history of external excitation is given, which is here the variation of axial forces of cables due to traffic. The function $N(t)$ is assumed to be described by a piecewise

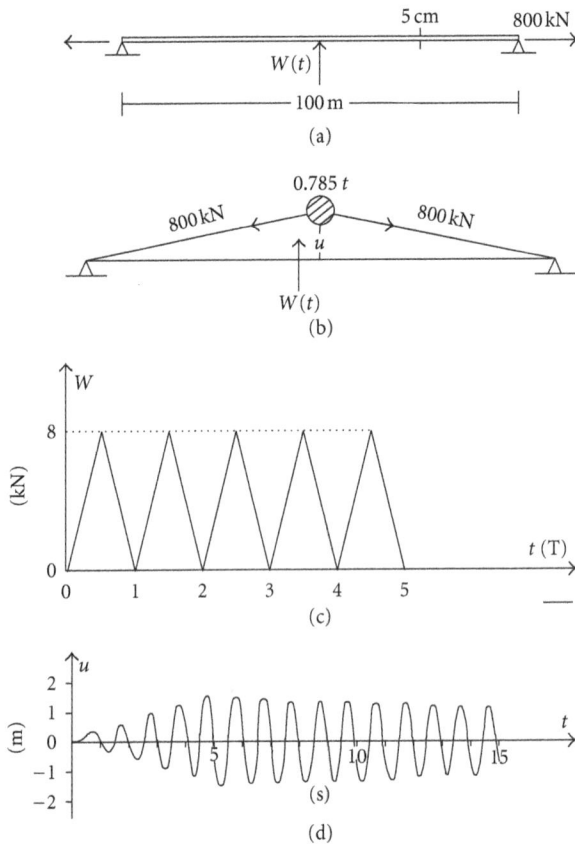

FIGURE 6: Fourth application: isolated cable subject to wind. (a) Given geometry and loading. (b) Parameters of dynamic analysis. (c) Given time-history of wind drag force. (d) Resulting time-history of displacement of center of cable.

linear curve, as shown in Figure 2(c). And within each time interval, between two successive nodes, a linear interpolation is performed, in order to find, from a specific time instant t, the corresponding axial force N of the cables.

2.6. Initial Value Problem. A state vector is introduced:

$$\mathbf{y} = \{\mathbf{u}\,\mathbf{v}\,\mathbf{c}\}, \qquad (8)$$

consisting of the vertical displacements $\mathbf{u} = \{u_o\,u_u\}$ (Figure 2(a)) and velocities $\mathbf{v} = \{\dot{u}_o\,\dot{u}_u\}$ of the centers of upper and lower cables, respectively, as well as of the constitutive variables $\mathbf{c} = \{\varepsilon_{opl}\,\varepsilon_{upl}\}$, which are the plastic strains (Figure 2(b)) of upper and lower cross-ties, respectively.

By combining all the previous equations, (1) up to (8), a system of first-order ordinary nonlinear differential equations is obtained:

$$\dot{\mathbf{y}} = \mathbf{q}(t,\mathbf{y}), \qquad (9a)$$

which, along with the initial value of the state vector

$$\mathbf{y}(0) = \mathbf{y_0} \qquad (9b)$$

for time $t = 0$, and with sought function the time-history of the state vector $\mathbf{y}(t)$, constitutes an initial value problem.

2.7. Proposed Algorithm. For the step-by-step dynamic analysis (direct time integration) of the previous initial value problem of (9a) and (9b), the algorithm of trapezoidal rule is proposed:

$$\mathbf{y}_{n+1} = \mathbf{y}_n + \frac{1}{2}\left[\mathbf{q}(t_n\,\mathbf{y}_n) + \mathbf{q}(t_{n+1}\,\mathbf{y}_{n+1})\right]\Delta t, \qquad (10)$$

where n and $n+1$ are two successive steps of the algorithm. This coincides with the algorithm of constant average acceleration of Newmark's group of algorithms for step-by step dynamic analysis.

The aforementioned algorithm is combined with a predictor-corrector technique, with two corrections per step, $PE(CE)^2$, where, in this symbol, P means prediction and C correction of the state vector $\mathbf{y}$, whereas E means evaluation of the function $\mathbf{q}(t,\mathbf{y})$ of (9a). In more detail, the proposed predictor-corrector technique can be written, within any nth step of the algorithm, as follows:

Prediction

$$\mathbf{y}_{n+1}^P = \mathbf{y}_n + \mathbf{q}(t_n\,\mathbf{y}_n)\Delta t, \qquad (11a)$$

First correction

$$\mathbf{y}_{n+1}^1 = \mathbf{y}_n + \frac{1}{2}\left[\mathbf{q}(t_n\,\mathbf{y}_n) + \mathbf{q}\left(t_{n+1}\,\mathbf{y}_{n+1}^P\right)\right]\Delta t, \qquad (11b)$$

Second and final correction

$$\mathbf{y}_{n+1} = \mathbf{y}_n + \frac{1}{2}\left[\mathbf{q}(t_n\,\mathbf{y}_n) + \mathbf{q}\left(t_{n+1}\,\mathbf{y}_{n+1}^1\right)\right]\Delta t. \qquad (11c)$$

Thanks to the aforementioned predictor-corrector technique, no solving of algebraic system is needed, within each step of the algorithm.

The stability criterion of the proposed algorithm is [9]

$$\omega_{\max}\Delta t < 2.0\,\text{rad}, \qquad (12)$$

that is, $\Delta t < T_{\min}/\pi$; otherwise a divergent solution results, whereas the accuracy criterion is at least

$$\omega_{\max}\Delta t < 0.5\,\text{rad}, \qquad (13)$$

that is, $\Delta t < T_{\min}/4\pi = T_{\min}/12.56$; otherwise a significant accumulated truncation error appears, which is expressed as amplitude decay of the vibration, as well as period elongation.

3. Computer Program

Based on the proposed algorithm of previous Section 2.7, a simple and very short computer program has been developed, with only 115 Fortran instructions, consisting of the MAIN program (79 instructions) which performs the algorithm of step-by-step dynamic analysis, and of three subroutines: (1) Subroutine EVAL (17 instructions) which evaluates the present strain and stress state of the cable structure under consideration, (2) subroutine SE (9 instructions)

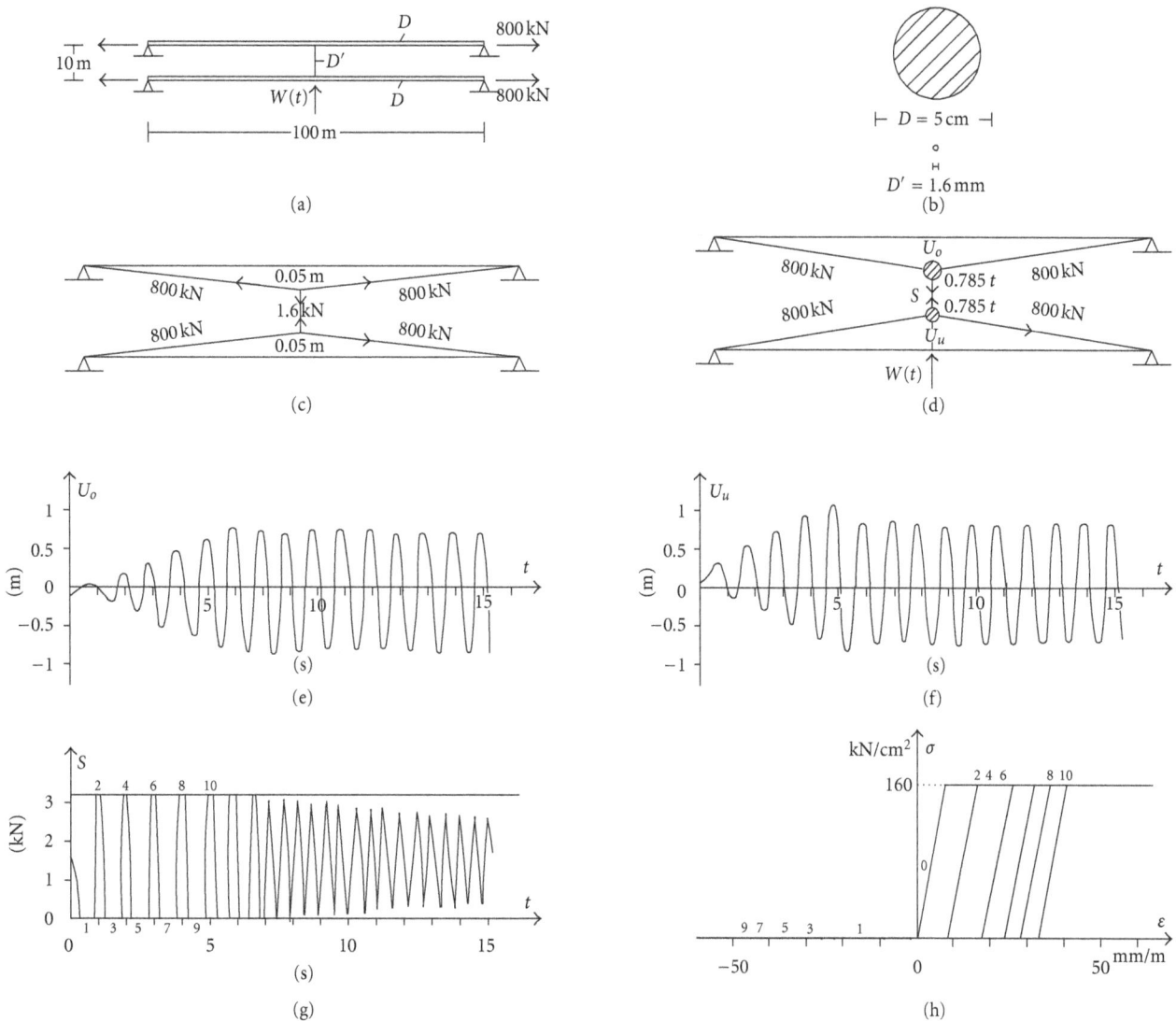

FIGURE 7: Fifth application: couple of interconnected cables subject to wind. (a) Given geometry and loading. (b) Cross sections of main cables and cross-tie. (c) Initial static analysis. (d) Parameters of dynamic analysis. (e), (f) Resulting time-histories of displacements of two main cables. (g) Resulting time-history of axial force of cross-tie. (h) Resulting hysteresis stress-strain loops of the cross-tie.

which describes the nonlinear uniaxial stress-strain law of a cross-tie, and (3) subroutine NHIST (10 instructions) describing the given time-history of the external excitation, which is, here, the variation, with respect to time, of the axial force of cables due to traffic.

A full documentation of the previous computer program is presented as Appendix, consisting of the description of program line by line in Section A.1, of the complete list of Fortran instructions in Section A.2, and the variables explanation in Section A.3. The documentation of the computer program is completed by the series of seven applications in Section 4. The program is particularly oriented to the specific third application of Section 4.3, as already mentioned in the equations of problem in Section 2. However, by simple and obvious modifications of the computer program, all the other numerical experiments, in Section 4, can be treated, too.

4. Applications (Numerical Experiments)

Seven applications (numerical experiments) follow, on the dynamic analysis of isolated pretensioned cables of a cable-stayed bridge or couples of parallel cables connected to each other and possibly with the deck of the bridge by very thin pretensioned single wires (cross-ties). Three of these cable structures are subjected to variation of axial forces of cables due to traffic (parametric excitation) and four of them are subjected to successive pulses of drag force due to a strong wind.

As already mentioned, the previously presented algorithm, in Section 2, is oriented only to the specific third application of Section 4.3. However, by simple and obvious modifications of this algorithm, all the other applications present in Section 4, which are presented herein after, have been analysed, too.

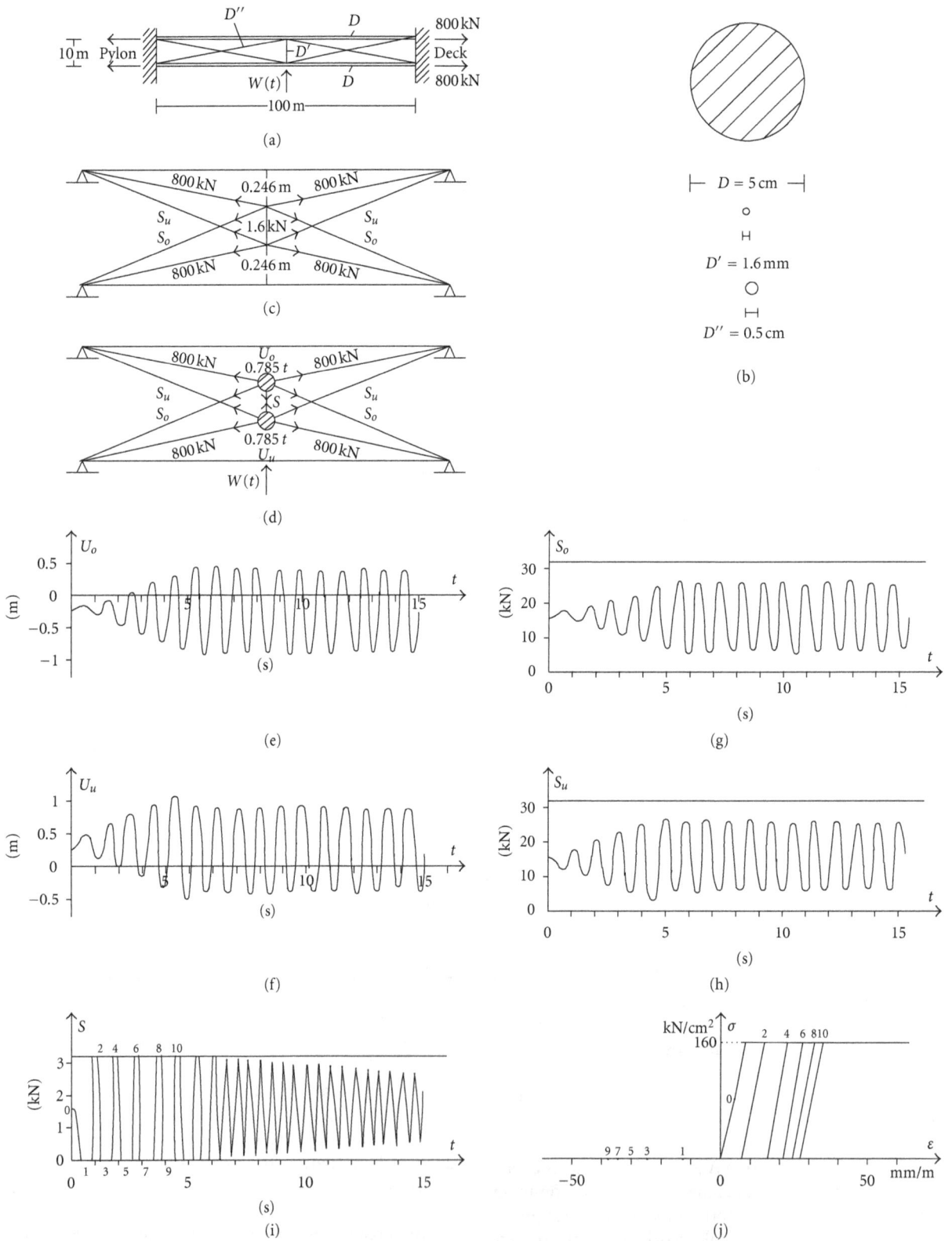

FIGURE 8: Sixth application: couple of cables interconnected by cross-tie, additionally connected by diagonals to pylon and deck, subject to wind. (a) Given geometry and loading. (b) Cross sections of main cables, cross-tie and diagonals. (c) Initial static analysis. (d) Parameters of dynamic analysis. (e), (f) Resulting time-histories of displacements of two main cables. (g), (h) Resulting time-histories of axial forces of diagonal bars. (i) Resulting time-history of axial force of cross-tie. (j) Resulting stress-strain loops of the cross-tie.

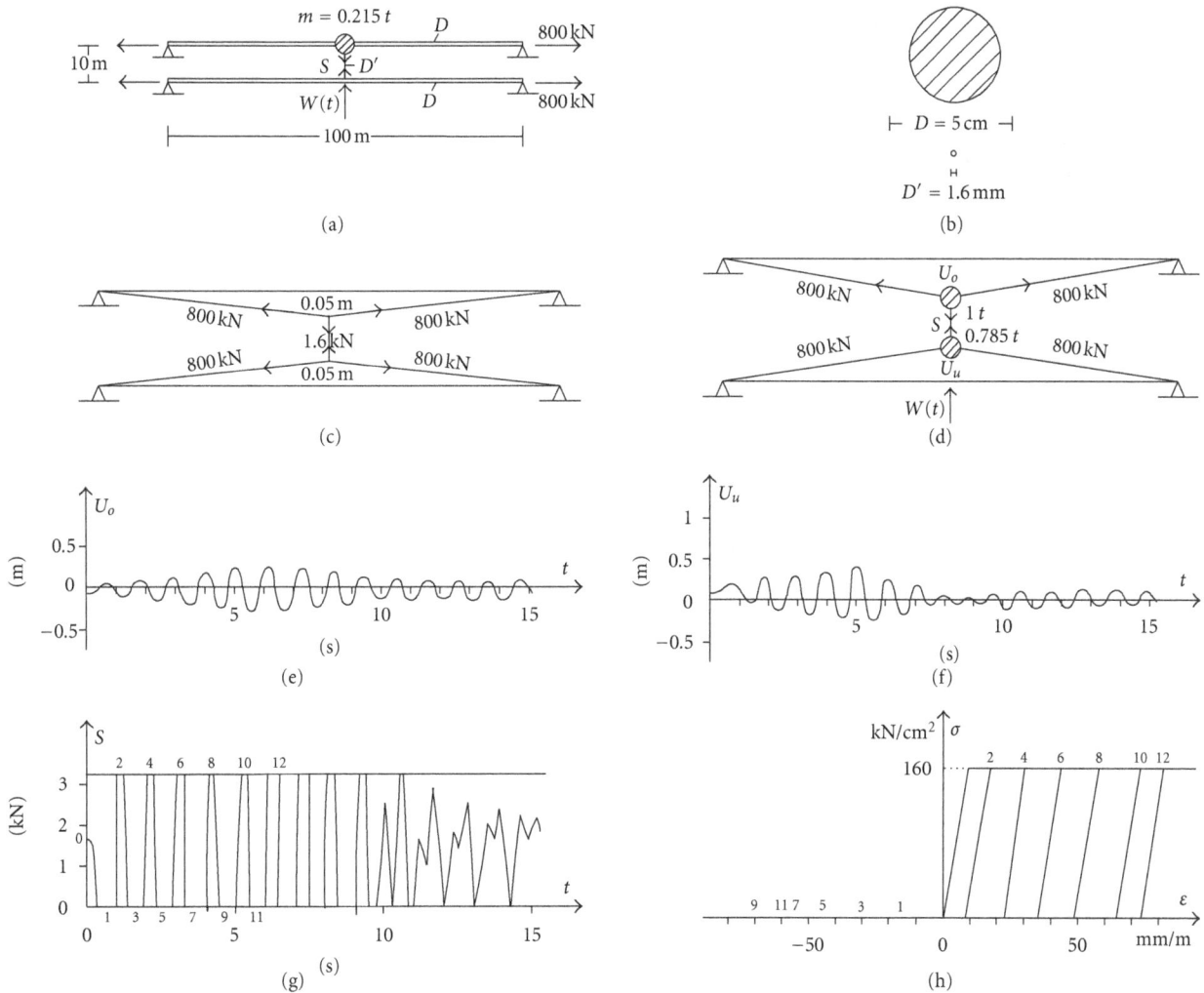

FIGURE 9: (a) Seventh application: Couple of inter-connected cables subject to wind with additional small mass on one cable. (a) Given geometry, loading and additional small mass. (b) Comparison of cross-sections of cables and cross-tie. (c) Initial static analysis. (d) Parameters of dynamic analysis. (e), (f) Resulting time-histories of displacements of two cables. (g) Resulting time-history of axial force of cross-tie. (h) Resulting hysteresis stress-strain loops of the cross-tie.

4.1. First Application: Isolated Cable Subject to Traffic.
As shown in Figure 3, an isolated cable is subjected to a periodic variation of its axial force with a period equal to its fundamental one. Up to the fifth cycle that the excitation lasts, the vibration amplitude of the cable increases up to about 1.6 m and then remains constant; only a slight algorithmic damping is observed.

4.2. Second Application: Couple of Interconnected Cables Subject to Traffic.
The cable of first application is, in Figure 4, connected to another shorter parallel cable by a thin cross-tie. Both cables are subjected to a periodic variation of their axial force with a period equal to the fundamental one of the cable system. Up to the fifth cycle that the excitation lasts, the vibration amplitude of the upper cable increases up to 1.4 m and that of the shorter lower cable up to 0.7 m. Then, both amplitudes are gradually reduced, the upper one up to 0.4 m and the lower one up to 0.3, in 10 sec. Because of the different lengths of the two cables, thus different geometric stiffness, masses, and weights, too, large differences of displacements at the ends of the cross-tie are obtained, thus large stress-strain loops with a total width $\Delta\varepsilon \approx 76$ mm/m, which are responsible for the significant hysteresis damping which is achieved.

4.3. Third Application: Couple of Cables Connected to Each Other and to Deck, Subject to Traffic.
The cable system of the second application is, in Figure 5, supplied by one more thin cross-tie connecting the lower cable with the deck of the bridge. Because of the stiffness of this additional cross-tie, the maximum vibration amplitudes of both cables are significantly reduced, that of the upper cable to 0.7 m and that of the lower cable to 0.3 m. However, at same time, this reduction of displacements has a consequence less wide stress-strain loops in the upper cross-tie with total width $\Delta\varepsilon \approx 38$ mm/m and very thin stress-strain loops in the lower

```
1        COMMON/A/AW, HO, HO0, DHO, EO, SO, TO, HU, HU0, DHU, *EU, SU, TU, WO, WU, N,
            LO, LU, FO, FU, MO, MU
2        COMMON/B/EY, ELAST
3        COMMON/C/NN, TK, NK
4        REAL LO, LU, MO, MU, N, NK
5        DIMENSION TK(20), NK(20)
6        OPEN(5,FILE = "TIEIN.TXT")
7        OPEN(6,FILE="TIEOUT.TXT")
8        READ(5,1)LO,LU,HO,HU,DC,DW,DENS
9    1   FORMAT(7F10.0)
10       AC = 3.14159*DC**2/4.
11       MO = DENS*AC*LO/2./10000.
12       WO = −MO*10.
13       MU = DENS*AC*LU/2./10000.
14       WU = −MU*10.
15       AW = 3.14159*DW**2/4./10000.
16       READ(5,1)SY,ELAST,SC0,SW0
17       EY = SY/ELAST
18       N = SC0*AC
19       TO = SW0*AW
20       TU = SW0*AW
21       READ(5, 2)NN,TMAX
22   2   FORMAT(I5, F10.0)
23       READ(5,3)(TK(I), NK(I), I = 1,NN)
24   3   FORMAT(40F5.2)
25       STIF1 = 2. *N/(LO/2.) + ELAST*AW/HO
26       STIF2 = 2. *N/(LU/2.) + ELAST*AW/HO+ELAST*AW/HU
27       STIF12= −ELAST*AW/HO
28       A = MO*MU
29       B = −(STIF1*MU + STIF2*MO)
30       C = STIF1*STIF2-STIF12**2
31       D = SQRT(B**2 − 4. *A*C)
32       W1=SQRT((−B + D)/(2. *A))
33       W2 = SQRT((−B + D)/(2. * A))
34       T1 = 2. * 3.14159/W1
35       T2 = 2. * 3.14159/W2
36       DO 4 I = 1, NN
37   4   TK(1) = TK(1) * T2
38       DT=T2/(4. * 3.14159)
39       T = 0.
40       UO = (WO−TO)/(2. *N/(LO/2.))
41       UU = (WU + TO − TU)/(2. *N/(LU/2.))
42       VO = 0.
43       VU = 0.
44       EOPL = 0.
45       EUPL = 0.
46       EW0 = SW0/ELAST
47       HO0 = (HO + UO − UU)/(1. + EW0)
48       HU0 = (HU + UU)/(1. + EW0)
49       CALL EVAL(UO,UU,EOPL,EUPL,GO,GU)
50   5   T = T + DT
51       CALL NHIST(T, N)
52       UOP = UO + VO*DT
53       UUP = UU + VU*DT
54       VOP = VO + GO*DT
55       VUP = VU + GU*DT
56       EOPLP = EOPL
57       EUPLP = EUPL
58       CALL EVAL (UOP, UUP, EOPLP, EUPLP, GOP, GUP)
59       UO1 = UO + (VO + VOP)/2. *DT
60       UU1= UU + (VU + VUP)/2. *DT
```

ALGORITHM 1: Continued.

```
61        VO1 = VO + (GO + GOP)/2. * DT
62        VU1 = VU + (GU + GUP)/2. *DT
63        EOPL1 = EOPL
64        EUPL1 = EUPL
65        CALL EVAL (UO1, UU1, EOPL1, EUPL1, GO1, GU1)
66        UO =UO + (VO+VO1)/2. *DT
67        UU = UU + (VU + VU1)/2. *DT
68        VO = VO + (GO+GO1)/2. *DT
69        VU = VU + (GU + GU1)/2. *DT
70        CALL EVAL (UO, UU, EOPL, EUPL, GO, GU)
71        WRITE (6,6)T,N,UO,UU,TO,TU
72        WRITE (6,6)EO,SO,EU,SU
73    6   FORMAT(1X,5(E10.4,1X))
74        IF(T.GT.TMAX) GO TO 7
75        GO TO 5
76    7   CLOSE(5)
77        CLOSE(6)
78        STOP
79        END
```

ALGORITHM 1: MAIN program.

```
1     SUBROUTINE EVAL(UO,UU,EOPL,EUPL,GO,GU)
2     COMMON/A/AW,HO,HO0,DHO,EO,SO,TO,HU,HU0,DHU,EU, *SU,TU, WO, WU,
      N, LO, LU, FO, FU, MO, MU
3     REAL LO, LU, MO, MU, N
4     DHO = HO + UO − UU − HO0
5     EO = DHO/HO0
6     CALL SE(EO,EOPL,SO)
7     TO = SO*AW
8     DHU = HU + UU − HU0
9     EU = DHU/HU0
10    CALL SE(EU,EUPL,SU)
11    TU = SU*AW
12    FO = WO − 2. *N/(LO/2.)*UO − TO
13    FU = WU − 2. *N/(LU/2.)*UU + TO − TU
14    GO = FO/MO
15    GU = FU/MU
16    RETURN
17    END
```

ALGORITHM 2: Subroutine EVAL.

```
1     SUBROUTINE SE(E, EPL, S)
2     COMMON/B/EY, ELAST
3     IF(E,GT.EPL + EY)EPL=E-EY
4     IF(E.LT.EPL)EPL=E
5     IF(EPL.LT.0.)EPL=0.
6     S = ELAST*(E − EPL)
7     IF(S.LT.0.)S = 0.
8     RETURN
9     END
```

ALGORITHM 3: Subroutine SE.

```
1        SUBROUTINE NHIST(T,N)
2        COMMON/C/NN, TK, NK
3        DIMENSION TK(20), NK(20)
4        REAL N, NK
5        DO 1 I = 1, NN−1
6        IF((T− TK(1))∗(T− TK(I + 1)).GT.0.) GO TO 1
7        N = NK(I) + (NK(I + 1)−NK(I))/(TK(I + 1)−TK(I))∗(T−TK(I))
8        RETURN
9    1   CONTINUE
10       END
```

ALGORITHM 4: Subroutine NHIST.

cross-tie with only $\Delta\varepsilon \approx 6$ mm/m, resulting in low values of hysteretic damping.

4.4. Fourth Application: Isolated Cable Subject to Wind.
As shown in Figure 6, an isolated cable is subjected to a resonant periodic wind drag force. Up to the fifth cycle that the excitation lasts, the vibration amplitude increases up to about 3.0 m and then remains constant; only a slight algorithmic damping is observed.

4.5. Fifth Application: Couple of Interconnected Cables Subject to Wind.
Two identical parallel cables are, in Figure 7, interconnected by a thin cross-tie. A wind drag force acts on one cable only; initially, this cable exhibits larger displacements, but gradually the movement is transferred to the other cable, too. So, the displacements are divided by two, compared with those of previous fourth application. During the initial stage of displacements transfer from one cable to the other, stress-strain loops of the cross-tie with a total width of medium size $\Delta\varepsilon \approx 33$ mm/m appear. From this point on, as the two cables are identical and perform similar movements, no more yielding of the cross-tie appears, as shown in Figure 7(g), thus no more stress-strain loops and hysteretic damping, too.

4.6. Sixth Application: Couple of Interconnected Cables, Additionally Connected by Diagonals to Pylon and Deck, Subject to Wind.
The cable system of the fifth application is, in Figure 8, supplied with diagonal ties connected with the pylon and the deck of the bridge. These diagonal ties offer a small additional stiffness, perpendicularly to the cables, which slightly reduces their displacements. However, at same time, this restriction of displacements further reduces the total width of stress-strain loops of the cross-tie to $\Delta\varepsilon \approx 27$ mm/m, thus reducing the hysteretic damping.

4.7. Seventh Application: Couple of Interconnected Cables Subject to Wind, with Small Additional Mass on One Cable.
The cable system of fifth application is in Figure 9, supplied by a small additional mass to the one of the two cables. So, the two cables have now different dynamic characteristics and they perform significantly different movements. As a consequence, large differences of displacements at the ends of the cross-tie result, thus wide stress-strain loops with a total

width $\Delta\varepsilon \approx 73$ mm/m, which implies significant hysteretic damping. The displacement amplitudes of the cables are now only about one-fourth of those of the fifth application.

5. Conclusions

Cable vibrations of cable-stayed bridges have been examined. Either isolated cables or couples of parallel cables, connected to each other and possibly with the deck of the bridge, by a very thin pretensioned wire (cross-tie), have been considered. External excitation is either traffic, which causes displacements of cable ends on deck and pylon, thus variation of axial forces, geometric stiffnesses and sags of cables, too (parametric excitation), or successive pulses of drag force due to a strong wind, perpendicularly to a vertical cables' plane at one side of the bridge.

The proposed analytical model is on the one hand simplified, as an SDOF oscillator is adopted for every individual cable, approximating its fundamental vibration mode. However, on the other hand, the proposed analytical model is accurate, as it takes into account the geometric nonlinearity of the cables by their geometric stiffness; also it includes the material nonlinearity of the cross-ties by their compressive loosening, tensile yielding, and hysteretic stress-strain loops.

The equations of the problem of dynamic analysis, oriented to a specific cable structure, have been written, consisting of the geometric, constitutive, static, and dynamic ones, as well as of the given time-history of the external excitation. By combining these equations, an initial value problem is obtained. For the step-by-step dynamic analysis of this problem, the algorithm of trapezoidal rule is proposed, combined with a predictor-corrector technique, with two corrections per step. So, no solving of algebraic system is required within each step of the algorithm.

Based on the proposed algorithm, a short computer program has been developed, with only 115 Fortran instructions, consisting of the main program and three subroutines. A full documentation is given for this program, which means transparency of computation.

Seven numerical experiments have been performed by the aforementioned program, three with variation of axial forces of cables due to traffic (parametric excitation) and four with successive pulses of drag force due to a strong wind.

On the basis of previous series of numerical experiments some observations with practical usefulness are made. (These are not strict theoretical conclusions, but simple observations based on the results of numerical experiments.)

It is confirmed by the series of numerical experiments, the great advantage of pretensioned cross-ties, that although they are very thin, with ratio of cross-section area of a cable to that of a cross-tie of magnitude order 1000, however, they possess an axial elastic stiffness comparable in magnitude to the geometric stiffness of cables, with magnitude order 50 kN/m, along the same direction, that is perpendicularly to cables axes.

The in-plane cross-ties (within a vertical cables plane) are intended to suppress cables' variations from parametric excitation due to traffic, whereas the out-of-plane cross-ties (transverse ones connecting cables at two sides of bridge) are intended to suppress cables vibrations from successive pulses of drag force due to a strong wind.

General observation from all numerical experiments: in a couple of parallel cables connected to each other and possibly with the deck of bridge by cross-ties, even a single cross-tie proves effective by its hysteresis damping (due to stress-strain loops) in suppressing large amplitude cable vibrations under the following circumstances: if the two cables have different dynamic characteristics, for example, different lengths which imply different masses, weights, and geometric stiffnesses, too, or if one of them has a small additional mass.

Appendices

A. Documentation of the Proposed Computer Program

In this appendix, documentation is given for the proposed computer program, for the step-by-step dynamic analysis of the third application, that of a couple of parallel cables connected to each other and to the deck of bridge by a thin cross-tie and subject to a variation of their axial forces.

A.1. Description of Program Line by Line. The description refers to the complete numbered list of Fortran instructions of Algorithms 1, 2, 3, 4.

MAIN Program. The first seven lines include nonexecutable statements. Particularly, in the three first lines, the COMMON instructions connect the MAIN program with the three subroutines, by their common variables.

In the next 17 lines, 8 up to 24, the input data are read: geometric data and density of steel in lines 8-9, the parameters of σ-ε law of steel along with the pretension stresses of cables and cross-ties in line 16, and the time-history of axial forces of cables given by the coordinates of nodes of piecewise linear curve N-t in lines 21–24. In lines 10–15 and 17–20, some simple preliminary calculations are performed to determine cross-section areas and pretension forces of cables and cross-ties, as well as masses and weights of cables and yield strain of steel. In lines 25–35, the initial characteristic equation of the cable structure is solved, so

that to find its natural frequencies and periods. In lines 36-37, the time scale of the given time history of external excitation is expanded so that to obtain a period equal to the fundamental one T_{max} of the structure, in order to cause resonance, whereas in line 38, the minimum natural period T_{min} of the structure dictates the time-step length Δt of the algorithm, so that to assure accuracy of computation.

In lines 39–49, the initial conditions are established: time $t = 0$ in line 39, determination of initial static displacements of cables in lines 40-41, evaluation of undeformed lengths of cross-ties in lines 42–44, zero initial velocities of cables in lines 45-46, zero initial plastic strains of cross-ties in lines 47-48, and evaluation of initial strain and stress state of structure by calling subroutine EVAL in line 49.

In line 50, any step of algorithm begins by increasing time t by Δt. In line 51, by calling subroutine NHIST, the present value of axial force of cables is determined. In lines 52–55, the prediction of values of displacements and velocities is performed, and in lines 56–58 by calling subroutine EVAL the corresponding plastic strains and accelerations are found. So in lines 52–58 the prediction of state vector within a step of algorithm is performed. In lines 59–65 the first correction of value of state vector is made by use of trapezoidal rule. And in lines 66–70, the second and final correction. In lines 71–73, the output data of present step of algorithm are written (time t, axial force N of cables, displacements u_o and u_u of cables, axial forces S_o and S_u of cross-ties, strains and stresses ε_o-σ_o and ε_u-σ_u of cross-ties).

In lines 74–79 if a maximum time has been exhausted, the algorithm is interrupted. Otherwise, we continue to the next step of the algorithm.

Subroutine EVAL. Lines 1–3 are nonexecutable statements. In lines 4–7, from the displacements of cables, the elongation, strain, stress by calling subroutine SE, and axial force of upper cross-tie are determined. In lines 8–11, the corresponding quantities are found for the lower cross-tie. In lines 12-13, the vertical nodal forces on centers of upper and lower cables are determined and in lines 14-15 the corresponding accelerations.

Subroutine SE. In lines 3–5, the new plastic strain of the cross-tie is found. In lines 6-7, the stress of the cross-tie is determined.

Subroutine NHIST. In line 5 is found the time interval where present time t is included. In line 6, a linear interpolation is performed between the two end-nodes of the previous time interval, in order to find the axial force N of cables corresponding to present time t.

A.2. List of Fortran Instructions

MAIN Program.
   See Algorithms 1, 2, 3, and 4.

A.3. Variables Explanation

MAIN Program.

$A = m_o \times m_u$ coefficient of characteristic equation

AC: cross-section area of a cable

AW: cross-section area of a wire (cross-tie)

$B = -(K_{11}m_u + K_{22}m_o)$ coefficient of characteristic equation

$C = K_{11} \times K_{22} - K_{12}^2$ coefficient of characteristic equation

$D = (b^2 - 4ac)^{1/2}$ coefficient of characteristic equation

DC: cross-section diameter of a cable

DENS: density of steel

DHO: elongation of upper tie

DHU: elongation of lower tie

DW: cross-section diameter of wire (cross-tie)

$DT = \Delta t$, time steplength of algorithm

ELAST: initial elasticity (Young) modulus

EO: strain of upper tie

EOPL: plastic strain of upper tie

EOPLP: prediction of EOPL

EOPL1: first correction of EOPL

EU: strain of lower tie

EUPL: plastic strain of lower tie

EUPLP: prediction of EUPL

EUPL1: first correction of EUPL

EVAL: subroutine for evaluation of strain and stress state of the structure

EW0: pretension strain of wires (cross-ties)

EY: yield strain of steel

FO: vertical nodal force at center of upper cable

FU: vertical nodal force at center of lower cable

GO: vertical acceleration at center of upper cable

GOP: prediction of GO

GO1: first correction of GO

GU: vertical acceleration at center of lower cable

GUP: prediction of GU

GU1: first correction of GU

HO: design (nominal) length (height) of upper cross-tie

HO0: undeformed length (height) of upper cross-tie

HU: design (nominal) length (height) of lower cross-tie

HU0: undeformed length (height) of lower cross-tie

LO: length of upper cable

LU: length of lower cable

MO: mass of upper cable

MU: mass of lower cable

N: axial force of a cable

NHIST: subroutine for given time-history of N

NK: ordinate of a node of piecewise linear curve N-T

NN: number of nodes of piecewise linear curve N-T

SO: stress of upper cross-tie

STIF1: K_{11} elements of stiffness matrix

STIF2: K_{22} elements of stiffness matrix

STIF12: K_{12} elements of stiffness matrix

SU: stress of lower cross-tie

SW0: pretension stress of wires (cross-ties)

SY: yield stress of steel

T: time

TIEIN: input file

TIEOUT: output file

TK: abscissa of a node of piecewise linear curve N-T

TMAX: t_{max}, maximum time of observation

TO: axial force of upper cross-tie

TU: axial force of lower cross-tie

$T1$: T_{min} extreme natural periods of structure ¡list-item¿¡label/¿

$T2$: T_{max} extreme natural periods of structure

UO: vertical displacement of center of upper cable

UOP: prediction of UO

UO1: first correction of UO

UU: vertical displacement of center of lower cable

UUP: prediction of UU

UU1: first correction of UU

VO: vertical velocity of center of upper cable

VOP: prediction of VO

VO1: first correction of VO

VU: vertical displacement of center of lower cable

VUP: prediction of VU

VU1: first correction of VU

WO: weight at center of upper cable

WU: weight at center of lower cable

W1: ω_{max} extreme natural frequencies of the structure

W2: ω_{min} extreme natural frequencies of the structure.

Subroutine EVAL. Only the following variable is different from those of MAIN program:

SE: subroutine for stress-strain law of a cross-tie.

Subroutine SE. Only the following variables are different from those of MAIN program:

E: strain of a cross-tie

EPL: plastic strain of a cross-tie

S: stress of a cross-tie.

Subroutine NHIST. All the variables are the same as in MAIN program.

References

[1] R. Walther, B. Houriet, W. Isler, P. Moia, and J. F. Klein, *Cable-Stayed Bridges*, Th. Telford, London, UK, 2nd edition, 1999.

[2] E. Caetano, "Cable vibrations in cable-stayed bridges," IABSE-Structural Engineering Documents, 2007.

[3] L. Caracoglia and D. Zuo, "Effectiveness of cable networks of various configurations in suppressing stay-cable vibration," *Engineering Structures*, vol. 31, no. 12, pp. 2851–2864, 2009.

[4] L. Caracoglia and N. P. Jones, "Passive hybrid technique for the vibration mitigation of systems of interconnected stays," *Journal of Sound and Vibration*, vol. 307, no. 3–5, pp. 849–864, 2007.

[5] L. Caracoglia and N. P. Jones, "In-plane dynamic behavior of cable networks. Part 1: formulation and basic solutions," *Journal of Sound and Vibration*, vol. 279, no. 3–5, pp. 969–991, 2005.

[6] L. Caracoglia and N. P. Jones, "In-plane dynamic behavior of cable networks. Part 2: prototype prediction and validation," *Journal of Sound and Vibration*, vol. 279, no. 3–5, pp. 993–1014, 2005.

[7] P. G. Papadopoulos, A. Diamantopoulos, P. Lazaridis, H. Xenidis, and C. Karayiannis, "Simplified numerical experiments on the effect of hysteretic damping of cross-ties on cable oscillations," in *9th International Conference on Computational Structures Technology*, Athens, Greece, September 2008.

[8] P. G. Papadopoulos, J. Arethas, P. Lazaridis, E. Mitsopoulou, and J. Tegos, "A simple method using a truss model for in-plane nonlinear static analysis of a cable-stayed bridge with a plate deck section," *Engineering Structures*, vol. 30, no. 1, pp. 42–53, 2008.

[9] P. G. Papadopoulos, "A simple algorithm for the nonlinear dynamic analysis of networks," *Computers and Structures*, vol. 18, no. 1, pp. 1–8, 1984.

11

Behavior of FRP Link Slabs in Jointless Bridge Decks

Aziz Saber and Ashok Reddy Aleti

Department of Civil Engineering, Louisiana Tech University, 600 West Arizona Avenue, Ruston, LA 71272, USA

Correspondence should be addressed to Aziz Saber, saber@latech.edu

Academic Editor: Sami W. Tabsh

The paper investigated the use of fiberglass-reinforced plastic (FRP) grid for reinforcement in link slabs for jointless bridge decks. The design concept of link slab was examined based on the ductility of the fiberglass-reinforced plastic grid to accommodate bridge deck deformations. The implementation of hybrid simulation assisted in combining the experimental results and the theoretical work. The numerical analyses and the experimental work investigated the behavior of the link slab and confirmed its feasibility. The results indicated that the technique would allow simultaneous achievement of structural need, lower flexural stiffness of the link slab approaching the behavior of a hinge, and sustainability need of the link slab. The outcome of the study supports the contention that jointless concrete bridge decks may be designed and constructed with fiberglass-reinforced plastic grid link slabs. This concept would also provide a solution to a number of deterioration problems associated with bridge deck joints and can be used during new construction of bridge decks. The federal highway administration provided funds to Louisiana Department of Transportation through the innovative bridge research and development program to implement the use of FRP grid as link slab.

1. Introduction

Thousands of bridges in the United States are constructed as simple spans. The bridges require the use of expansion joints over piers. The joints create short-term and long-term problems. Some examples of these problems are leaks through the joints deteriorating the supporting girders and the piers and debris accumulating in the joints which prevents them from functioning properly. These problems lead to massive direct and indirect costs (Saber et al. [1, 2]). Therefore there is a need for reducing or eliminating expansion joints in bridge decks. The objectives of this study are to develop and evaluate a new technique using advancement in materials and current technology. An innovative system is proposed for this study. The new system replaces expansion joints with a link slab. The link slab joins decks of adjacent spans without imposing any continuity in the bridge girders. The link slab is subjected to tensile forces and stresses due to the negative moment developed at the joint. Fiberglass-reinforced plastic (FRP) reinforcement is used to carry the tension forces (Saber [3]) and its corrosion resistance.

The most common type of reinforcement used in bridge construction is steel rods. The deterioration of steel caused by corrosion has been plaguing these structures across the nation, decreasing their service life, and increasing the cost of repair and maintenance. Many investigations were conducted to resolve the problems associated with corrosion by such methods as decreasing the porosity of concrete, coating steel bars with a protective outer layer, and increasing the reinforcement cover. But these methods only extend the time it takes for corrosion to take place.

For more than three decades, researchers have investigated the use of FRP (fiberglass reinforced polymers) as an alternate to steel reinforcement in concrete structures. In recent years, the use of FRP rods for structural applications has been gaining acceptance around the world. Recently FRP grids have been used for reinforcement of concretes beams and slabs (Dutta et al. [4]). A grid is a latticework of rigid, interconnecting ribs in two, three, or four groups and directions. Such grid reinforcement enhances the energy absorption capability and the overall ductility of the structure is improved. This leads to an increase in the ultimate load carrying capacity of concrete beams and slabs. When the opening of grids is filled with concrete, the combined structure derives its shear rigidity from the concrete filler and the concrete prevents the ribs from buckling. FRP composite grids provide a mechanical anchorage within the concrete due to the interlocking elements (cross-ribs), and therefore no bond is necessary for proper load transfer.

Although there have been a number of studies on the use of FRP-grid-reinforced concrete beams or slabs, there is currently a lack of information on the use of FRP-grid-reinforced concrete link slabs for the replacement of expansion joint. Because the link slab will be subjected to a negative bending moment and thermal stress, it is expected that the design and performance will be different from conventional beams or slabs, which is primarily subjected to a positive bending moment and transverse shear force. Therefore, there is a need to conduct experimental testing and theoretical modeling analysis of FRP-grid-reinforced concrete link slabs for the replacement of expansion joints.

2. Experimental Work

A test program was conducted to determine the behavior and strength of jointless bridge decks under static loading. The jointless decks could be achieved by replacing expansion joints with a link slab that could join bridge decks of adjacent spans without imposing any continuity in the bridge girders. The link slab would be subjected to tensile forces due to negative moment developed at the location of the joint. The link slab panel was cut into beam specimens to determine the strength of the link slab against tensile forces. The test program included specimens with two layers of FRP grids. The specimens were tested under the same support conditions. Loads, deflections, strains, and load carrying capacity were measured for each test specimen.

2.1. Test Specimens. The specimens were designed as per ASTM C 78, ACI 318, and ACI 440 guidelines, [5, 6]. Since there was no design code for FRP-grid-reinforced concrete beams, the existing design equations in ACI 440 for FRP rebar-reinforced concrete beams were modified and used. The cross section of the specimens was rectangular in shape with a width of 300 mm (1 ft), 200 mm (8 in) deep, and 2.4 m (8 ft) long. The FRP grids were placed in the center 1.2 m (4 ft) of the beams. The first specimen, beam 1, contained two layers of FRP grid; each is 25 mm (1 in) deep, 1.2 m (4 ft) long, and 225 mm (9 in) wide. The clear spacing between the two FRP grids was 25 mm (1 in). Shear reinforcement was not provided to the beams since the depth of the beam did not exceed the requirements of ACI 318 [5]. Also, three number 13 (number 4) rebars were placed in the specimens for handling. The dimensions and cross-section details of the beam 1 were shown in Figures 1 and 2, respectively.

The second beam, beam 2, contained two layers of 31 mm (1.25 in) deep FRP grids, 1200 mm (4 ft) long, and 225 mm (9 in) wide. The dimensions and cross-section details of beam 2 were similar to beam 1, as shown in Figure 3. The two rectangular beams were cast from the batch delivered by a ready mix truck to the Structural and Materials Laboratory at Louisiana Tech University. To simulate field conditions, the beams were cured in dry air conditions for 28 days before they were tested.

The specimens were tested under the same set-up as shown in Figure 1. The applied loads and reactions were symmetrical with respect to the center of the beam. The specimen was placed on a high reaction stand of a stiffened steel section. At each reaction point, a roller support was placed between the specimen and the steel section. Load was applied by an MTS hydraulic jack at load points. A steel section was used between the hydraulic jack and the beam specimen to apply the load equally at the load locations. At the load points, roller supports were provided to disperse the load from the steel section to the specimen. The jack was activated by a single automatic MTS electric pump.

2.2. Instrumentation Plan. The instrumentation used for the testing of each beam included a deflectometer, a twenty-four channel data acquisition system, and 50 mm (2 in) long strain gages installed at locations on the FRP grids where the shear forces and bending moments were high. The strain gages were installed on the outer surface along the longitudinal direction. The top grid was designated as layer 1 and the bottom grid was designated as layer 2. Layer 1 strain gages were designated as L1G1 through L1G8 from left end to the right end of the grid. Similarly, layer 2 strain gages were designated as L2G1 through L2G8 from left end to the right end of the grid. The deflection of each beam was measured during the test by a deflectometer placed at the midspan of the beam.

2.3. Test Procedure. A four-point bending test was conducted, and the test load was applied in such a way that a negative bending moment was produced in the beam at the FRP grid locations. The beams were loaded continuously at a constant rate of 8.9 kN/min (2 kip/min) until failure. The four-point bending tests were conducted using the MTS machine. The data collection system stored the strain and load data for every quarter second. For each load increment, data for the FRP strains and loads were collected. The applied loads and corresponding deflections at midspan for each beam were measured during the tests.

2.4. Material Characteristics. The concrete mix constituents were shown in Table 4. The concrete cylinders were cast from the same batch delivered by a local ready mix truck to the Structural and Materials Laboratory at Louisiana Tech University.

The concrete cylinders 100×200 mm (4×8 in) were cured in accordance with ASTM C511, and the compressive strength was determined in accordance with ASTM C39. The average compressive strength of three cylinders was recorded for each test day and the strength development over time is shown in Figure 4. When the beam specimens were tested at 28 days, the compressive strength of the concrete was 36.4 MPa (5277 psi).

The material properties of FRP grid were obtained from the manufacturer (Fibergrate, Composite Structures) and were listed in Table 1.

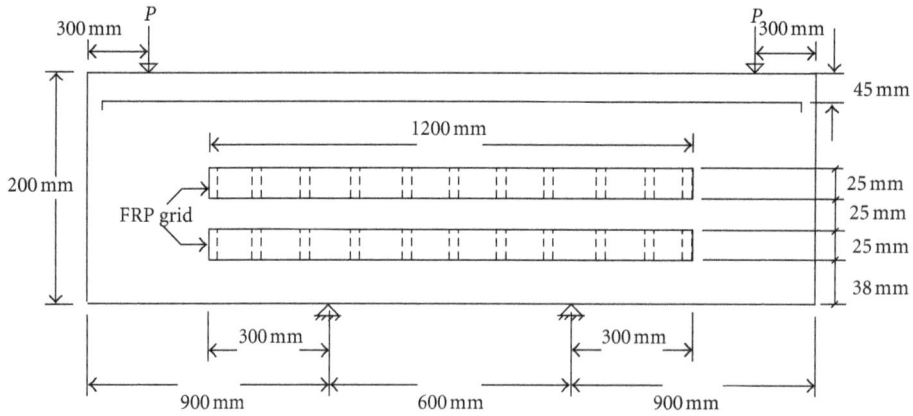

FIGURE 1: Beam 1 dimensions (not to scale).

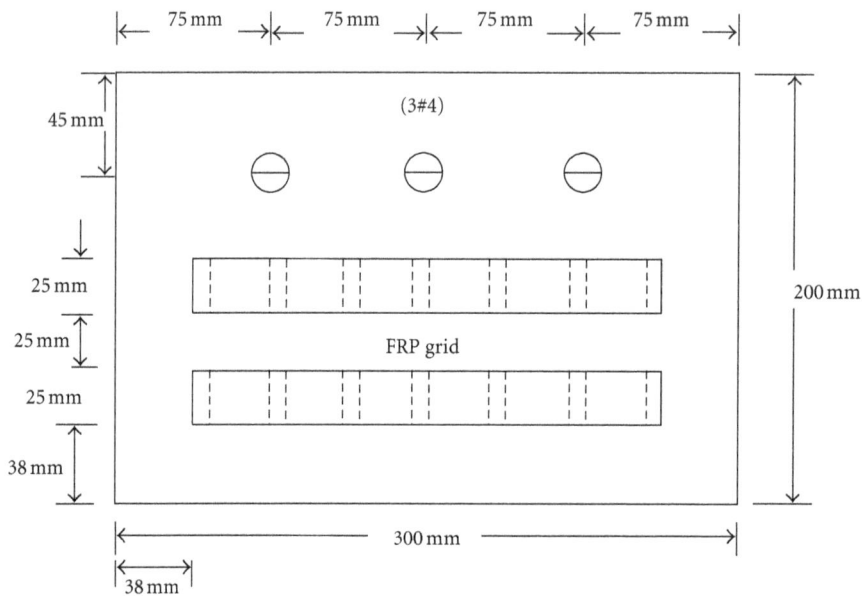

FIGURE 2: Beam 1 cross-section details.

TABLE 1: Concrete mix proportions.

Cement	222 kg/m³ (489 lb/yd³)
Fly Ash	55 kg/m³ (122 lb/yd³)
Coarse aggregate pea gravel	849 kg/m³ (1870 lb/yd³)
Natural Sand	602 kg/m³ (1325 lb/yd³)
Admixture (900 P0Y-5)	0.53 liter/m³ (18 Oz/yd³)
Air content	0.05
Slump	125 mm (5 inch)
Water	112 liter/m³ (29.5 gal/yd³)

2.5. Experimental Results. The specimens were designed to be underreinforced so that large strains in FRP grids preceded the crushing of the concrete in compression. The discussion will be given on strain responses up to failure, the overall load/deflection, and the mode of failure of the specimens. The beams were designed to have ductile failure at the ultimate load, as would be the case for existing bridge decks in service. The flexural cracks formed in the constant moment region extended vertically and became wider and then progressed towards the load points in a diagonal fashion. The beam then collapsed as shown in Figure 5.

2.6. Beam 1 Failure. The longitudinal strains in the FRP grids due to the applied loads were recorded. The strain data in the cantilever section indicated that the longitudinal strain distribution followed the bending moment diagram. In Figures 6 and 7, the data obtained from the strain gages indicated that at higher loads the longitudinal strains in the shear spans increased above those of a linear variation. This showed that strains were not proportional to the applied moment at these locations. At ultimate conditions, the axial

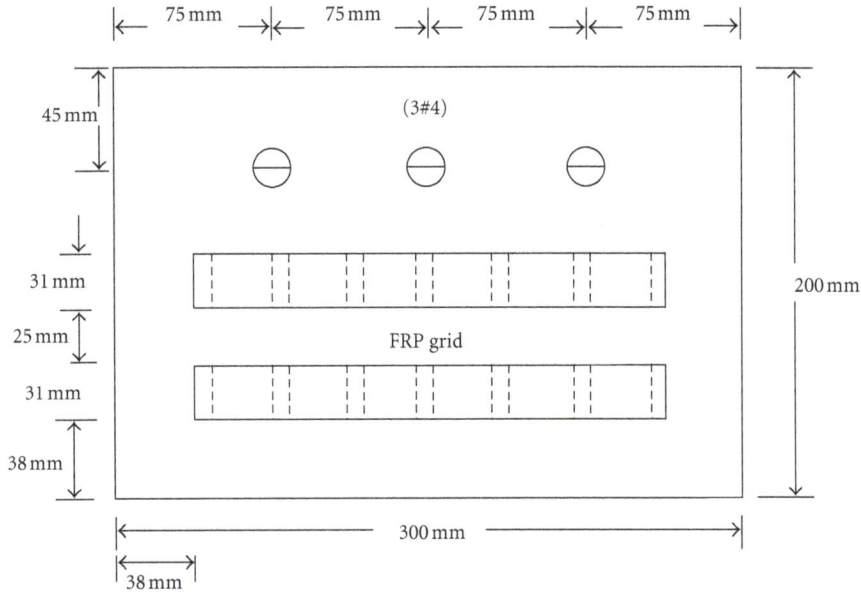

FIGURE 3: Beam 2 cross-section details.

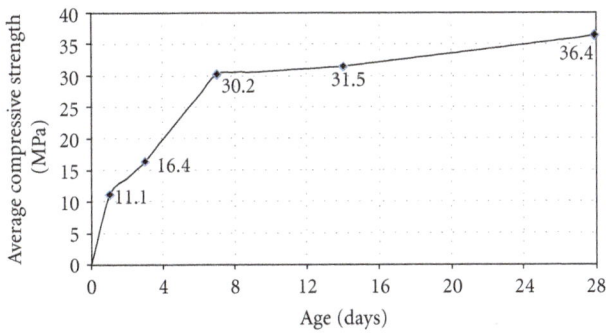

FIGURE 4: Concrete average compressive strength.

FIGURE 5: Beam 2 at collapse.

TABLE 2: FRP grid material properties provided by manufacturer.

Material characteristics	Value
Tensile stress, LW	207 MPa (30,000 psi)
Tensile modulus, LW	17.2 GPa (2.5×10^6 psi)
Compressive stress, LW	207 MPa (30,000 psi)
Compressive modulus, LW	17.2 GPa (2.5×10^6 psi)
Flexural stress, LW	207 MPa (30,000 psi)
Flexural modulus, LW	12.4 GPa (1.8×10^6 psi)
Shear modulus	3.1 GPa (0.45×10^6 psi)
Short beam shear	31 MPa (4,500 psi)
Punch shear	68.9 MPa (10,000 psi)
Bearing stress, LW	207 MPa (30,000 psi)
Area of 1 inch deep FRP per 9 inch width per layer	894 mm^2 (1.43 in^2)
Area of 1.25 inch deep FRP per 9 inch width per layer	1112 mm^2 (1.78 in^2)

6 and 7 indicated that the variations in the strain with the load at the beam center were slightly higher than those close to the load point, but the two curves were of similar form. As the applied load increased, the rate of change in the strains in the shear span was higher than that in the constant moment region. The higher rates demonstrated the initiation and progress of cracking in the region close to the support. The high level of strains in the shear span explained the flexural/shear cracking in the collapse mechanism for the beam.

2.7. Strains in Beam 1 Layer 1. A total of eight strain gages were installed to monitor the strain distribution. The strains measured were tensile strains in all of the gages

strain in the FRP grid varied linearly along the end of the FRP gird and at the point of load. Based on the previous discussion, it was concluded that the bond between the FRP grid and concrete is uniform. Moreover, the data in Figures

TABLE 3: Maximum flexural stresses for bottom elements for bridge girders.

Girder number	Open joint bridge (OJB) maximum flexural stress (MPa)	Link slab bridge (LSB) maximum flexural stress (MPa)	Percentage of decrease in girder stresses due to link slab
S1G1	2.13	1.75	18%
S1G2	2.55	2.07	19%
S1G3	2.53	2.12	16%
S1G4	2.16	1.69	22%
S2G1	0.82	0.54	34%
S2G2	0.82	0.55	33%
S2G3	0.83	0.57	32%
S2G4	0.80	0.54	32%
S3G1	0.63	0.56	12%
S3G2	0.65	0.59	10%
S3G3	0.65	0.59	9%
S3G4	0.65	0.56	14%

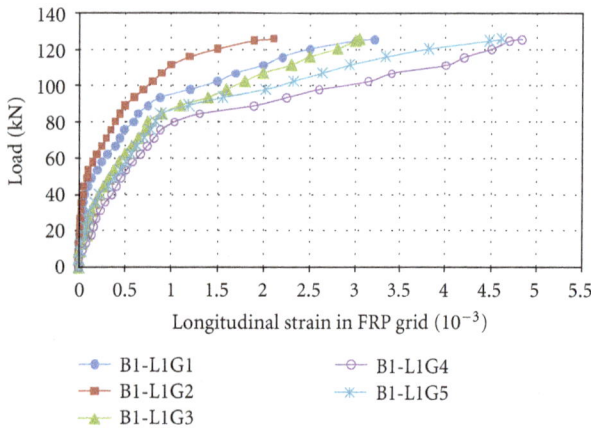

FIGURE 6: Load/strain along FRP grid for layer 1 in beam 1.

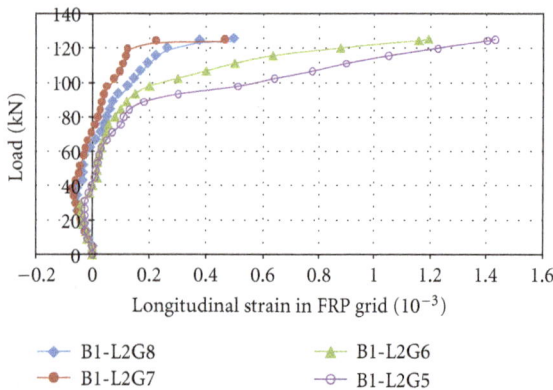

FIGURE 7: Load/strain along FRP grid for layer 2 in beam 1.

The maximum strain was 4.8 milli strains at the ultimate load of 125.5 kN (28.2 kips). The tensile modulus of the grid was 17.2 GPa (2.5×10^3 ksi). Therefore, the tensile stress corresponding to maximum tensile strain was 82.7 MPa (12.0 ksi) which is 40% of the maximum tensile stress recommended by the manufacturer, as shown in Table 1. The load-strain relationship was linear up to the load level of 75.6 kN (17 kip) when the beam began to yield. The load-strain distribution of gage 5 in layer 1 (B1-L1G5) located at the center of grid and beam indicated that the changes in the strains were low up to the load level of 84.6 kN (19 kip), and after that, change in strains were higher until the ultimate load was reached. The strain distribution for layer 1 of beam 1 indicated that as the applied load increases towards its maximum value, the distribution of strain in the FRP grid became unsymmetrical.

2.8. Strains in Beam 1 Layer 2. The strains measured were compressive strains in all the gages up to an applied load of 40 kN (9 kips). Then the measured strains were changed to tensile strains for the ultimate load test. These measurements indicated that the grid was in compression until the applied load reached a value of 40 kN (9 kip); then the grid was in tension. Among all the gages, the maximum compressive strain was found in gage 7 (B1-L2G7) located at 1.65 m (66 in) from the left end of the beam at an applied load of 40 kN (9 kips). The load-strain distribution of gage 7 in layer 2 (B1-L2G7) indicated that the maximum compressive strain was −0.074 millistrains. The compressive modulus of the grid was 17.2 GPa (2.5×10^3 ksi). Therefore, the compressive stress corresponding to maximum compressive strain was 1.24 MPa (0.18 ksi) which is 0.6% of the maximum compressive stress recommended by the manufacturer. The maximum tensile strain was found in gage 4 (B1-L2G4) which was located at just right to the left support. The maximum strain was 1.6 millistrains at the ultimate load 125.5 kN (28.2 kips). The tensile modulus of the grid was 17.2 GPa (2.5×10^3 ksi). Therefore, the

at different applied loads for the ultimate load test. These measurements indicated that the grid was in tension. Among all the gages, maximum tensile strain was found in gage 4 (B1-L1G4), which was located just right of the left support.

tensile stress corresponding to maximum tensile strain was 27.4 MPa (3.98 ksi) which is 13.3% of the maximum tensile stress. The strain distribution for layer 1 of beam 2 indicated that as the applied load increased towards its maximum value, the distribution of strain in the FRP grid became unsymmetrical.

2.9. Beam 2 Failure. The same discussion presented previously applies to the behavior for beam 2 with two 31 mm (1.25 in) FRP grids. The strain distribution in the FRP grid in layer 1 of the specimen beam 2 is shown in Figure 8.

2.10. Strains in Beam 2 Layer 1. A total of eight strain gages were installed to monitor the strain distribution. The strains measured were tensile strains in all the gages at different applied loads for the ultimate load test. These measurements indicated that the grid was in tension. Among all the gages, maximum tensile strain was found in gage 4 (B2-L1G4), which was located at just right to the left support. The maximum strain was 4.0 millistrains at the ultimate load of 113.9 kN (25.6 kips). The tensile modulus of the grid was 17.2 GPa (2.5×10^3 ksi). Therefore, the tensile stress corresponding to maximum tensile strain was 69.6 MPa (10.1 ksi) which is 34% of the maximum tensile stress recommended by the manufacturer. The load-strain distribution of gage 5 in layer 1 (B2-L1G5) located at center of grid and beam indicated that at higher loads, strain varied linearly with the applied loads.

2.11. Strains in Beam 2 Layer 2. The strains measured were compressive strains in all eight gages up to an applied load of 62.3 kN (14 kips); after that, the measured strains were changed to tensile strains for the ultimate load test. These measurements indicated that the grid was in compression till the applied load reached a value of 62.3 kN (14 kips); then the grid was in tension. Among all the gages, the maximum compressive strain was found in gage 7 (B2-L2G7) located at right end of the grid at an applied load of 31.2 kN (7 kips). The load-strain distribution of gage 7 in layer 2 (B2-L2G7) indicated that the maximum compressive strain was −0.058 millistrains. The compressive modulus of the grid was 17.2 GPa (2.5×10^3 ksi). Therefore, the compressive stress corresponding to maximum compressive strain was 1 MPa (0.15 ksi) which is 0.5% of the maximum compressive stress recommended by the manufacturer. The maximum tensile strain was also found in gage 7 (B2-L2G7). The maximum strain was 0.21 millistrains at the ultimate load 114 kN (25.6 kips). The tensile modulus of the grid was 2.5×10^3 ksi. Therefore, the tensile stress corresponding to maximum tensile strain was 3.7 MPa (0.53 ksi) which is 1.8% of the maximum tensile stress.

2.12. Load-Deflection Behavior. All specimens were tested in a four-point bending configuration. The ultimate loads and corresponding deflections for both beams were measured during the tests. The load carrying capacity of the beam 1 was 125.5 kN (28.2 kips) and that for beam 2 was 114 kN (25.6 kips). The deflection of each beam at collapse was

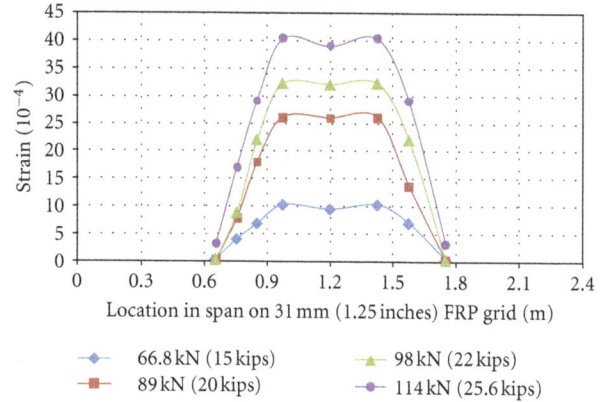

FIGURE 8: Longitudinal strain along FRP grid for layer 1 in beam 2.

substantial (L/240) accompanied by excessive cracking. The load deflection response of the specimens exhibited three regions of behavior. At low applied loads the stiffness of the reinforced concrete beam was relatively high, indicating that the concrete behaved in a linear elastic manner. As the load increased, the bending stress in the extreme fibers increased until the tensile strength at the top of the section of the concrete was reached. This caused flexural cracks to form, first in the constant moment region, then through the beam cantilever section. As the flexural cracks developed in the span, the member stiffness was reduced and the response after the cracking load was approximately linear due to the postcracking stiffness. After the concrete in the tension zone cracked, the FRP grid carried the tensile forces due to applied loads. As the applied load increased, the tensile stress increased the, beam stiffness was decreasing due to the loss of material stiffness, and the ability of the section to support the tensile stress was reduced. The yield plateau in the slope-deflection curve for beam 2 was longer than that of beam 1, which indicated that beam 2 was more ductile than beam 1, even though the areas of the FRP grids in beam 2 were greater than beam 1.

3. Theoretical Work

In this study, finite element models are developed to investigate the behavior of the bridge with link slabs. Two models are considered, one with open joints and another with the joints closed over the supports. The results of the models are compared and used to evaluate the structural behavior of the FRP-grid-reinforced link slab.

3.1. Bridge Model Description. The bridge models were developed using the software ANSYS. A typical three-span bridge was considered. In each span, four AASHTO Type III girders, end and intermediate diaphragms were modeled. The dimensions of the deck in the z-axis were 18 m (60 ft) long, 9 m (30 ft) wide in the x-axis, and 200 mm (8 in) thick in the y-axis. A gap of 25 mm (1 in) and 150 mm (6 in) was considered between two adjacent decks (open joint) and two girders in adjacent spans, respectively. The center-to-center distance between adjacent girders in a span was 2.6 m (104 in). The end diaphragms were provided between two adjacent girders, from the middepth of girder to the bottom of the top flange. The intermediate diaphragms were placed from the top of bottom flange to the bottom of the top flange (Saber et al. [7]) The thickness of the intermediate and end diaphragms was 175 mm (7 in). The link slab was modeled for a distance of 600 mm (2 ft) at both adjacent ends of the open joint. The link slab length was based on the theoretical studies which showed that the load-deflection behavior of the structure would not be affected by a debonding length of up to 5% of the girder span length (Zia et al. [8]). The girders were restrained at supports and both extreme ends of the decks were restrained in x, y, and z directions (translations). The HL-93 loads were applied to the bridge models in such a way that they produced maximum negative moment and tensile force in the link slab.

3.2. Elements Used in Modeling.
The FRP layers were meshed using SOLID46 element and material properties (Modulus of elasticity, Poisson's ratio, and density) were assigned during the process. The element edge length was 150 mm (6 in). A small size element was chosen because the depth of FRP was just 25 mm (1 in). The bridge decks and diaphragms were meshed using SOLID65 element and material properties (Modulus of elasticity, Poisson's ratio, and density) were assigned during the process. The element edge length of concrete element was 600 mm (24 in). The mesh was refined twice at the girder supports to restrain the girders properly over the piers (Saber et al. [9]).

3.3. Boundary Conditions and Loading System.
In the model, the interface area between girders and substructure was restrained in the x and y directions (translations). Both extreme ends of the decks (area along the depth) were restrained in the x, y, and z directions (translations). LRFD Bridge Design [10] load combinations were considered, and the corresponding load factors were applied to the model. The vehicular live load and the live load surcharge were applied on the bridge. The truck load was applied to produce maximum negative moments in the link slab.

3.4. Comparison between Open Joint Bridge and Link Slab Bridge.
The details for the finite element models are shown in Figure 9. The four girders in the first span of the bridge were designated as S1G1, S1G2, S1G3, and S1G4. Similarly, girders in the second span of the bridge were designated as S2G1, S2G2, S2G3, and S2G4 and girders in the third span of the bridge were designated as S3G1, S3G2, S3G3, and S3G4. Among all the girders, the maximum flexural stresses (tensile) were found in the second girder of the first span.

3.5. Girder Stresses

3.5.1. Span 1.
The flexural stresses (tensile) for the bottom elements along the length of the second girder in the first span for the two bridge models were shown in Figure 10.

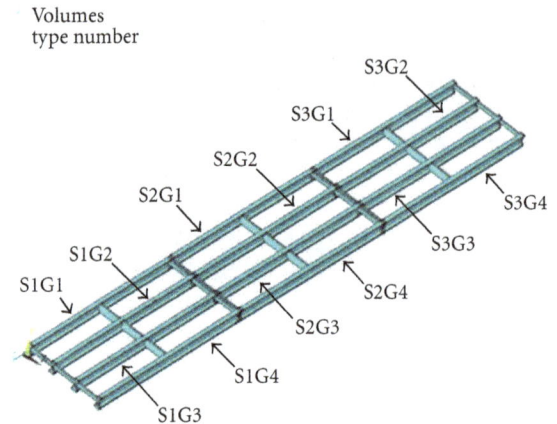

FIGURE 9: Model with the girders.

The flexural stresses were higher in the open joint bridge than those in the link slab bridge at most of the locations. A maximum flexural stress difference of 1 MPa (150 psi) was observed between the two girders, at a distance of 14.9 m (49 ft-8 in) from the left support. The flexural stresses were almost the same for a length of 6 m (20 ft) from the left support for both cases, but after that, the stresses in the open joint bridge were much higher. It can be inferred from the figure that the continuity in the decks reduces the flexural stresses in the girders.

3.5.2. Span 2.
The flexural stresses (tensile) for the bottom elements along the length of the second girder in the second span for two bridge models were shown in Figure 11. The flexural stresses were higher in the open joint bridge than those in the link slab bridge at all locations. A maximum flexural stress difference of 0.28 MPa (40 psi) was observed between two girders, at a distance of 13.1 m (43 ft-9 in) from the left support.

3.5.3. Span 3.
The flexural stresses (tensile) for the bottom elements along the length of the second girder in the third span for two bridge models were shown in Figure 12. A maximum flexural stress difference of 0.21 MPa (31 psi) was observed between two girders, at a distance of 3.7 m (12 ft-4 in) from the left support. The flexural stresses were higher in the open joint bridge up to 12 m (40 ft) from the left support and after that the flexural stresses were high in the link slab bridge.

3.6. Maximum Flexural Stresses in Girders.
The maximum flexural stresses in the twelve girders of the open joint bridge, the link slab bridge, and the percentage change in stresses of the open joint bridge compared with link slab bridge were shown in Table 2. The stresses were higher in girders of the open joint bridge. The maximum decrease was 34% found in the girders of span 2 of the bridge, and the minimum decrease was 9% found in span 3. The maximum effects in span 1 where truck load was applied were in the range of 16% to 22%.

TABLE 4: Stresses in link slabs at the top and the bottom of bridge deck.

Results	Link slab 1 stress (MPa)		link slab 2 stress (MPa)	
Transverse stress (S_x)	Max.	0.53	Max.	0.23
	Min.	−0.09	Min.	−0.04
Longitudinal stress (S_z)	Max.	2.29	Max.	1.05
	Min.	−1.01	Min.	−0.20
Shear stress (S_{yz})	Max.	0.05	Max.	0.07
	Min.	−0.03	Min.	−0.01

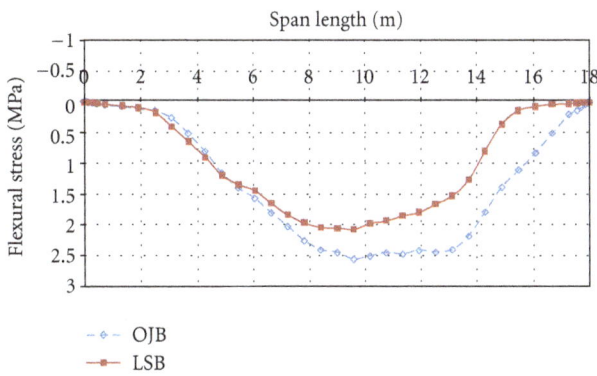

FIGURE 10: Flexural stresses for bottom elements of 2nd girder in 1st span (S1G2).

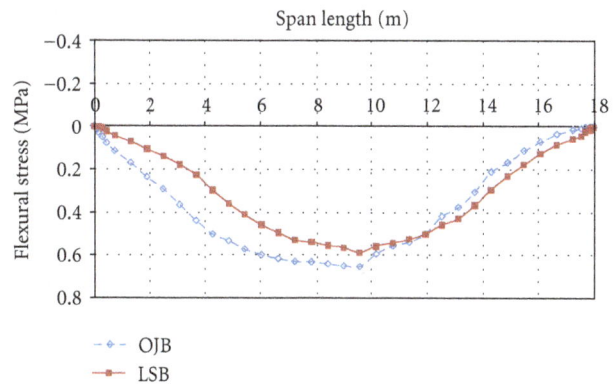

FIGURE 12: Flexural atresses for bottom elements of 2nd girder in 3rd span (S3G2).

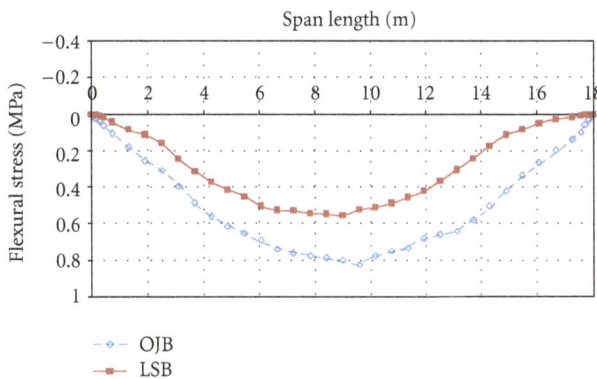

FIGURE 11: Flexural stresses for bottom elements of 2nd girder in 2nd span (S2G2).

3.7. Stresses in Bridge Decks. In bridge decks, the maximum and minimum transverse, longitudinal, and shear stresses were found in the first deck of the open joint bridge or the link slab bridge, since the load was applied on the first span of the bridge. The use of link slab reduced the bridge deck stresses. The transverse stresses were reduced by more than 13%, the longitudinal stresses were reduced by more than 36%, and the shear stresses were reduced by more than 43%. Based on these results, the use of the link slab will improve the performance of the bridge decks.

3.8. Stresses in Link Slabs. The maximum and minimum transverse, longitudinal, and shear stresses in two link slabs were shown in Table 3. Slabs 1 and 2 were joined by link slab 1, and slabs 2 and 3 were joined by link slab 2. The stresses were higher in link slab 1 than those in the link slab 2 because the truck was placed on the first span of the bridge. Maximum and minimum stresses were either at the top surface or at the bottom surface of the link slab.

The longitudinal stresses along the depth or thickness of the link slab 1 for the bottom element and at the top element were −1.01 MPa (−146.7 psi) and 2.29 MPa (332.8 psi), respectively. The longitudinal stresses along the depth for the bottom element and at the top element of the link slab 2 were −0.20 MPa (−28.4 psi) and 1.05 MPa (151.9 psi), respectively. The longitudinal stresses varied from compression to tension from the bottom to the top elements of both link slabs. The maximum longitudinal stresses along the length of the link slabs for the top elements were 1.21 MPa (176.1 psi) and 0.62 MPa (89.6 psi) for link slabs 1 and 2, respectively. Along the length of the link slab, all top elements for both link slabs were in tension. The maximum and minimum longitudinal stresses were higher in the link slab 1 than those in the link slab 2 because the truck load was placed in the first span of the bridge and the link slab 1 was connecting the span 1 and the span 2 decks of the bridge. The maximum longitudinal stresses along the length of the link slabs for the bottom elements were 0.29 MPa (−42.1 psi) and 0.02 MPa (2.6 psi) for link slabs 1 and 2, respectively. Along the length of the link slab, the bottom elements of link slab 2 were in tension.

4. Conclusions

The ductility of the FRP grid material was utilized to accommodate bridge deck deformations imposed by girder deflection, concrete shrinkage, and temperature variations. It would also provide a cost-effective solution to a number of deterioration problems associated with bridge deck joints.

The structural behavior of two types of FRP-grid-reinforced concrete slabs was investigated. Scaled-up beam specimens simulating the actual deck joint were prepared and tested. The design concept of link slabs was then examined to form the basis of design for FRP grid link slabs. Improved design of FRP grid link slab/concrete deck slab interface was confirmed in the numerical analysis.

The results indicated that the technique would allow simultaneous achievement of structural need, lower flexural stiffness of the link slab approaching the behavior of a hinge, and durability need of the link slab. The overall investigation supports the contention that durable jointless concrete bridge decks may be designed and constructed with FRP grid link slabs. It is recommended that the link slab technique be used during new construction of bridge decks. Also, it is recommended that the advantages of using the FRP grid link slab technique in repair and retrofit of bridge decks are considered along with the amount of intrusive field work required to develop the required mechanical properties at the bridge deck joints. The Louisiana Transportation Research Center received funds from the federal highway administration through the Innovative Bridge Research and Development program to support the implementation of the use of FRP grid as link slab.

Acknowledgments

Support of this work was provided by Louisiana Transportation Research Center under research Project no. 06-2ST and State Project no. 736-99-1391. Also, the fiberglass grid used in this study was provided by the Fibergrate Composite Structures (http://www.fibergrate.com/). The contents of this study reflect the views of the authors who are responsible for the facts and the accuracy of the data presented herein. The contents do not necessarily reflect the official views or policies of the Louisiana Department of Transportation, or the Louisiana Transportation Research Center, or the Fibergrate Composite Structures. This paper does not constitute a standard, specification, or regulation.

References

[1] A. Saber, F. Roberts, J. Toups, and W. Alaywan, "Effects of continuity diaphragm for skewed continuous span precast prestressed concrete girder bridges," *The Precast/Prestressed Concrete Institute Journa*, vol. 52, no. 2, pp. 108–114, 2007.

[2] A. Saber, J. Toups, and W. Alaywan, "Effects of continuity on load transfer in prestressed concrete skewed bridges," in *Proceedings of the 3rd International Structural Engineering and Construction*, 2005.

[3] A. Saber, "'Failure Behavior of RC T-Beams retrofitted with Carbon FRP Sheets.' Creative Systems in Structural and Construction Engineering," in *Proceeding of the 1st International Structural Engineering and Construction*, Honolulu, Hawaii, USA, 2001.

[4] K. P. Dutta, M. D. Bailey, R. J. Hayes, W. D. Jensen, and W. S. Tsai, "Composite grids for reinforcement of concrete structures," *Construction Productivity Advancement Research Program*, Final report, 1998.

[5] American Concrete Institute, "Building Code Requirements for Structural Concrete," ACI 318-05.

[6] American Concrete Institute, "Report on fiber reinforced plastic, reinforcement for concrete structure," ACI 440R, 2002.

[7] A. Saber, J. Toups, and A. Tayebi, "Continuity diaphragms for aashto type II girders," in *Proceedings of the 2nd International Structural Engineering and Construction*, Rome, Italy, August 2003.

[8] P. Zia, C. Alp., and K. E. S. Adel, *Jointless Bridge Decks*, North Carolina Department of Transportation, 1995.

[9] A. Saber, F. Roberts, X. Zhou, and W. R. Alaywan, "Impact of higher truck loads on remaining safe life of Louisiana bridge decks," in *Proceedings of the 9th International Conference, Applications of Advanced Technology in Transportation*, Chicago, Ill, USA, August 2006.

[10] American Association of State and Highway Transportation Officials, *LRFD Bridge Design Specifications*, U.S. Customary Units, 3rd edition, 2004.

12

Sensitivity Analysis of the Influence of Structural Parameters on Dynamic Behaviour of Highly Redundant Cable-Stayed Bridges

B. Asgari,[1] S. A. Osman,[1] and A. Adnan[2]

[1] *Department of Civil & Structural Engineering, Faculty of Engineering & Built Environment, National University of Malaysia (UKM), 43600 Bangi, Selangor, Malaysia*
[2] *Department of Civil & Structural Engineering, Faculty of Engineering & Built Environment, UTM University of Malaysia, 81310 Skudai, Johor, Malaysia*

Correspondence should be addressed to B. Asgari; basgari1360@yahoo.com

Academic Editor: John Mander

The model tuning through sensitivity analysis is a prominent procedure to assess the structural behavior and dynamic characteristics of cable-stayed bridges. Most of the previous sensitivity-based model tuning methods are automatic iterative processes; however, the results of recent studies show that the most reasonable results are achievable by applying the manual methods to update the analytical model of cable-stayed bridges. This paper presents a model updating algorithm for highly redundant cable-stayed bridges that can be used as an iterative manual procedure. The updating parameters are selected through the sensitivity analysis which helps to better understand the structural behavior of the bridge. The finite element model of Tatara Bridge is considered for the numerical studies. The results of the simulations indicate the efficiency and applicability of the presented manual tuning method for updating the finite element model of cable-stayed bridges. The new aspects regarding effective material and structural parameters and model tuning procedure presented in this paper will be useful for analyzing and model updating of cable-stayed bridges.

1. Introduction

In the past decade the construction of cable-stayed bridges has increased and their span lengths are growing due to improvements in design and analysis technologies. However, the complex structural characteristics of long-span cable-stayed bridges cause difficulties in understanding their dynamic behavior and make them vulnerable to dynamic loadings from phenomena such as wind or earthquakes. In recent years, many experimental and analytical investigations have studied effective factors on the dynamic behavior of cable-stayed bridges, such as natural periods, mode shapes, and damping properties [1–5]. The sensitivity analysis is a promising way to provide a tuned analytical model and assess the actual dynamic characteristics of superstructures such as cable-stayed bridges.

Sensitivity analysis is a technique to determine the influence of different properties, such as boundary conditions, damping properties, material constants, and geometrical parameters, on the structural responses. A number

of sensitivity methods for model updating purposes have been proposed for different structures [6–12]; there have been successful applications of sensitivity-based updating technology for bridges in recent years. Cantieni [13] and Pavic et al. [14] were among the first to investigate the model updating of bridges using the sensitivity method. Mackie and Stojadinović [15] conducted a sensitivity study to identify the effect of abutment mass and stiffness on seismic demand for short- and medium-length bridges. For a complex structure with high degrees of indeterminacy, such as cable-supported bridges, model updating becomes difficult because it may inevitably involve uncertainties in many parameters, for example, material and geometrical properties, and boundary conditions. Most of the recent studies have applied iterative solutions for sensitivity-based model updating of cable-stayed bridges. Zhang et al. [16] implemented finite element (FE) model updating for a 430 m main span cable-stayed bridge in Hong Kong. The updating method was an iterative eigenvalue sensitivity-based approach with lower and upper bounds for the parameter values and the degrees of uncertainty. Brownjohn and Xia [17] successfully applied iterative sensitivity-based model updating to the dynamic assessment of a curved cable-stayed bridge in Singapore. The results of previous studies demonstrate that during an iterative procedure, it is impossible to obtain a desirable result after only one or two iterations, and it is necessary to frequently adjust the tuning strategies for a successful updating. Although most of the proposed tuning methods are automatic processes, the reasonable approximations obtained by manual methods (i.e., engineering judgment) are required to obtain the best results. Daniell and Macdonald [18] successfully applied manual tuning to a 3D FE model of one balanced cantilever section of the Second Severn Crossing (SSC) cable-stayed bridge, using data measured from ambient vibration tests during its construction. Benedettini and Gentile [19] investigated the sensitivity of the natural frequencies to changes in uncertain parameters of a cable-stayed bridge in Italy with a 70 m main span through a manual sensitivity analysis. However, the application of the manual sensitivity-based model updating technology to long-span cable-stayed bridges is still a challenge to the civil engineering community.

This paper presents a manual iterative model tuning algorithm for updating the analytical model of long-span cable-stayed bridges. The sensitivity analysis is applied to identify the effect of different parameters on dynamic characteristics of the bridge, which provides a better understanding of the nature of the structural behavior of highly redundant cable-stayed bridges. The FE model of Tatara cable-stayed bridge in Japan is generated in ANSYS 12 [20], and the sensitivity of the bridge to structural and material parameters is investigated to understand their effect on the dynamic behavior of the structure. The new and important aspects of this study can be useful as a guide for the analysis and model updating of long-span cable-stayed bridges.

2. Model Updating Procedure Based on Sensitivity Analysis

Sensitivity analysis is a common way to find effective parameters for structural responses of cable-stayed bridges, such as static displacements, mode shapes, natural frequencies, or correlation values such as modal assurance criterion (MAC) values. Conducting a sensitivity study helps to update the FE model of existing bridges and understand the structural behavior for future design of cable-stayed bridges. In recent years, different methods have been proposed to apply sensitivity-based model updating to bridges. Most of the proposed methods are automatic solutions that iteratively update selected structural and material properties to improve the correlation of model responses and test results of the bridge. However, new investigations show that during the process, it is necessary to adjust the tuning strategies frequently based on engineering judgment for a successful updating. The automatic sensitivity-based updating algorithm for cable-stayed bridges was described previously by Zhang et al. [16]. This paper investigates a manual model updating algorithm based on sensitivity study to assess the dynamic behavior of the long-span cable-stayed bridges. The natural frequencies

of selected modes are considered to compare the results. The preference for the proposed method over other model updating methods results from the possibility of using manual tuning to obtain the most reasonable results. However, the automatic model updating is also possible using the presented algorithm by applying the iteration loops in ANSYS.

The order and the main features of the updating procedure are represented as follows.

The tuning procedure requires correct selection of updating parameters and reference responses to obtain the best match between FE and measured results. The following parameters ($P_i, i = 1, \ldots, n$) are selected for sensitivity analysis:

(i) material properties: Young's modulus, Poisson's ratio, and mass density of the deck, towers, and cables;

(ii) geometrical element properties: spring stiffness and beam cross-sectional properties;

(iii) lumped properties: lumped stiffness (boundary conditions) and lumped masses.

The choice of sensitive parameters from the mentioned updating parameters is the most important aspect of the process, which requires wise engineering insights. The uncertain parameters must be selected to prevent meaningless results in the FE simulation. Furthermore, the number of updating parameters should be kept small, and such parameters should be chosen with the aim of correcting recognized uncertainty in the model and ensuring the data sensitivity to them.

The sensitivity study is a promising way to find the right parameters to update. A comprehensive sensitivity study of the parameters is presented in this paper using sensitivity coefficients. The sensitivity coefficient (S_{ji}) is defined as the rate of response change with respect to a parameter adjustment:

$$S_{ji} = \frac{\partial R_j}{\partial P_i}, \tag{1}$$

where R_j and P_i represent structural response and parameter, respectively. The subscripts are $i = 1, 2, \ldots, n$ for n parameters and $j = 1, 2, \ldots, m$ for m responses. Although some researchers [19] have considered the sensitivity coefficient as the change percentage in mode frequency per 100% change in the parameter, the changes in parameters should be small and restricted to lower and higher values to achieve reasonable responses and prevent physically meaningless updated results ($L < \delta P_i < H$).

The sensitivity of the natural frequencies to variations in different parameters of a long-span cable-stayed bridge is investigated in this study. The effective parameters (P^k) are chosen considering the sensitivity coefficients. As mentioned before, the change in parameters should be restricted to lower and higher values ($L < \delta P_i^k < H$). The tolerance (ε) is defined as the difference between the analyzed natural frequencies in each iteration (f_a^k) and the measured frequencies (f_m). The algorithm of the described procedure is represented in Figure 1. In the manual model tuning, the different bounds for parameters can be adjusted in each iteration,

Sensitivity Analysis of the Influence of Structural Parameters on Dynamic Behaviour of Highly
Redundant Cable-Stayed Bridges

115

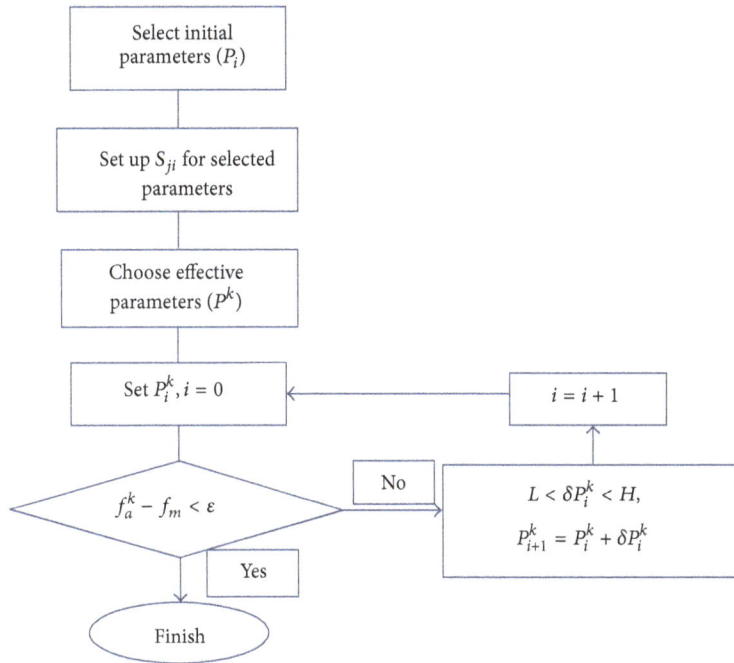

FIGURE 1: FE model tuning algorithm.

FIGURE 2: Tatara bridge.

based on engineering judgment. Moreover, choosing the changing parameters may have different priorities based on the parameter importance. The try- and -error procedure can be used for iterations. In the automatic procedure, the FE model updating starts with an iteration loop that considers reasonable tolerance for the results to converge. The manual model tuning results in better understanding of structural behavior and considers engineering considerations in long-span cable-stayed bridges.

The updated analytical models using the mentioned procedure can be used in the future for tasks such as health monitoring of the bridges.

3. Numerical Study

3.1. Bridge Description. The analytical FE model of Tatara Bridge is considered for simulations in this study. The Tatara

Bridge (Figure 2) is completed in 1999 in Japan, with a main span of 890 m, and it is now the 4th longest cable-stayed bridge in the world. The bridge comprises a steel box deck 2.7 m deep with prestressed concrete (PC) girders serving as counterweights in side spans (Figure 3).

The bridge has two diamond-shaped steel towers with heights of 220 m. Figure 4 shows the schematic elevation view of the bridge. All 21 stay cables are arranged in a two-plane multifan shape and are stressed at the anchor points of the bridge deck. The stay cables have a polyethylene cable coating to resist rain vibration. Figure 5 shows the arrangement of the cables on the towers.

The bridge is modeled in ANSYS software based on the design information of the bridge. The effect of different structural and material parameters on the dynamic properties of the bridge is investigated from the analysis of a three-dimensional FE model.

3.2. Three-Dimensional FE Model of the Bridge. The geometric and inertial parameters, connections, and boundary conditions of the bridge are simulated in ANSYS software. The simplified three-dimensional FE model of the bridge is developed using elastic beam and link elements. The bridge deck is modeled by a single central spine with offset rigid links to accommodate cable anchor points (fishbone model). The BEAM4 elements are used to model the deck, towers, heads, and struts of the towers. The MPC184 elements are applied to model the rigid links, and MASS21 elements are used to include the mass of nonstructural members. The cables are modeled by tension-only LINK10 elements that have a stress stiffening capability. The cable pretensions are considered to ensure small deformations under the deck self-weight. Table 1

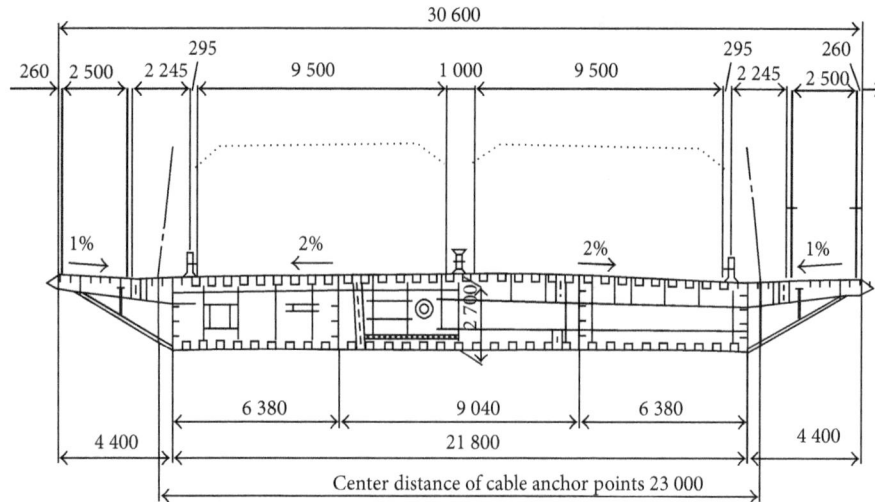

(a)   Cross section of steel girder

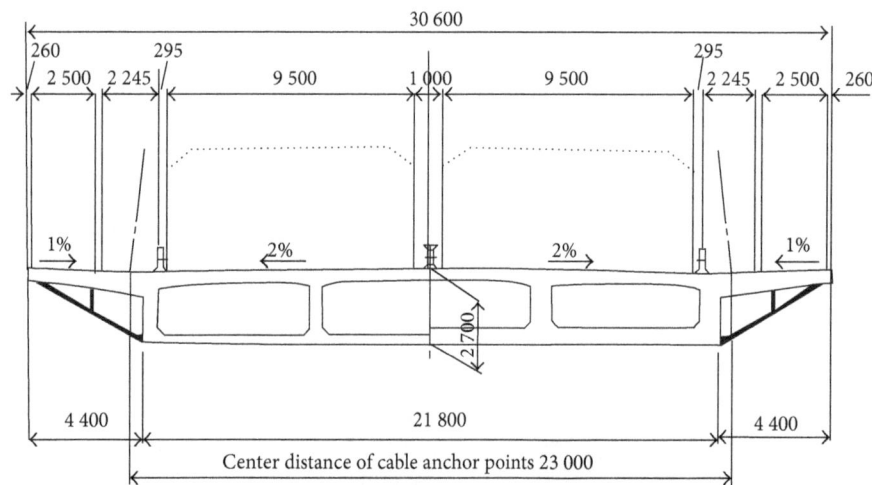

(b)   Cross section of concrete girder

FIGURE 3: Main girder sections [21].

TABLE 1: Properties of structural members.

Structural members	Element type	Material	E (MPa)	ρ (kg/m^3)
Tower	BEAM4, MPC184	Steel	2.10×10^5	7850
Rigid link of towers		—	—	
Girder	BEAM4	Steel	2.10×10^5	7850
Rigid link of girders	BEAM4	Concrete	3.00×10^4	2400
	MPC184	—	—	
Cable	Link10	Steel	2.00×10^5	7850

shows the properties and element types of the structural members that are used in the FE model of the Tatara Bridge.

The boundary conditions at the piers, at the base of the towers and at the connection of each abutment, are simulated in the FE model. The tower bases are considered as being fixed in all degrees of freedom. The end connections permit the end of the deck to rotate freely about the vertical and transverse axes. Rotation about the longitudinal axis (x) and two translational degrees of freedom at each abutment are fixed. Elastic bearings were used for vertical bearings on the tower-to-deck connections. Thus, the spring constant about 3.92×10^6 (N/m) is adopted for limit girder displacement in the direction of the bridge axis based on design information about the bridge. Figure 6 represents the boundary conditions assigned to the FE model of the bridge.

One of the important features of cable-stayed bridges is the dead load influence on the stiffness of the bridge. To include this influence, the static analysis under self-weight

Sensitivity Analysis of the Influence of Structural Parameters on Dynamic Behaviour of Highly
Redundant Cable-Stayed Bridges

117

TABLE 2: Comparison of natural frequencies (Hz) between field tests and FE modeling.

Mode no.	FE frequencies (HZ)	Deck component of motions			Tower component of motions			Measured frequencies	Frequency error (%)
		Vertical	Lateral	Torsional	Longitudinal	Lateral	Torsional		
V1	0.216	✓			✓			0.225	4.0
TL1	0.120		✓	✓		✓	✓	0.097	23.0
V2	0.270	✓			✓			0.263	2.7
TL2	0.320		✓	✓		✓	✓	0.248	29.0
V3	0.367	✓			✓			0.365	0.5
TL3	0.564		✓	✓		✓	✓	0.470	20.0

V indicates vertical modes; TL indicates torsional-lateral modes.

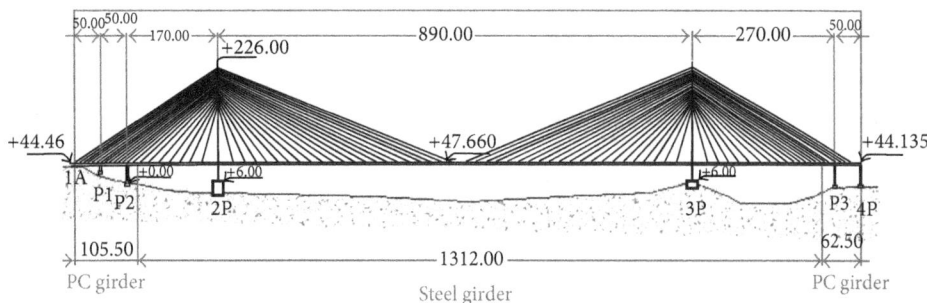

FIGURE 4: General arrangement of the Tatara cable-stayed bridge.

and cable pretension, in which all the structural members are prestressed, is performed before the modal analysis. The sag effect, the most important nonlinear effect, is considered to include geometrical nonlinearity in the static analysis. The results of recent studies show that nonlinearities are more significant in static analysis than dynamic analysis of the cable-stayed bridges and the most effective nonlinearity is the sag effect [4, 5, 22–25]. Following a nonlinear static analysis, the linear prestressed modal analysis is conducted to extract the natural frequencies and mode shapes of the initial FE model of the bridge.

3.3. Verification of FE Simulation Results with Field Test Results of the Bridge. The natural frequencies and mode shapes of the FE modal analysis should be verified with field test results to ensure that the structural parameters used in the FE modeling reflect the real dynamic characteristics of the bridge.

The field forced vibration test of the Tatara Bridge had been performed in 1998 when the construction of pavement was almost complete. The exciters vibrated the girder in the vertical direction to investigate three vertical bending vibration modes and in the horizontal direction to investigate three bending vibration modes in the normal direction of the bridge axis [26]. In the case of the Tatara Bridge, the test results are available only for first-mode frequencies. Thus, the results of FE modeling are verified with a few measured mode frequencies. However, the results of recent studies show that when the purpose of parametric updating is the response estimation of a long-span bridge under wind excitation, the verification of the lowest few vertical and lateral-torsional deck modes would be sufficient. It is conceived that the

response of a bridge can be quite accurately spanned by the lower modes [27]. Table 2 shows the comparison of FE calculated frequencies and identified frequencies from test results.

The other method available to compare the results of FE modeling with test results involves using the MAC value, which compares the ordinates of mode shapes and gives a value of unity for perfect correlation, while returning a value of zero for uncorrelated modes. The MAC values are not compared in this study with respect to lack of information about actual mode shapes from test results. However, the comparison of natural frequencies is considered to assess the sensitivity of the dynamic characteristics of the bridge to different parameters.

4. Model Updating Results

4.1. Sensitivity Analysis of the Parameters. Having established the basic model, the sensitivity of the natural frequencies is investigated for variations in material and structural properties of the deck, towers, and cables of the modeled bridge. Based on the natural characteristics of the bridge a maximum of 30% is considered for lower and higher change values of parameters.

The following properties are considered to investigate the sensitivity of the modal responses of the bridge:

(i) Young's modulus, self-weight (which is relative to mass density), and section properties of steel girders $(E_{d1}, \rho_{d1}, A_{d1}, I_{xx(d1)}, I_{zz(d1)}, I_{yy(d1)})$ and PC girders $(E_{d2}, \rho_{d2}, A_{d2}, I_{xx(d2)}, I_{zz(d2)}, I_{yy(d2)})$;

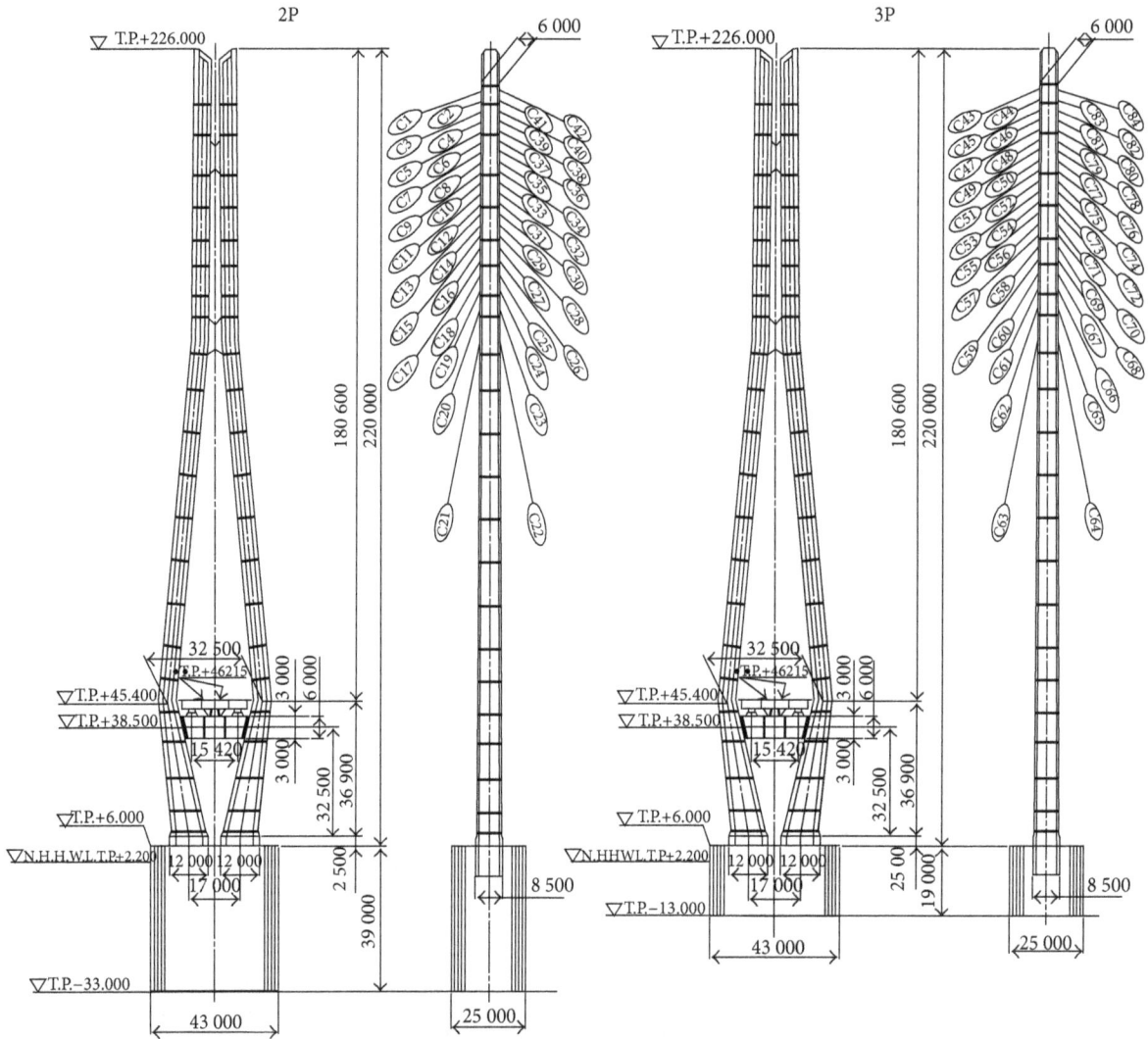

FIGURE 5: Main tower general arrangement [21].

FIGURE 6: Boundary conditions of the Tatara cable-stayed bridge.

(ii) the mass density and Young's modulus of the towers (ρ_t, E_t) and cables (ρ_c, E_c);

(iii) the constant of transverse springs in the deck-to-tower connections (K_z).

The sensitivity coefficients are defined as $(\partial R_j /\partial P_i)(P_i/R_j)$, and they represent the change in natural frequencies for an approximately 60% change in parameters (30% in each direction). The sensitivity of the model to change in the deck, tower, and cable properties is investigated in the prestressed modal analysis to compare the natural frequencies. The tensions in the cables are adjusted manually to correspond with the adjustments in deck mass.

The elastic moduli of cables are modified automatically in ANSYS to consider the sag effect.

Young's modulus and the weight of steel girders affect the bridge responses significantly, while the same parameters of PC girders have almost no effect on natural frequencies (Tables 3 and 4). It can be justified that the PC girders are much smaller than the steel girders of the main span; the PC girders perform only as counterweights and do not have any significant effect on the dynamic characteristics of the bridge. Thus, the properties of steel girders are considered for the sensitivity analysis of the deck $(E_{\text{deck}}, \rho_{\text{deck}}, A_{\text{deck}}, I_{xx}, I_{zz}, I_{yy})$. The inspection of the sensitivity coefficients of the first vertical and torsional-lateral modes for the parameters of steel girders (as shown in Figure 7) clearly reveals the following.

(i) The E_{deck} and ρ_{deck} change the natural frequencies inversely, while ρ_{deck} changes the first vertical modes more significantly. The change of E_{deck} primarily affects the first torsional-lateral modes of the deck.

Sensitivity Analysis of the Influence of Structural Parameters on Dynamic Behaviour of Highly
Redundant Cable-Stayed Bridges

119

TABLE 3: Comparison of natural frequencies (Hz) with changes in Young's modulus of steel girders (E_{d1}) and PC girders (E_{d2}).

E_{deck}	TL1	V1	V2	TL2	V3	TL3
$0.7E_{d1}, E_{d2}$	0.103	0.209	0.26	0.276	0.355	0.535
$0.8E_{d1}, E_{d2}$	0.109	0.212	0.264	0.293	0.359	0.547
$0.9E_{d1}, E_{d2}$	0.115	0.214	0.267	0.308	0.363	0.555
$1.1E_{d1}, E_{d2}$	0.126	0.217	0.273	0.327	0.370	0.571
$1.2E_{d1}, E_{d2}$	0.130	0.218	0.275	0.331	0.373	0.578
$1.3E_{d1}, E_{d2}$	0.135	0.220	0.277	0.333	0.376	0.585
$\mathbf{E_{d1}, E_{d2}}$	**0.120**	**0.216**	**0.270**	**0.32**	**0.367**	**0.564**
$0.7E_{d2}, E_{d1}$	0.120	0.215	0.269	0.320	0.367	0.564
$0.8E_{d2}, E_{d1}$	0.120	0.215	0.269	0.320	0.367	0.564
$0.9E_{d2}, E_{d1}$	0.120	0.215	0.270	0.320	0.367	0.564
$1.1E_{d2}, E_{d1}$	0.120	0.216	0.270	0.320	0.367	0.564
$1.2E_{d2}, E_{d1}$	0.120	0.216	0.270	0.320	0.368	0.564
$1.3E_{d2}, E_{d1}$	0.120	0.216	0.270	0.320	0.368	0.564

TABLE 4: Comparison of natural frequencies (Hz) with changes in mass density of steel girders (ρ_{d1}) and PC girders (ρ_{d2}).

ρ_{deck}	TL1	V1	V2	TL2	V3	TL3
$0.7\rho_{d1}, \rho_{d2}$	0.136	0.243	0.305	0.333	0.374	0.577
$0.8\rho_{d1}, \rho_{d2}$	0.130	0.233	0.292	0.331	0.369	0.571
$0.9\rho_{d1}, \rho_{d2}$	0.125	0.224	0.28	0.327	0.365	0.569
$1.1\rho_{d1}, \rho_{d2}$	0.116	0.208	0.261	0.311	0.354	0.546
$1.2\rho_{d1}, \rho_{d2}$	0.112	0.202	0.252	0.303	0.343	0.529
$1.3\rho_{d1}, \rho_{d2}$	0.109	0.196	0.245	0.294	0.336	0.514
$\mathbf{\rho_{d1}, \rho_{d2}}$	**0.120**	**0.216**	**0.270**	**0.320**	**0.367**	**0.564**
$0.7\rho_{d2}, \rho_{d1}$	0.121	0.216	0.270	0.320	0.367	0.564
$0.8\rho_{d2}, \rho_{d1}$	0.120	0.216	0.270	0.320	0.367	0.564
$0.9\rho_{d2}, \rho_{d1}$	0.120	0.216	0.270	0.320	0.367	0.564
$1.1\rho_{d2}, \rho_{d1}$	0.120	0.216	0.270	0.320	0.367	0.564
$1.2\rho_{d2}, \rho_{d1}$	0.120	0.216	0.270	0.320	0.367	0.564
$1.3\rho_{d2}, \rho_{d1}$	0.120	0.216	0.270	0.320	0.367	0.564

TABLE 5: Comparison of natural frequencies (Hz) with changes in the transverse stiffness (K_z) of bearings.

K_z	TL1	V1	V2	TL2	V3	TL3	TOW1	TOW2
$K_z = K_x$	0.067	0.216	0.270	0.14	0.367	0.238	0.342	0.434
$K_z = 2K_x$	0.075	0.216	0.270	0.154	0.367	0.248	0.342	0.434
$K_z = 10K_x$	0.100	0.216	0.270	0.224	0.367	0.352	0.342	0.434
$K_z = 15K_x$	0.105	0.216	0.270	0.248	0.367	0.435	0.342	0.430
$K = \infty$	0.120	0.216	0.270	0.320	0.367	0.564	0.342	0.427

TOW indicates tower modes.

(ii) The A_{deck} slightly affects both the vertical and torsional-lateral modes of the bridge.

(iii) The changes of I_{xx}, I_{zz} have no significant effect on mode frequencies, while the change of I_{yy} has an increasing effect on the first two torsional-lateral modes.

The sensitivity coefficients of the first deck and tower modes to the parameters of steel towers and cables are shown

TABLE 6: Effective parameters on the first mode responses of the bridge.

Mode	Effective parameters
TL1	$E_{\text{deck}}, I_{yy}, \rho_{\text{deck}}, A_{\text{deck}}, K_z$
TL2	$E_{\text{deck}}, \rho_{\text{deck}}, K_z, I_{yy}, E_{\text{tower}}, \rho_{\text{tower}}$
TL3	$E_{\text{deck}}, \rho_{\text{deck}}, A_{\text{deck}}, K_z$
V1	$\rho_{\text{deck}}, E_{\text{cable}}, A_{\text{deck}}, E_{\text{deck}}$
V2	$\rho_{\text{deck}}, E_{\text{cable}}, A_{\text{deck}}, E_{\text{deck}}$
V3	$E_{\text{cable}}, \rho_{\text{deck}}, A_{\text{deck}}, E_{\text{deck}}$
TOW1	$\rho_{\text{tower}}, E_{\text{tower}}, I_{yy}$
TOW2	$\rho_{\text{tower}}, E_{\text{tower}}, I_{yy}$

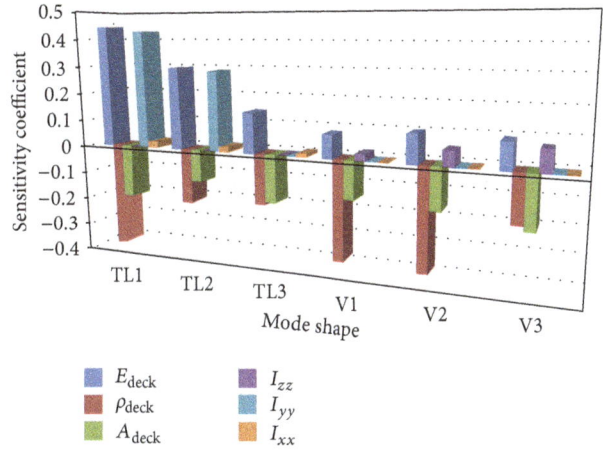

FIGURE 7: Sensitivity of modes to the parameters of steel deck.

in Figure 8. The tower properties (E_{tower} and ρ_{tower}) change the vertical and transverse deck modes slightly; however, they have a significant effect on the following tower modes. The cross-sectional areas of the towers (A_{tower}) are not considered for updating because they are parameters that cannot be updated simply. Increasing Young's modulus of the cables (E_{cable}) will considerably increase the natural frequencies of the first modes, while the mass density of cables (ρ_{cable}) does not have a significant effect on the natural frequencies.

The dynamic characteristics of the recent cable-stayed bridges prove that the deck-to-tower bearings act in all directions and using uniaxial spring elements cannot represent the dynamic behavior of the cable-stayed bridge accurately. Therefore, the transverse stiffness of the springs (K_z) is also considered to represent the actual behavior of bearings in the deck-to-tower connections. Longitudinal and transverse springs (K_x, K_z) representing the stiffness of bearings in deck-to-tower connections mainly affect the deck torsional-lateral modes. Table 5 indicates that increasing the transverse spring stiffness (K_z) results in an enhancement of the natural frequencies. The spring parameters can be freely updated, and no variation bounds are assumed because they are quite uncertain in the first application.

The results of the sensitivity analysis clearly reveal that the dynamic characteristic of a cable-stayed bridge model can be improved by changing the structural and material

TABLE 7: Comparison between calculated FE and measured frequencies.

Modes	Measured frequencies (Hz)	FE calculated frequencies (Hz)						Final frequency error (%)
		Initial FE frequency (HZ)	Estimated frequencies					
			First	Second	Third	Forth		
TL1	0.097	0.120	0.120	0.110	0.100	**0.100**		3
TL2	0.248	0.320	0.269	0.260	0.255	**0.250**		3
TL3	0.470	0.564	0.390	0.410	0.450	**0.463**		1
V1	0.225	0.216	0.220	0.218	0.218	**0.218**		3
V2	0.263	0.270	0.269	0.269	0.267	**0.267**		1
V3	0.365	0.367	0.367	0.370	0.370	**0.370**		1

TABLE 8: Updated parameters of the bridge.

Parameters	Initial FE parameters	Updated FE parameters	Change (%)
E_{deck} (MPa)	2.10×10^5	2.15×10^5	2.38
ρ_{deck} (kg/m^3)	7850	8242.5	5
A_{deck} (m^2)	1.367	1.367	0
I_{yy} (m^4)	140	133	-5
E_{cable} (MPa)	2.00×10^5	2.05×10^5	2.5
K_z (N/m)	—	3.92×10^7	—

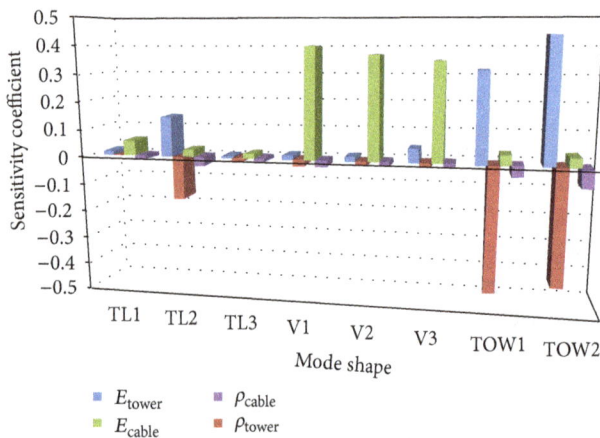

FIGURE 8: Sensitivity of modes to the parameters of towers and cables (TOW indicates tower modes).

parameters. In the case of the Tatara cable-stayed bridge, the most effective parameters on the first frequency modes are shown in Table 6. Based on the results of sensitivity studies, material and structural parameters can be updated to achieve the closest natural frequencies to the experimental tests of the bridge.

4.2. Manual FE Model Tuning of the Bridge. Based on the sensitivity analysis results, the FE model of the bridge is updated using the algorithm that is presented in Figure 1. The manual model tuning is considered rather than automatic methods because of the following.

(i) In manual model tuning, the selection of parameters for updating is more meaningful than other proposed

methods. In the case of the modeled cable-stayed bridge, the homogeneous properties of steel main deck and towers permit fewer changes in the mass density and cross-sectional parameters of the deck and towers in the updating procedure in comparison with concrete ones.

(ii) In the manual tuning method the restricted lower and higher values of parameters can be different during the procedure with respect to engineering judgment. The different variations in the higher and lower bounds of parameters increase the accuracy of the procedure. The deck-to-tower connection parameters, which act as bearings, can be updated according to trials and errors with no variation bounds, while other sectional properties should vary in a restricted bound to prevent meaningless results. A maximum 10% variation is considered for the elastic moduli, mass densities, and cross-sectional areas of all components. For the moments of inertia of the deck, a maximum 20% variation is considered due to the relative complexity compared with the other components.

(iii) The number of updating candidates of parameters is lower than that of other methods due to elimination of some parameters based on engineering judgment and natural characteristics of the bridge.

Despite all the advantages, the disadvantage of the manual tuning method is that the variation in selected stiffness and mass parameters cannot be very refined without an impracticable number of model variations [8]. However, the manual method is considered in this study for model updating of a long-span cable-stayed bridge considering all the mentioned advantages. Table 7 shows the different attempts at model updating of the bridge. The tuning procedure will be stopped when the differences between the measured and the calculated frequencies become less than 3% for the selected first vertical and lateral-torsional modes of the bridge. The values of updated parameters are shown in Table 8.

Using the presented manual tuning method, material and structural parameters are updated to achieve the closest natural frequencies to the experimental tests of the bridge. The updated mode shapes of the bridge are shown in Figure 9. In the case of the updated bridge model, it seems that the combinations of updated parameters represent the dynamic

Sensitivity Analysis of the Influence of Structural Parameters on Dynamic Behaviour of Highly
Redundant Cable-Stayed Bridges

121

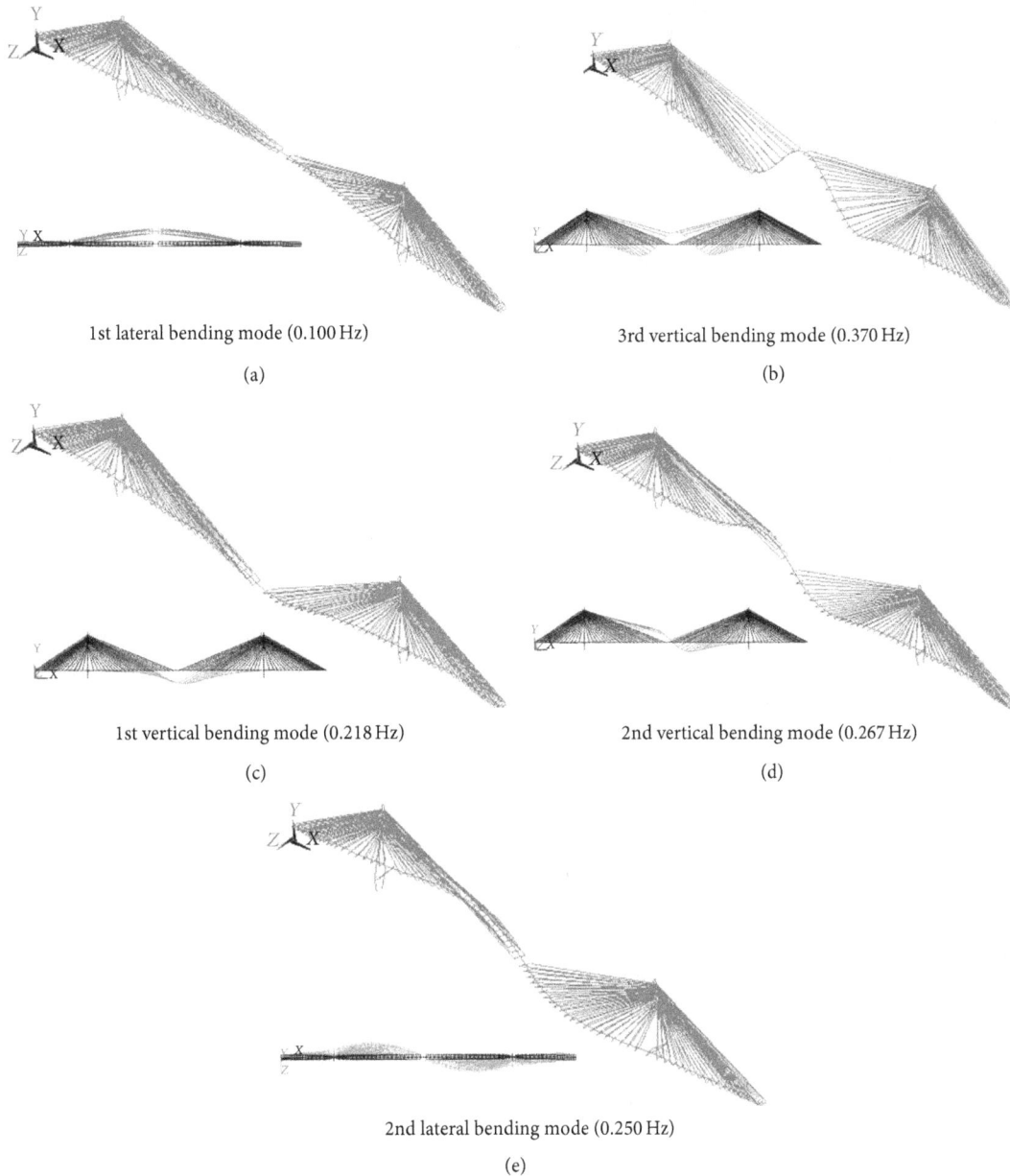

1st lateral bending mode (0.100 Hz)

(a)

3rd vertical bending mode (0.370 Hz)

(b)

1st vertical bending mode (0.218 Hz)

(c)

2nd vertical bending mode (0.267 Hz)

(d)

2nd lateral bending mode (0.250 Hz)

(e)

FIGURE 9: First mode shapes from the updated FE model.

behavior of the bridge accurately, and further changes would not be effective in modeling bridge responses. The dynamic and modal characteristics of the FE model show good agreement with the field test results, and the updated model can be used as a baseline model for future uses such as health monitoring of the bridge.

5. Conclusions

A manual updating algorithm is presented in this paper for model tuning of highly redundant cable-stayed bridges based on sensitivity analysis. The following aspects are concluded from the proposed model updating procedure.

(i) Sensitivity analysis is a proper way to identify the effects of structural and material parameters on the dynamic behavior of cable-stayed bridges. The results of the sensitivity analysis on the FE model of the simulated cable-stayed bridge show that the vertical modes of the deck are mostly affected by ρ_{deck}, A_{deck}, and E_{cable}, while the lateral-torsional modes are affected by E_{deck}, $I_{yy(\text{deck})}$ and the stiffness of deck-to-tower connections. The results of the mentioned sensitivity analysis can be helpful for understanding the dynamic behavior of long-span cable-stayed bridges and future health monitoring of these superstructures.

(ii) The verification of simulation results in this study shows that in FE models of cable-stayed bridges, the vertical deck modes are calculated more accurately than the lateral-torsional ones, which could be the result of errors in parameters across the width of the deck, assumptions made in initial modeling simplifications of nonstructural members, or defined boundary conditions and connections of the bridge. The parameters that influence lateral-torsional deck modes are the first parameters to be updated in FE models of cable-stayed bridges.

(iii) The proposed manual tuning algorithm can successfully update the FE model of long-span cable-stayed bridges. In comparison with other proposed automatic model updating methods the manual method makes more reasonable adjustments in parameters. Furthermore, selecting fewer updating parameters and assigning different lower and higher variations of parameters increase the accuracy of the tuning procedure.

Conflict of Interests

The authors do not have any conflict of interests with regard to the content of the paper.

Acknowledgments

The authors would like to thank all previous researchers that their studies have been reviewed in this paper. The research reported in this paper is sponsored by remarked GRA research grants (UKM-HEJIM-INDUSTRI-07-2010) funded by the National University of Malaysia (UKM) and (FRGS/1/2011/TK/UKM/02/13) founded by Ministry of Higher Education, Malaysia.

References

[1] J. F. Fleming and E. A. Egeseli, "Dynamic behavior of a cable-stayed bridge," *Earthquake Engineering & Structural Dynamics*, vol. 8, no. 1, pp. 1–16, 1980.

[2] A. S. Nazmy and A. M. Abdel-Ghaffar, "Nonlinear earthquake-response analysis of long-span cable-stayed bridges: theory," *Earthquake Engineering & Structural Dynamics*, vol. 19, no. 1, pp. 45–62, 1990.

[3] J. C. Wilson and W. Gravelle, "Modeling of a cable-stayed bridge for dynamic analysis," *Earthquake Engineering & Structural Dynamics*, vol. 20, no. 8, pp. 707–721, 1991.

[4] W. X. Ren, X. L. Peng, and Y. Q. Lin, "Experimental and analytical studies on dynamic characteristics of a large span cable-stayed bridge," *Engineering Structures*, vol. 27, no. 4, pp. 535–548, 2005.

[5] P. H. Wang and C. G. Yang, "Parametric studies on cable-stayed bridges," *Computers & Structures*, vol. 60, no. 2, pp. 243–260, 1995.

[6] M. I. Friswell and J. E. Mottershead, *Finite Element Model Updating in Structural Dynamics*, Kluwer Academic Publishers, Dordrecht, The Netherlands, 1995.

[7] H. Ahmadian, J. E. Mottershead, and M. I. Friswell, "Regularisation methods for finite element model updating," *Mechanical Systems and Signal Processing*, vol. 12, no. 1, pp. 47–64, 1998.

[8] S. Ziaei-Rad and M. Imregun, "On the use of regularization techniques for finite element model updating," *Inverse Problems in Engineering*, vol. 7, no. 5, pp. 471–503, 1999.

[9] J. K. Sinha and M. I. Friswell, "Model updating: a tool for reliable modelling, design modification and diagnosis," *The Shock and Vibration Digest*, vol. 34, no. 1, pp. 27–35, 2002.

[10] I. Kreja, T. Mikulski, and C. Szymczak, "Adjoint approach sensitivity analysis of thin-walled beams and frames," *Journal of Civil Engineering and Management*, vol. 11, no. 1, pp. 57–64, 2005.

[11] R. Baušys, G. Dundulis, R. Kačianauskas et al., "Sensitivity of dynamic behaviour of the FE model: case study for the ignalina NPP reactor building," *Journal of Civil Engineering and Management*, vol. 14, no. 2, pp. 121–129, 2008.

[12] R. M. Ferreira, "Implications on RC structure performance of model parameter sensitivity: effect of chlorides," *Journal of Civil Engineering and Management*, vol. 16, no. 4, pp. 561–566, 2010.

[13] R. Cantieni, "Updating of analytical models of existing large structures based on modal testing," in *Proceedings of the US-Europe Workshop on Bridge Engineering: Evaluation, Management and Repair*, pp. 153–177, ASCE, Reston, Va, USA, 1996.

[14] A. Pavic, M. J. Hartley, and P. Waldron, "Updating of the analytical models of two footbridges based on modal testing of full scale structures," in *Proceedings of the 23rd International Seminar on Modal Analysis*, pp. 1111–1118, Society for Experimental Mechanics, Bethel, NY, USA, 1998.

[15] K. Mackie and B. Stojadinović, "Probabilistic seismic demand model for California highway bridges," *Journal of Bridge Engineering*, vol. 6, no. 6, pp. 468–481, 2001.

[16] Q. W. Zhang, C. C. Chang, and T. Y. P. Chang, "Finite element model updating for structures with parametric constraints," *Earthquake Engineering & Structural Dynamics*, vol. 29, no. 7, pp. 927–944, 2000.

[17] J. M. W. Brownjohn and P.-Q. Xia, "Dynamic assessment of curved cable-stayed bridge by model updating," *Journal of Structural Engineering*, vol. 126, no. 2, pp. 252–260, 2000.

[18] W. E. Daniell and J. H. G. Macdonald, "Improved finite element modelling of a cable-stayed bridge through systematic manual tuning," *Engineering Structures*, vol. 29, no. 3, pp. 358–371, 2007.

[19] F. Benedettini and C. Gentile, "Operational modal testing and FE model tuning of a cable-stayed bridge," *Engineering Structures*, vol. 33, no. 6, pp. 2063–2073, 2011.

[20] ANSYS, User's manual, revision12-0-1, Swanson Analysis System, USA, 2009.

[21] A. Wilson, "A critical analysis of Tatara Bridge, Japan," in *Proceedings of Bridge Engineering*, p. 37, University of Bath, Bath, UK, April 2009.

[22] R. Karoumi, "Some modeling aspects in the nonlinear finite element analysis of cable supported bridges," *Computers & Structures*, vol. 71, no. 4, pp. 397–412, 1999.

[23] A. M. S. Freire, J. H. O. Negrão, and A. V. Lopes, "Geometrical nonlinearities on the static analysis of highly flexible steel cable-stayed bridges," *Computers & Structures*, vol. 84, no. 31-32, pp. 2128–2140, 2006.

[24] J. Hu, I. E. Harik, S. W. Smith, J. Gagel, J. E. Campbel, and R. C. Graves, "Baseline modeling of the Owensboro cable-stayed bridge over the Ohio River," Kentuky Transport System Report KTC-64-04, 2006.

[25] Y. Y. Lin and Y. L. Lieu, "Geometrically non-linear analysis of cable-stayed bridges subject to wind excitations," *Journal of the Chinese Institute of Engineers*, vol. 26, no. 4, pp. 503–511, 2003.

Sensitivity Analysis of the Influence of Structural Parameters on Dynamic Behaviour of Highly
Redundant Cable-Stayed Bridges

123

[26] K. Yamagushi, "Vibration test of Tatara Bridge," *Science Links Japan*, vol. 3694, pp. 493–500, 2000.

[27] Q. W. Zhang, T. Y. P. Chang, and C. C. Chang, "Finite-element model updating for the Kap Shui Mun cable-stayed bridge," *Journal of Bridge Engineering*, vol. 6, no. 4, pp. 285–293, 2001.

13

Nutrient Release from Disturbance of Infiltration System Soils during Construction

Daniel P. Treese,[1] Shirley E. Clark,[1] and Katherine H. Baker[2]

[1] *Environmental Engineering Program, Penn State Harrisburg, Middletown, PA, USA*
[2] *Life Sciences Program, Penn State Harrisburg, Middletown, PA, USA*

Correspondence should be addressed to Shirley E. Clark, seclark@psu.edu

Academic Editor: Cumaraswamy Vipulanandan

Subsurface infiltration and surface bioretention systems composed of engineered and/or native soils are preferred tools for stormwater management. However, the disturbance of native soils, especially during the process of adding amendments to improve infiltration rates and pollutant removal, may result in releases of nutrients in the early life of these systems. This project investigated the nutrient release from two soils, one disturbed and one undisturbed. The disturbed soil was collected intact, but had to be air-dried, and the columns repacked when soil shrinkage caused bypassing of water along the walls of the column. The undisturbed soil was collected and used intact, with no repacking. The disturbed soil showed elevated releases of nitrogen and phosphorus compared to the undisturbed soil for approximately 0.4 and 0.8 m of runoff loading, respectively. For the undisturbed soil, the nitrogen release was delayed, indicating that the soil disturbance accelerated the release of nitrogen into a very short time period. Leaving the soil undisturbed resulted in lower but still elevated effluent nitrogen concentrations over a longer period of time. For phosphorus, these results confirm prior research which demonstrated that the soil, if shown to be phosphorus-deficient during fertility testing, can remove phosphorus from runoff even when disturbed.

1. Introduction

To decrease the volume of stormwater runoff reaching already-degraded urban streams, many localities in the US are either mandating or encouraging the use of green infrastructure. Infiltration is a primary component of green infrastructure/low-impact development because it restores some of the natural hydrologic function to urbanized areas by introducing water back to the groundwater, either through surface or subsurface devices. Infiltration systems also have the potential to remove some of the pollutants transported in urban runoff and reduce their discharge to surface receiving waters through the interaction of pollutants and the infiltration media. Many state guidance documents describe the ideal media characteristics for this pollutant removal.

One concern with the heavy reliance on infiltration systems for pollutant removal is the potential for groundwater contamination. Papers of Pitt et al. [1, 2] and Clark et al. [3] contain extensive literature reviews on known and modeled impacts of stormwater infiltration on groundwater quality. Clark and Pitt [4] illustrate two levels of modeling that can be performed to evaluate whether groundwater contamination is a concern, and, if so, how long before pollutants are estimated to reach the groundwater.

The focus of much of the research on groundwater contamination from stormwater infiltration has been on the fate of stormwater pollutants. Little attention has been paid to the components of the media mix itself. Guidance documents often specify that the native soil be incorporated into the media mixture. First, this assumes, or requires that testing demonstrate, that the native soil is not contaminated. Second, it assumes that disturbing the soil to incorporate more organic matter and/or sand for improved removal and hydraulic stability will not have negative impacts on the water passing through the filter.

Leaching of nutrients has been observed from newly constructed infiltration devices [5–7], as well as from engineered filter media [8]. This leaching has been assumed to be related to the increased organic matter typically added to the native soil (such as the addition of compost to glacial till resulting in increased phosphorus export, as reported in Pitt et al. [9]). However, a literature review using the Agricola database on soil nutrient release shows that the disturbance of native soils also can release nutrients. For example, a reduction of nutrient pools was observed in tilled soils [10] with the reduction linked to the destruction of soil chemical bonds [11]. The destruction of these bonds between soil aggregates reduces soil macropores, increases bulk density, and reduces hydraulic conductivity, with effects that can exist for decades [12, 13]. Since increased movement of water through soil and removal of pollutants from influent stormwater are the priorities of infiltration system designers, the effects of disturbance should be of concern. This research is designed to address two of those concerns: the magnitude of the pollutant release from the media and the temporality of the release.

2. Materials and Methods

The soil selected for testing was a Wharton silt loam from central Pennsylvania. In the field, 21 ten-centimeter diameter columns of the soil were encased in 0.8 m length PVC pipe and removed intact from the sampling site. The collection location was a sloped field with a shallow soil that is less than 1 m to bedrock. Currently, the land is maintained as a lawn but there has been agricultural activity in the past and plowing may have occurred. The visible O horizon was 3 cm deep but was exaggerated to 7.5 cm in the O horizon columns to keep the soil intact; the final O horizon consisted of the visible O horizon and the transition to the A horizon. The A and B horizons were moderately rocky. Once the columns were returned to the laboratory, the soil profile was separated into layers by slicing off a portion of the top or bottom of the encased soil, depending on the horizon desired for testing. Five columns were used for each horizon group (O, A, AB, and entire profile) testing with one column used as the control or pretesting soil condition. Vegetation that was extracted with the columns was cut at the level of the soil surface and removed. The vegetation was not weeded because of the concern for disturbing the soil.

Within two days of returning the soils to the laboratory and separating the columns into the specified horizons, it was observed that the soil had shrunk away from the walls of the pipe in all columns. New samples were collected at the same location; however, even though they were covered to maintain moisture in the soil profile, shrinkage was observed. Therefore, a second local soil of similar quality for pH and organic content, as reported in the USDA/NRCS soil surveys, with similar geographic location and accessibility, was selected for comparison with the silt loam. The soil selected was a Leetonia loamy sand, again collected from central Pennsylvania (Table 1). Both soils are listed as moderately well drained to well drained. The Leetonia loamy sand columns were collected from a wooded section of state gameland, about 15 meters from a timber harvest landing and 100 meters from abandoned strip mines. The O horizon

was 10 cm deep and consisted almost entirely of deciduous forest canopy leaf litter with the root mass of a forest meadow grass species. After removing the leaf litter, the visible O horizon was approximately 7-8 cm deep. The A horizon was very sandy and mostly rock free. Impenetrable compaction was encountered at a depth of about 30 centimeters.

As noted in Table 1, the CECs are different between the two soils with the loamy sand having a higher sand content, which should reduce shrinkage. Because the concern for this investigation was nutrient leaching, it was more important that the organic content be similar since the organic content should be the primary source of nitrogen and phosphorus leaching. When these samples were returned to the laboratory, each soil was analyzed by horizon to confirm the information found in the USDA Soil Survey. These results also are included in Table 1.

For the silt loam soil, the laboratory disturbance consisted of extracting the soil from the column, separating it into 7.5-cm layers, air drying, and repacking without compaction except from the weight of the soil above any layer. While this procedure is more rigorous in terms of not compacting the soil than the field construction of infiltration systems, it is similar in its intent.

The test water for this project was stormwater runoff collected from the Penn State Harrisburg campus. Approximately once a week, 600 mL (equivalent to 75 mm of runoff on the soil surface) was distributed into each column. Given that most infiltration systems are designed at a 5 : 1 or 10 : 1 loading ratio, this 75 mm of runoff on the soil surface is equivalent to 15 mm or 7.5 mm of runoff from a drainage area. These "events" are much smaller than a typical design runoff event; this small loading was selected in order to evaluate the change in nutrient release over much smaller time steps to determine the length of time (measured as a water loading) for which nutrient release could be expected. Infiltration through the columns was by gravity only; no artificial pressure was applied to either the top or bottom of the columns. Hydraulic head was maintained between 2.5 and 7.5 cm. Each soil type received a total of 40 simulated storm events over the course of one year.

Samples of the influent and effluent from each column were collected weekly and the effluent volumes recorded. Water quality tests included pH and conductivity, total hardness (calcium/magnesium) by titration, and turbidity, color, total nitrogen, total phosphorus (phosphate), potassium, and sulfur (sulfate). All samples were collected and analyzed according to approved US EPA protocols and/or *Standard Methods for the Examination of Water and Wastewater* [14].

At the start and end of the project, plus four times throughout the project, a column of each representative test group (OAB, O, A, AB) was sacrificed for soil testing at the Penn State College of Agricultural Sciences Agricultural Analytical Services Lab. Each sacrificed column was subdivided into 7.5 cm segments and tested for soil pH, soluble salts, total carbon and total nitrogen through combustion, and phosphorus, potassium, magnesium, calcium, zinc, copper, and sulfur by Mehlich 3 extraction and ICP analysis.

The data below are presented as ratios of the effluent to influent concentration. C_E/C_0 values greater than 1 indicate

TABLE 1: Comparison of silt loam and loamy sand from USDA Soil Survey and analytical testing.

	Silt loam	Loamy sand
USDA Soil Survey Information		
Soil pH	4.0–5.0	3.6–5.0
Organic content	1–4%	1–5%
Cation Exchange Capacity [CEC] (meq/100 g)	3.8–8.0	0.6–2.0
Results of soil fertility analysis		
Soil pH	O horizon: 4.5	O horizon: 4.7
	AB horizon: 5.7	AB horizon: 4.7
Organic content	O horizon: 5.5%	O horizon: 9.5%
	AB horizon: 1.8%	AB horizon: 1.4%
Cation Exchange Capacity [CEC] (meq/100 g)	O horizon: 19	O horizon: 15
	AB horizon: 12	AB horizon: 11
Total nitrogen (mg/kg)	O horizon: 2,900	O horizon: 4,700
	AB horizon: 1,000	AB horizon: 700
Total phosphorus (mg/kg)	O horizon: 35	O horizon: 16
	AB horizon: 5	AB horizon: 2

that the soil is releasing nutrients, whereas values less than 1 indicate removal from the influent water. Because these are two different soils in terms of USDA textural class and of soil chemical characterization, they cannot be compared statistically.

3. Results and Discussion

This paper focuses on the nutrient release from each of the two soil types and from the organic (O) and mineral (AB) horizons since nutrient release is the issue of concern for both surface and groundwater contamination. The full data set may be found in Treese [15]. Because the two soils had different extraction depths, the graphs and discussion below are based on common locations in the soil profile. The O horizon data corresponds to the results obtained from the 0–7.5 cm section of each soil, while the AB horizon corresponds to the results at the 15–22.5 cm section.

3.1. Total Phosphorus (TP). Figure 1(a) compares the trends in phosphorus effluent quality for the O horizons of the two soils. Initially, the organic horizon of both soils released phosphorus. However, the disturbed soil had a substantially higher release (2.5 to 3 times the influent) until approximately 0.4 m of stormwater had been applied. The undisturbed soil also had a higher initial phosphorus effluent concentration (approximately 1.5 times the influent) for the same 0.4 m of stormwater, but it was approximately half of the disturbed soil values. After the 0.4 m of stormwater has been applied, the disturbed and undisturbed organic horizons are indistinguishable in their performance. Excess phosphorus is washed rapidly from the soil profile. The soil fertility testing of these two horizons found that these soils contained an excess of phosphorus for typical crops. This indicates that there is a reservoir of phosphorus available for leaching. The initial soil P concentrations showed that

the disturbed soil had approximately twice as much as phosphorus as the undisturbed soil, so attributing the phosphorus release to only disturbance is not possible.

Figure 1(b) compares the two mineral horizons. Generally, the mineral layers of the two soils provided some removal of phosphorus, with the greater removal being seen with the undisturbed soil. The mineral layers removed approximately 50% of the total phosphorus applied for the undisturbed soil and 30% applied to the disturbed soil. Therefore, an infiltration system that incorporated both the organic and mineral layers would be expected to remove phosphorus slightly, even after disturbance. However, extracting and reusing only the organic and topsoil (A horizon where organic leaching has occurred) layers could result in either no removal or the release of phosphorus from the soil if it is not taken up by any vegetation. Because the soil concentration of extractable phosphorus was large, this initial release of phosphorus into the passing water was not visible as a decrease in soil phosphorus concentration in the soil analyses (Figure 2).

Generally, these results are in agreement with the literature on long-term phosphorus behavior in agricultural and forest soils, although no prior studies have investigated the short-term behavior of phosphorus at the resolution used in this study. The silt loam had higher initial phosphorus content than the loamy sand (twice as high), as would be expected since the loamy sand had more sand in the mixture. However, the loamy sand had a much higher initial organic content. For phosphorus, this higher initial concentration and the disturbance of the soil had a very limited impact (0.4 m of cumulative loading). This agrees with Boem et al. [16] who observed that there was no difference in total phosphorus, available phosphorus, or phosphorus sorption indexes in the upper 5 cm of tilled and no-till agricultural soils despite a 14% increase of total OM and 56% increase of particulate OM in no-till soils. This indicates that organic matter content may not impact phosphorus leaching.

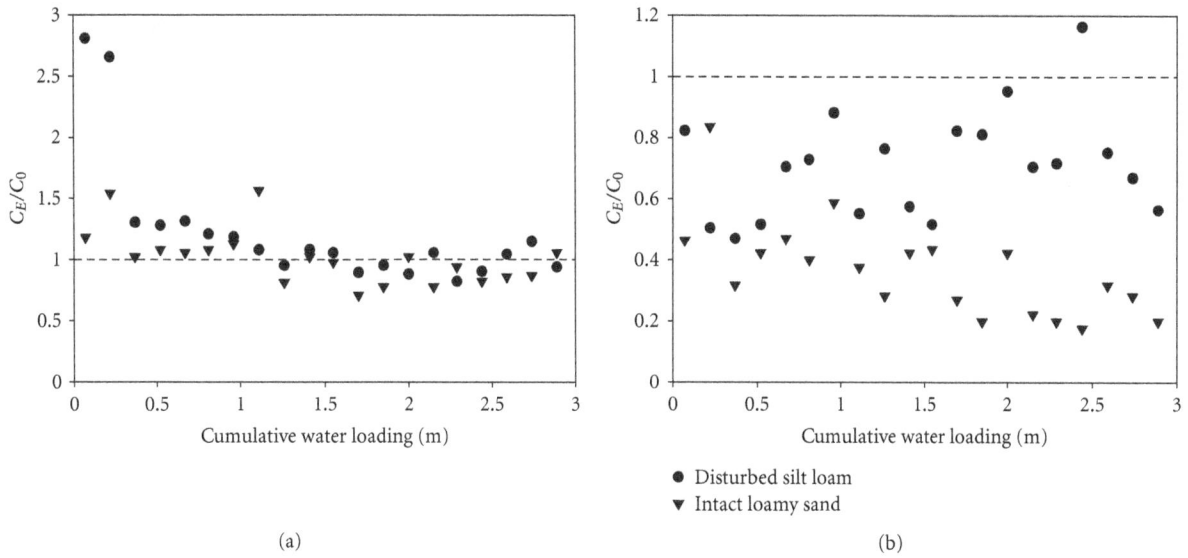

(a)

(b)

FIGURE 1: Phosphorus export from (a) organic horizon and (b) mineral horizon.

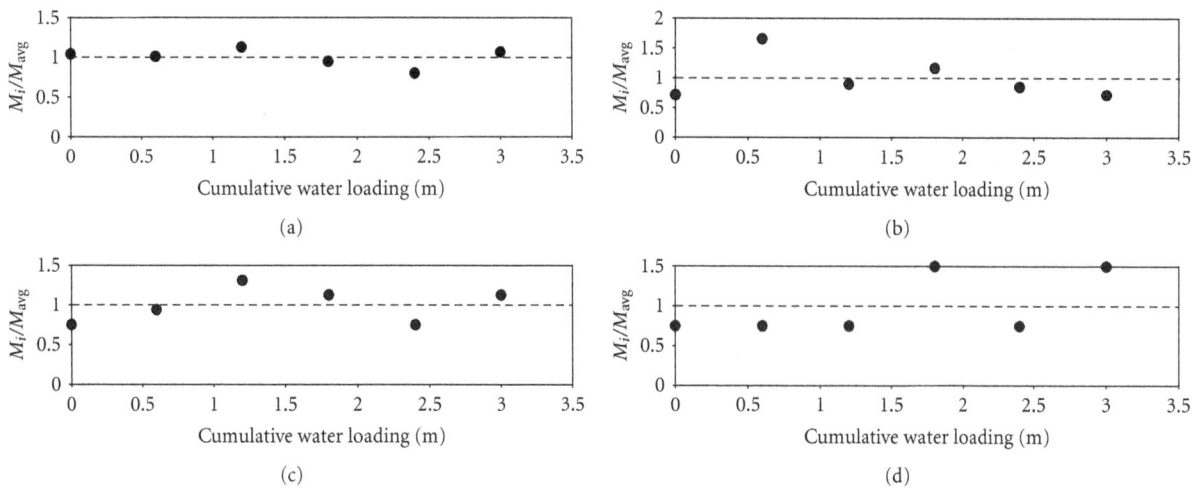

(a)

(b)

(c)

(d)

FIGURE 2: Ratio of sample soil results for phosphorus for (a) disturbed O horizon, (b) undisturbed O horizon, (c) disturbed AB horizons, and (d) undisturbed AB horizons.

These results also illustrated the impact of the initial phosphorus content of soils on phosphorus release. Both soils had an initial release from the organic layer, and a removal from the passing water in the mineral layers. Increased retention of phosphorus by subsurface soils has been noted before in both forest soils [17] and no-till agricultural soils. It was assumed to result from the fact that subsurface soils were not saturated with phosphorus [18]. This removal by the subsurface layers recently has been correlated with other factors. Differences in phosphorus retention by the mineral soil horizon have been correlated with soil mineralogy, with improved retention seen by aluminum-bearing minerals [19]. Calcium, magnesium, and aluminum have been shown to have a direct effect on phosphorus retention in the organic horizon with cation exchange capacity, pH, carbonate, organic carbon, sand content, silt content, and

clay content having indirect correlations [20]. This appears to be partly due to the formation of calcium-phosphate-clay cement bridges between soil aggregates [21]. Disturbance breaks up the larger soil clumps into smaller particles, increasing the surface area available for reaction with water. These reactions include hydration and ion exchange. These surface reactions change the surface chemistry, breaking bonds [11], which may result in excess release of calcium [10]. Specifically for engineered bioretention soils, Hunt et al. [7] recommended soils with a low P-index value, which translates into soils that are considered deficient in phosphorus in a soil fertility test. The organic layers of these soils were considered sufficient or high for phosphorus content for soil fertility while both mineral layers were considered deficient.

The rapid decrease of the initially elevated effluent phosphorus concentrations in this study correlates well

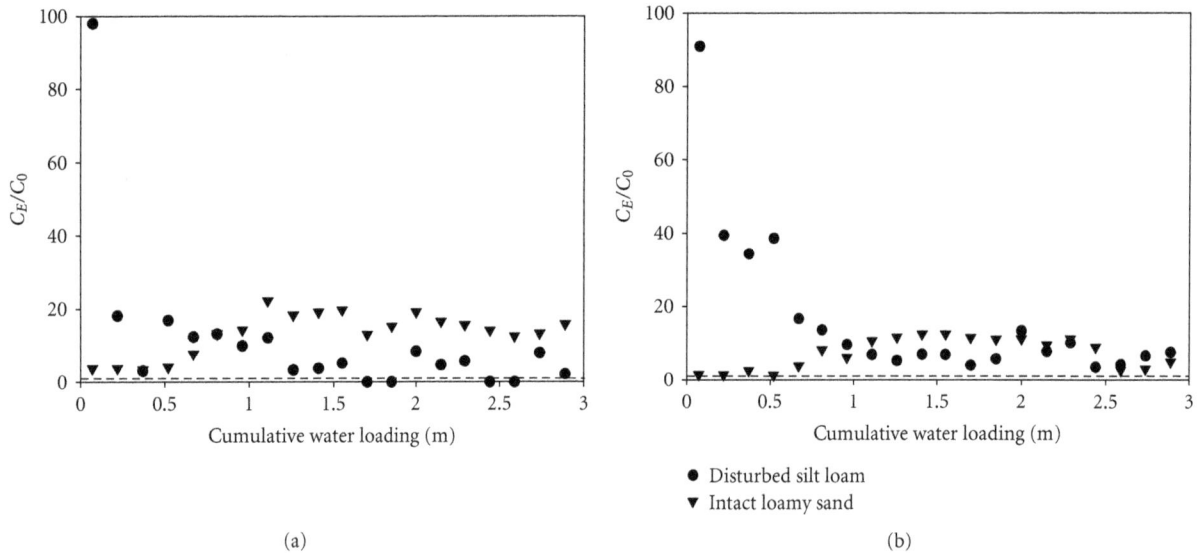

(a)   (b)

FIGURE 3: Nitrogen export from (a) organic horizon and (b) mineral horizon.

with the trends observed from other elements leached from the disturbed soil's organic horizon plus the trends in the aggregate ionic measurement of conductivity (data not shown). For phosphorus, it appears that disturbance may have no impact on phosphorus release by itself, but instead the initial phosphorus release results from initial soil concentrations in excess of plant needs. Initial releases are substantially higher in both soils; however, the 0.4 m of runoff loading would be less runoff than would be expected during a single large storm.

3.2. Total Nitrogen. Compared to the total phosphorus, the initial release of total nitrogen from both the organic and mineral horizons of the disturbed soil was very high (100 times the influent concentration) and lasted approximately twice as long (approximately 0.8 m of runoff loading), despite the lower initial concentration of nitrogen in the disturbed soil (Figure 3). In contrast to the total phosphorus results, the undisturbed soil saw a delayed elevated release of total nitrogen. For the phosphorus, the release decreased or remained constant over time, whereas the total nitrogen saw an increase after approximately 0.4 m of runoff loading. This delayed release indicates that there potentially was a reservoir of nitrogen available for release or uptake by plants, but its transport through the soil was retarded for undetermined reasons. The disturbance of the soil accelerated that release potentially by breaking any chemical bonds that detained the nitrogen in the soil profile. Once this available nitrogen was released, it rapidly exited the column and the nitrogen effluent concentration asymptotically approached the influent concentration. Not disturbing the soil appears to reduce the magnitude of this release but does not prevent it. Figure 4 illustrates the change in soil nitrogen concentration as a function of loading. For the disturbed soils, there appears to be no trend in the soil nitrogen concentration; however, for the undisturbed soil, it appears that there could be a slight reduction in soil nitrogen over time that corresponds

to the delayed elevated nitrogen release from this soil. Given the lack of soil analysis replication at the individual water loadings, additional analysis would be required to confirm this trend and document its magnitude.

This higher effluent concentration from the disturbed soil columns occurs despite the initial O horizon nitrogen concentration and total organic content of the disturbed silt loam soil being approximately 60% of the concentration in the undisturbed loamy sand. A release of nitrogen from soils which were dried, sieved, and then repacked has been published and linked to bound nitrogen (12–27% contribution) in the upper 2 cm of tilled and no-till soils with greater release from the no-till soils of higher organic content [22]. It would appear that the disturbance caused an immediate and very high release of the bound nitrogen in the silt loam. The organic horizon of the undisturbed loamy sand only gradually began to release nitrogen as either the structure degraded [23], or, more likely, as any remaining leaf litter decomposed. These results are in contrast with those of Hsieh and Davis [6], who found that nitrate removal increased with increased organic matter in the media mix. These results showed that the nitrogen release was substantially greater in the organic layer of both the disturbed and undisturbed soils, which had the higher initial nitrogen and organic content.

4. Conclusions

The initial substantial leaching of some tested parameters by the silt loam soil columns, which had to be air-dried and repacked, may resemble what occurs after construction of infiltration units. An initial release of nutrients from infiltration system media has been observed before the establishment of vegetation [24] and should be recognized as a concern, even though it appears that this release is of short term. The nitrogen release of the disturbed silt loam soil columns quickly declined but the initial elevated

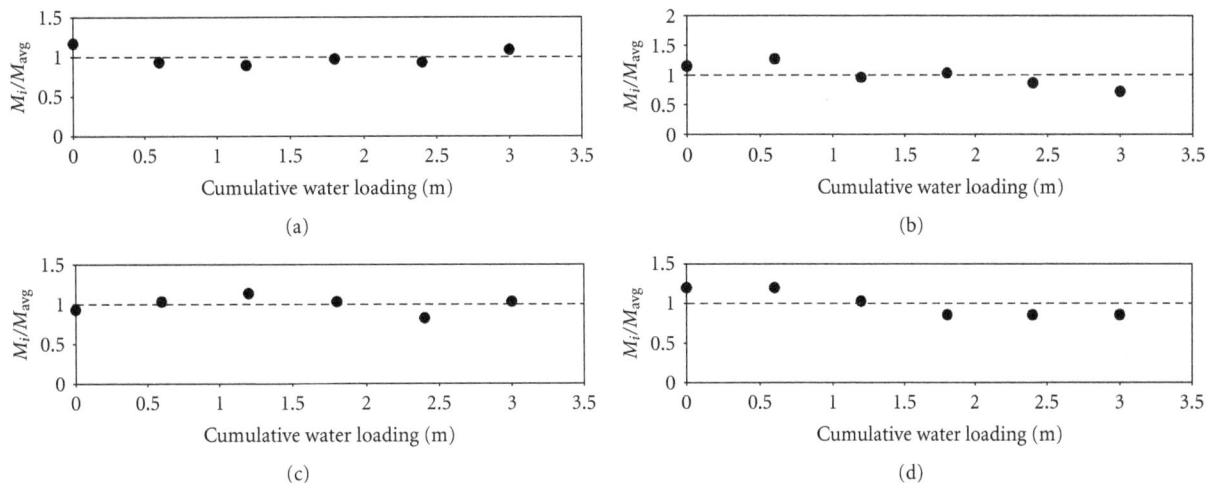

FIGURE 4: Ratio of sample soil results for nitrogen for (a) disturbed O horizon, (b) undisturbed O horizon, (c) disturbed AB horizons, and (d) undisturbed AB horizons.

concentrations are well above the U.S. Environmental Protection Agency's drinking water standards. Vegetation may abate these concerns, even during the plant establishment phase, at least to the depth of the root zone. However, this release would be expected for subsurface infiltration systems, such as dry wells and any soil disturbed below devices such as porous pavement.

Phosphorus leaching has been a problem in bioretention systems with underdrains and assumed to be due to disturbance [5] or initially high phosphorus content in organic media [7]. These results indicate that the organic horizon is a source of leaching phosphorus, but lower soil horizons exhibit removal even with disturbance. Given the higher initial phosphorus content in the disturbed soil, it appears that the initial phosphorus release may be primarily a function of initial concentration available for release. The lack of removal and only slightly elevated effluent concentrations from the O horizon, when combined with the removal of phosphorus by lower horizons, suggest that phosphorus removal is independent of organic matter and is dependent on one or more components of the mineral matter.

This study, with its focus on the early life of soil media, indicates that there are potential concerns with nutrient releases during the early storm events. For infiltration in native soils, this study reinforces the need to evaluate whether soil disturbance is required or whether an area with good infiltration should be left undisturbed. When combined with the results of Clark and Pitt [25] on the nutrient release from the organic matter if the organic part of the media goes anaerobic, these results indicate that the organic content of the infiltration system media should be limited to that needed for plant growth. For subsurface systems, it should be limited to the amount needed to provide the desired pollutant removal for the lifespan of the device before maintenance and media replacement. These results indicate that the organic content of the organic horizon is large enough for nitrogen release to occur and that,

for unvegetated systems, the organic content should be minimized.

References

[1] R. Pitt, S. Clark, K. Parmer, and R. Field, "Groundwater contamination from stormwater infiltration," U.S. EPA Report, 1995.

[2] R. Pitt, S. Clark, K. Parmer, and R. Field, *Groundwater Contamination from Stormwater Infiltration*, Ann Arbor, Chelsea, Mich, USA, 1996.

[3] S. E. Clark, K. H. Baker, D. P. Treese, J. B. Mikula, C. Y. S. Siu, and C. S. Burkardt, *Sustainable Stormwater Management: Infiltration vs. Surface Treatment Strategies*, Water Environment Research Foundation, 2010, Project Number 04-SW-3.

[4] S. E. Clark and R. Pitt, "Influencing factors and a proposed evaluation methodology for predicting groundwater contamination potential from stormwater infiltration activities," *Water Environment Research*, vol. 79, no. 1, pp. 29–36, 2007.

[5] M. E. Dietz and J. C. Clausen, "A field evaluation of rain garden flow and pollutant treatment," *Water, Air, and Soil Pollution*, vol. 167, no. 1–4, pp. 123–138, 2005.

[6] C. Hsieh and A. P. Davis, "Evaluation and optimization of bioretention media for treatment of urban storm water runoff," *Journal of Environmental Engineering*, vol. 131, no. 11, pp. 1521–1531, 2005.

[7] W. F. Hunt, A. R. Jarrett, J. T. Smith, and L. J. Sharkey, "Evaluating bioretention hydrology and nutrient removal at three field sites in North Carolina," *Journal of Irrigation and Drainage Engineering*, vol. 132, no. 6, pp. 600–608, 2006.

[8] S. E. Clark, *Urban stormwater filtration: Optimization of design parameters and a pilot-scale evaluation*, Ph.D. dissertation, The University of Alabama at Birmingham, 2000.

[9] R. Pitt, R. Harrison, J. Lantrip, and C. L. Henry, "Infiltration through disturbed urban soils and compost-amended soil effects on runoff quality and quantity," in *Proceedings for the Water Environment. Federation's 72nd Annual Conference and Exposition (WEFTEC '99)*, New Orleans, La, USA, 1999.

[10] B. R. Stinner, D. A. Crossley, E. P. Odum, and R. L. Todd, "Nutrient budgets and internal cycling of N, P, K, Ca and Mg in conventional tillage, no-tillage, and old-field ecosystems on

the Georgia Piedmont," *Ecology*, vol. 65, no. 2, pp. 354–369, 1984.

[11] D. H. Powers and E. L. Skidmore, "Soil structure as influenced by simulated tillage," *Soil Science Society of America Journal*, vol. 48, no. 4, pp. 879–884, 1984.

[12] R. Pitt, S. E. Chen, S. E. Clark, J. Swenson, and C. K. Ong, "Compaction's impacts on urban storm-water infiltration," *Journal of Irrigation and Drainage Engineering*, vol. 134, no. 5, pp. 652–658, 2008.

[13] K. N. Potter, F. S. Carter, and E. C. Doll, "Physical properties of constructed and undisturbed soils," *Soil Science Society of America Journal*, vol. 52, no. 5, pp. 1435–1438, 1988.

[14] APHA (American Public Health Association), *Standard Methods for the Examination of Water and Wastewater*, American Public Health Association, Washington, DC, USA, 17th edition, 1989.

[15] D. P. Treese, *Pollutant transport within the vadose zone of natural soils*, Master's of Environmental Engineering Paper, Penn State Harrisburg, 2009.

[16] F. H. G. Boem, C. R. Alvarez, M. J. Cabello et al., "Phosphorus retention on soil surface of tilled and no-tilled soils," *Soil Science Society of America Journal*, vol. 72, no. 4, pp. 1158–1162, 2008.

[17] H. Riekerk, "The mobility of phosphorus, potassium, and calcium in a forest soil," *Soil Science Society of America Journal*, vol. 35, pp. 350–356, 1971.

[18] E. A. Guertal, D. J. Eckert, S. J. Traina, and T. J. Logan, "Differential phosphorus retention in soil profiles under no-till crop production," *Soil Science Society of America Journal*, vol. 55, no. 2, pp. 410–413, 1991.

[19] C. J. Penn, G. L. Mullins, and L. W. Zelazny, "Mineralogy in relation to phosphorus sorption and dissolved phosphorus losses in runoff," *Soil Science Society of America Journal*, vol. 69, no. 5, pp. 1532–1540, 2005.

[20] D. V. Ige, O. O. Akinremi, and D. N. Flaten, "Direct and indirect effects of soil properties on phosphorus retention capacity," *Soil Science Society of America Journal*, vol. 71, no. 1, pp. 95–100, 2007.

[21] J. L. Ragland and W. A. Seay, "The effects of exchangeable calcium on the retention and fixation of phosphorus by clay fractions of soil," *Soil Science Society Proceedings*, vol. 21, article 261, 1957.

[22] H. L. Kristensen, G. W. McCarty, and J. J. Meisinger, "Effects of soil structure disturbance on mineralization of organic soil nitrogen," *Soil Science Society of America Journal*, vol. 64, no. 1, pp. 371–378, 2000.

[23] J. J. Schoenau and J. R. Bettany, "Organic matter leaching as a component of carbon, nitrogen, phosphorus, and sulfur cycles in a forest, grassland, and gleyed soil," *Soil Science Society of America Journal*, vol. 51, no. 3, pp. 646–651, 1987.

[24] W. C. Lucas and M. Greenway, "Nutrient retention in vegetated and nonvegetated bioretention mesocosms," *Journal of Irrigation and Drainage Engineering*, vol. 134, no. 5, pp. 613–623, 2008.

[25] S. E. Clark and R. Pitt, "Storm-water filter media pollutant retention under aerobic versus anaerobic conditions," *Journal of Environmental Engineering*, vol. 135, no. 5, pp. 367–371, 2009.

14

Polycyclic Aromatic Hydrocarbons in Urban Stream Sediments

Jejal Reddy Bathi,[1, 2] **Robert E. Pitt,**[1] **and Shirley E. Clark**[3]

[1] Department of Civil, Construction and Environmental Engineering, The University of Alabama, Tuscaloosa, AL 35487, USA
[2] Global Solutions International, LLC, P.O. Box 223, Mobile, AL 36652, USA
[3] Penn State Harrisburg, Environmental Engineering Program, Middletown, PA 17057, USA

Correspondence should be addressed to Jejal Reddy Bathi, jejalb@gmail.com

Academic Editor: Cumaraswamy Vipulanandan

Polycyclic aromatic hydrocarbons (PAHs) are persistent organic pollutants of high environmental concern with known carcinogenic activity. Although literature documents PAH fate in urban runoff, little is known about their distribution on sediment sizes, which is essential for determining their treatability and fate in receiving waters. This paper has quantified the concentrations of selected PAHs in urban creek sediments and examined possible relationships between sediment PAH content and sediment characteristics, such as particle size, volatile organic content (VOC), and sediment chemical oxygen demand (SCOD). SCOD, VOC, and PAH concentrations of sediments showed a bimodal distribution by particle size. The large diameter sediments had the highest VOC and also had the highest PAH concentrations. The spatial variation of PAH content by sediment sizes also was statistically significant; however, the mass of the PAH material was significantly affected by the relative abundance of the different particle size classes in the sediment mixtures.

1. Introduction

Polycyclic aromatic hydrocarbons (PAHs) are a class of frequently detected organic pollutants in urban stormwater runoff. According to Metre et al. [1], PAH levels in urban freshwater sediments in North America have increased over time, indicating additional discharges associated with industrialization and urbanization, including increased use of vehicles and wear and tear of asphalt [1–3]. For example, Stein et al. [4] found PAHs in storm fluxes ranged from $1.3\,\mathrm{g/km^2}$ for the largely undeveloped Arroyo Sequit watershed to $224\,\mathrm{g/km^2}$ for the highly urbanized Verdugo Wash watershed in California, USA. Similarly, according to Huston et al. [5], there is an increase in the PAHs and other contaminant flux in traffic and industrial areas compared to outer suburbs, implicating these developments positively influencing PAHs contribution in the runoff. The relative distribution of individual PAHs in stormwater runoff can indicate their originating source category, with high molecular weight PAHs indicating pyrogenic (combustion) sources, whereas low molecular weight PAHs indicating petrogenic sources. However, tracking sources in this manner becomes

questionable if the expected environmental biological or physical degradation processes change the relative abundance of the different PAHs in runoff. Regardless of the type of PAH source in urban areas, stormwater is a major delivery system of PAHs to receiving water bodies. It is important to understand the behavior of PAHs at their source and sink, that is, in stormwater and in nearby streams, in order to address their effective control and remediation. Since contaminated runoff also poses a risk to the aquatic environment, the presence of these compounds can result in regulatory demands for sediment assessment and remediation, or requirements for stormwater management controls or treatment [6].

PAHs in urban runoff can occur both in soluble and particulate-associated forms. Studies have identified particulate associated PAHs as the most abundant [7–10], which is expected based on their hydrophobic nature and low vapor pressure, especially for PAHs with more than three aromatic rings. Studies have also documented the impact of sediment texture and organic content affecting PAH associations with sediment [11]. Investigations have found high PAH concentrations associated with large organic material (leaf and other vegetation litter) trapped in stormwater floatable controls [12], further indicating that sediment composition affects the PAHs association with particulate matter, especially those solids high in organic content. Li et al. [13] also observed a strong positive relation between sediment total organic carbon or black carbon and its PAH concentrations. In addition, source areas may affect the distribution pattern of PAHs with sediment particle sizes. For example, Guggenberger et al. [14] noticed homogenous distributions of PAHs by particle sizes in rural soils, whereas Müller et al. [15] observed nonhomogeneous distributions of PAHs in urban soils. Biodegradation is one of the primary degradation processes for PAH reductions in sediments [16], and which might be affected by many factors including sediment size and hence the relative abundance of PAHs by sediment size. Relatively extensive data is available to confirm the major distribution fate of PAHs between dissolved and particulate form. However, limited information is available regarding distribution fate of the PAHs by particle sizes, which is the focus of this study.

Studies have looked at the particle size distribution in stormwater runoff and have attributed differences between studies to factors including source area, geographic location, and antecedent dry period. The transport energy available to move the larger particles through the urban drainage system is limited, with the large particles being deposited along the flow path or within the drainage system. Hence stormwater samples obtained near sediment sources contain more of the larger particles, whereas samples obtained at outfalls have fewer of the larger solids. The National Urban Runoff Pollution (NURP) study examined stormwater runoff data collected at outfalls from different locations in the USA and noted that more than 90% of the stormwater particulates (by volume and mass) were in the 1 to 100 μm range [17]. Because of their strong association with solids in stormwater runoff, the fate and transport of PAHs will be directly related to their association with particle size. Associations with smaller particles likely result in further transport, less sedimentation in the drainage system, and potential treatment using filtration systems. For PAHs associated with larger particles, it can be anticipated that during most storm conditions, they will settle closer to the source and sedimentation will be a very effective control practice. Hence, the sampling location relative to the source area may influence the abundance of particle sizes as well PAH mass in the sample, which makes it is critical to better understand the particle size influenced association of PAHs.

Due to the trace concentrations of selected PAHs on stormwater particles and need for large sample volumes to separate stormwater solids into aliquots for PAH analysis, large sample volumes are required, which is difficult to achieve in most stormwater sampling situations. A surrogate for the representation of PAH associations by particle size would be to analyze the surficial sediment in urban receiving waters at the stormwater outfall. Such sampling may not give the true representation of the source stormwater particle distribution, but it will be adequate to quantify the expected PAHs concentration on sediment particle sizes for the source areas draining to that outfall. Careful selection of the outfall

to isolate specific sources and landuses will allow for an improved understanding of source PAH generation and transport within the drainage system. This project used this approach to characterize PAH distribution by stormwater particle sizes and to test the influence of contributing source areas on such distribution. This project focused on collecting urban stream sediments located immediately downstream of stormwater outfalls in completely developed watersheds. The samples were obtained in a variety of urban streams in the Tuscaloosa and Northport, AL area, as described later. The range of sample characteristics is intended to represent the typical range of local urban stream conditions (particle sizes, contaminant concentrations, etc.). This study also examined the relationships between the urban creek sediment characteristics such as particle size distribution (psd), Volatile Organic Carbon (VOC), and Sediment Chemical Oxygen Demand (SCOD) with the PAH concentrations. Relationships such as these may allow stormwater managers to estimate the magnitude of PAH problems and associations based on more directly and easily measurable characteristics of the sediment.

This paper presents observed size distributions of sediment samples and PAH concentrations by particle size ranges. Results of tests for possible relationships between sediment PAH concentrations and respective sediment characteristics are also described in this paper.

2. Methodology

2.1. Source Area, Sample Collection, and Sample Processing. Fifteen separate sediment samples were collected from three different urban creeks (Cribbs Mill Creek, Hunter Creek, and Carroll Creek) in and around Tuscaloosa and Northport, AL, USA. Cribbs Mill Creek drained single-family, medium-to-high density residential areas, while Hunter Creek received stormwater runoff from a heavily trafficked road next to the creek, commercial areas (including automotive repair facilities), and runoff from a trailer park residential area. Carroll Creek was mostly affected by runoff from a high-density residential area on one side of creek and forested lands on the other side of the creek. These three creeks and the samples, therefore, represent a wide range of typical urban characteristics.

Sediment subsamples were obtained using a manual, polypropylene dipper sampler. Each subsample was collected from the top 2 cm of sediment (approximately 1000 g wet weight per sample). Five subsamples per creek were collected within a 100 meter reach for each sample period and composited for analyses. Samples were placed in aluminum trays, dried at about 100°C to remove moisture. It was assumed that, since the analytes of interests (selected PAHs, as described below) in this study have boiling points above 200°C, drying sediment samples at about 100°C for small residence time (approximately 12 hours) under normal pressure conditions will not cause any significant loss of analytes of interest from the sediment samples. Extended heating under pressurized conditions, though, could result desorption of analytes from solid matrices at a temperature below their boiling point. If there is any such loss in our study, it will be primarily for more volatile low molecular weight analytes such as naphthalene (boiling point 218°C). However, as all samples were treated in similar manner, such effect will be equally applicable to all sample results and will not affect comparison analyses of the sample results presented in this paper. Dried sediments were sieved through 45, 90, 180, 355, 710, 1.400, and 2,800 μm stainless steel sieves with mechanical shaking (100 rpm, 60 min) prior to PAH and other analyses. The mechanical shaker used has throw-action sieving where the vertical throwing motion is overlaid with slight circular motion, which helps the particles to distribute over the sieve and as well fall back to interact with the sieve mesh and pass through the opening if the particles are sufficiently small. Sediment separation using a mechanical shaker is a commonly employed procedure because it is believed to produce sediment separation that is reproducible and precise. Large organic materials (LOM; leaves and other debris >2.800 μm) were manually separated from the largest particle fraction and analyzed separately. Sediments collected on individual sieves and LOM fraction were weighed and stored in a refrigerator until they were analyzed.

2.2. Sediment Chemical Analysis. All sediment fractions were analyzed for thirteen PAHs (selected based on literature reports describing their common occurrence in urban stormwater and which were also noted to be highly toxic). The thirteen PAHs analyzed include Naphthalene, Fluorene, Phenanthrene, Anthracene, Fluoranthene, Pyrene, Benzo(a)anthracene, Chrysene, Benzo(b)fluoranthene, Benzo(a)pyrene, Indeno(1,2,3-cd) pyrene, Dibenz(a,h)anthracene, and Benzo(g,h,i)perylene. The sample preparation and analysis technique, reported in Bathi [18], used AutoDesorb (Scientific Instrument Services, Inc., Ringoes, NJ, USA) thermal extraction methods to extract the PAHs from the sediment for direct injection into a GC-MSD.

2.3. Sediment Material Composition and SCOD Analysis. The volatile organic content of the sediment fractions was determined using "Thermal Chromatography" (an expansion of the volatile solids analyses) techniques per Ray [19]. SCOD analysis of size fractionated sediments was conducted per reactor digestion HACH Method 8000 (http://www.hach.com/).

3. Results and Discussion

3.1. Sediment Particle Size Distribution. Figure 1 shows the observed particle size distribution, of the urban creek sediments from the three locations. Overall, most of the particles were distributed in the range of 90 to 710 μm, with the medians ranging from about 200 to 710 μm. Only about 10% of the sample mass, on average, was greater than 1,000 μm. Lack of small particles in the samples reflects the transport of fines during high creek flows. Cribbs Mill Creek had more large particles and Carroll Creek had more small particles. Increased large-size sediments at the Cribbs Mill Creek sampling location was thought to be caused

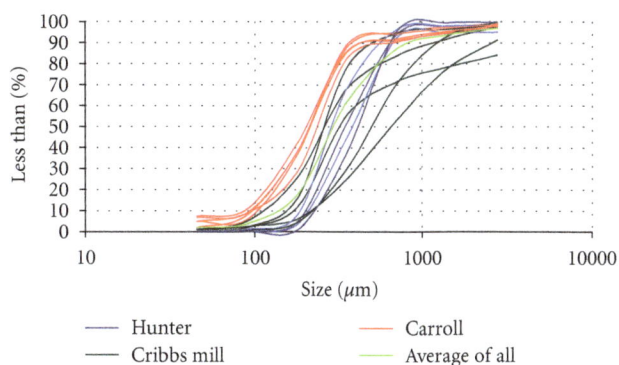

Figure 1: Observed creek sediment samples particle size distributions, by Mass.

by runoff from highly paved high-density urban areas as well the concrete lined creek bottom at the location that avoids dilution of the sediment particles with creek bottom erosion of smaller particles. In addition, the lined bottom also had considerable amounts of algae growth, which was included in larger size class of sediment samples. On the other hand, the Carroll Creek sample location's source area had large amounts of pervious areas. Hunter Creek was an intermediate location—it had a high percentage of paved areas in the drainage area, but with a natural creek bottom.

Compared to past studies of urban runoff samples (e.g., NURP), the creek sediment samples had lower percentages of smaller particles and higher percentages of larger particles, showing the likely deposition of the larger particles as creek sediment and the downstream transport of the smaller particles. These preferential removal mechanisms result in a shifted particle size distribution for the stream sediments compared to the particles in the stormwater discharges, as reported in the literature. As an example, the overall median particle sizes of the sampled stream sediments were between about 200 and 710 μm, which is about 50 to 100 times larger than the median particle sizes of most stormwater outfall particulates.

3.2. Sediment PAHs. The observed concentration of individual PAHs on size fractioned sediments with all samples combined from three locations are presented in Table 1. Observed PAH concentration within the sediment size fraction was highly variable (Table 1) and is similar to what one would expect for stormwater pollutants. Overall, the PAH analytical results showed a bimodal distribution of PAH concentrations with smaller (<90 μm) and larger (>710 μm) fractions having higher PAH concentrations, especially for the large organic matter (LOM). The trends are similar to what one would expect for semivolatile pollutant distributions in sediments given their low water solubility values and their propensity to be attracted to organic matter based on the literature, showing a strong positive correlation between the particles' organic content and their associated PAHs concentration. Based on the literature, it is expected that the smaller particles have a higher organic content in their mass [20], partly because of the native material (silts and clays) and

TABLE 1: Observed PAHs concentration by creek sediment particle sizes and associated standard deviation (all samples combined).

PAH	Mean concentration (μg/kg) (standard deviation)								
	<45 μm	45–90 μm	90–180 μm	180–355 μm	355–710 μm	710–1400 μm	1400–2800 μm	>2800 μm (w/o LOM*)	>2800 μm LOM
Naphthalene	255 (275)	177 (156)	163 (224)	94 (87)	124 (131)	790 (2046)	891 (2014)	124 (74)	2637 (2107)
Fluorene	257 (295)	189 (134)	225 (187)	125 (135)	139 (140)	196 (144)	293 (173)	216 (161)	1771 (945)
Phenanthrene	264 (278)	205 (211)	140 (158)	92 (92)	110 (85)	130 (136)	197 (230)	188 (164)	2007 (1422)
Anthracene	354 (397)	288 (273)	261 (253)	152 (150)	182 (182)	366 (314)	491 (614)	218 (152)	2255 (1089)
Fluoranthene	650 (868)	624 (753)	345 (372)	202 (242)	247 (336)	259 (237)	237 (197)	191 (173)	1520 (902)
Pyrene	653 (738)	519 (548)	412 (577)	175 (174)	240 (405)	207 (153)	192 (122)	172 (129)	2054 (954)
Benzo(a)anthracene	501 (595)	408 (537)	258 (286)	169 (171)	224 (229)	167 (134)	271 (252)	278 (371)	2164 (1045)
Chrysene	591 (618)	602 (689)	363 (363)	202 (199)	273 (268)	190 (125)	296 (242)	171 (130)	1810 (852)
Benzo(b)fluoranthene	597 (522)	517 (598)	358 (389)	402 (671)	227 (150)	316 (262)	375 (369)	329 (375)	2179 (1425)
Benzo(a)pyrene	1474 (2210)	1524 (3079)	662 (459)	434 (513)	351 (210)	502 (533)	1119 (2086)	392 (255)	2330 (1866)
Indeno(1,2,3-cd)pyrene	787 (544)	657 (538)	942 (794)	258 (187)	332 (189)	576 (774)	706 (917)	357 (424)	1774 (933)
Dibenz(a,h)anthracene	1267 (1864)	787 (1022)	675 (545)	276 (234)	355 (226)	687 (511)	835 (1254)	286 (191)	1492 (775)
Benzo(g,h,i)perylene	706 (686)	465 (451)	591 (691)	199 (226)	174 (116)	551 (567)	396 (299)	348 (229)	2236 (1728)

* Sediment size fraction organic material removed.

partly because of their ability to absorb dissolved organic matter in the water. The literature generally indicates a poor relationship between large mineral sediment sizes and PAH concentrations. This project, however, found that the large sediment particles, especially the LOM (large organic matter) samples that were of vegetation organic matter (leaves, grass clumps, etc.), had high PAH concentrations. These PAH concentrations correlated with the measured organic content of the sediment, which also showed the bimodal distribution.

The significant differences in PAHs concentration by particle size were verified using One-Way ANOVA analysis, and results of the test are presented in Table 2. Other than naphthalene at Cribbs Mill Creek and benzo(a)pyrene, indeno(1,2,3-cd)pyrene, and benzo(g,h,i)perylene at Hunter Creek, all PAHs analyzed at the three locations indicated significant differences in their concentration by sediment particle sizes (P value less than 0.05). Cluster analyses of sediment PAH concentrations by particle size showed a separate single group for the >2800 μm LOM fraction for most of the PAHs, indicating a difference exists between LOM PAH concentrations and other sediment size fractions PAHs, see Bathi 2008 for full statistical analyses [18]. The two-way ANOVA analysis of PAH concentrations by particle size and sample location indicated that concentrations were affected by sediment particle sizes and, except for naphthalene, fluorene, phenanthrene, and indeno(1,2,3-cd)pyrene, PAH concentrations were influenced by the sample location. With few exceptions, the interaction of sediment location and particle size statistically influenced the PAH concentrations (Table 3).

Figures 2(a) and 2(b) illustrate the mass load of the PAHs by sediment particle size for the creek sediment samples. As expected, most of the PAH mass was associated with small and intermediate size sediments, as LOM (0.6%) and the large mineral sediment sizes were only a small fraction of the total sediment. Since more than 90% of all stormwater particles are expected in the 1 to 100 μm range and, given the observed higher concentrations of PAHs in the smaller sediment particles, most of the stormwater runoff PAH load would be expected to be associated with the finer particles. This is similar to findings by Stein et al. [4], who observed a strong and consistent pattern of high PAH concentrations in dissolved and fine particulate matter. Stein also found that between 30 and 60% of the total PAH load was discharged in the first 20% of the runoff volume for the studied storms and location.

This association of PAHs with the small and large particle sizes affects their treatability. In an advanced stormwater treatment device, the Multi-Chambered Treatment Train (MCTT), Pitt et al. [21] found that very high PAH removals occurred with sedimentation processes designed to remove <5 μm particles. Postsedimentation media treatment using mixtures of sand and peat reduced the PAH concentrations to below detection limits. PAH control data is sparse, but it is expected that simple sedimentation in well-designed wet detention ponds typically results in moderate PAH removals. Media filtration can be very effective in removing particles in the range of 1 to 5 μm, indicating that filtration/biofiltration

TABLE 2: One-way ANOVA Analysis of PAHs concentration by creek sediment particle size.

PAH	One-way ANOVA P value		
	Cribbs Mill Creek	Hunter Creek	Carroll Creek
Naphthalene	0.324	0.000	0.000
Fluorene	0.000	0.000	0.000
Phenanthrene	0.000	0.000	0.000
Anthracene	0.000	0.000	0.000
Fluoranthene	0.000	0.000	0.000
Pyrene	0.000	0.000	0.000
Benzo(a)anthracene	0.000	0.000	0.000
Chrysene	0.000	0.001	0.000
Benzo(b)fluoranthene	0.000	0.011	0.000
Benzo(a)pyrene	0.039	0.060	0.000
Indeno(1,2,3-cd)pyrene	0.002	0.437	0.000
Dibenz(a,h)anthracene	0.024	0.010	0.000
Benzo(g,h,i)perylene	0.004	0.118	0.000

TABLE 3: Two-way ANOVA P values comparing PAH concentration by sediment particles size and sample location.

PAH	Two-way ANOVA P value		
	Particle size	Sample location	Size and location interaction
Naphthalene	0.000	0.088	0.116
Fluorene	0.000	0.721	0.481
Phenanthrene	0.000	0.389	0.043
Anthracene	0.000	0.032	0.821
Fluranthene	0.000	0.000	0.000
Pyrene	0.000	0.000	0.000
Benzo(a)anthracene	0.000	0.005	0.002
Chrysene	0.000	0.004	0.000
Benzo(b)fluoranthene	0.000	0.002	0.254
Benzo(a)pyrene	0.004	0.032	0.022
Indeno(1,2,3-cd)pyrene	0.000	0.284	0.250
Dibenz(a,h)anthracene	0.000	0.019	0.002
Benzo(g,h,i)perylene	0.000	0.041	0.493

systems are likely very effective at treating PAH particulate matter.

3.3. Relating Sediment Composition, COD, and PAH Content. The thermal chromatography technique showed that the sediment sample, when heated from 240 to 365°C, lost volatile organic content. The smaller (<90 μm) and larger (>710 μm) size fractions had higher percentage volatile organic content and COD than intermediate-sized particles (Table 4). The observed bimodal distribution of the volatile organic content and COD by sediment particle sizes was

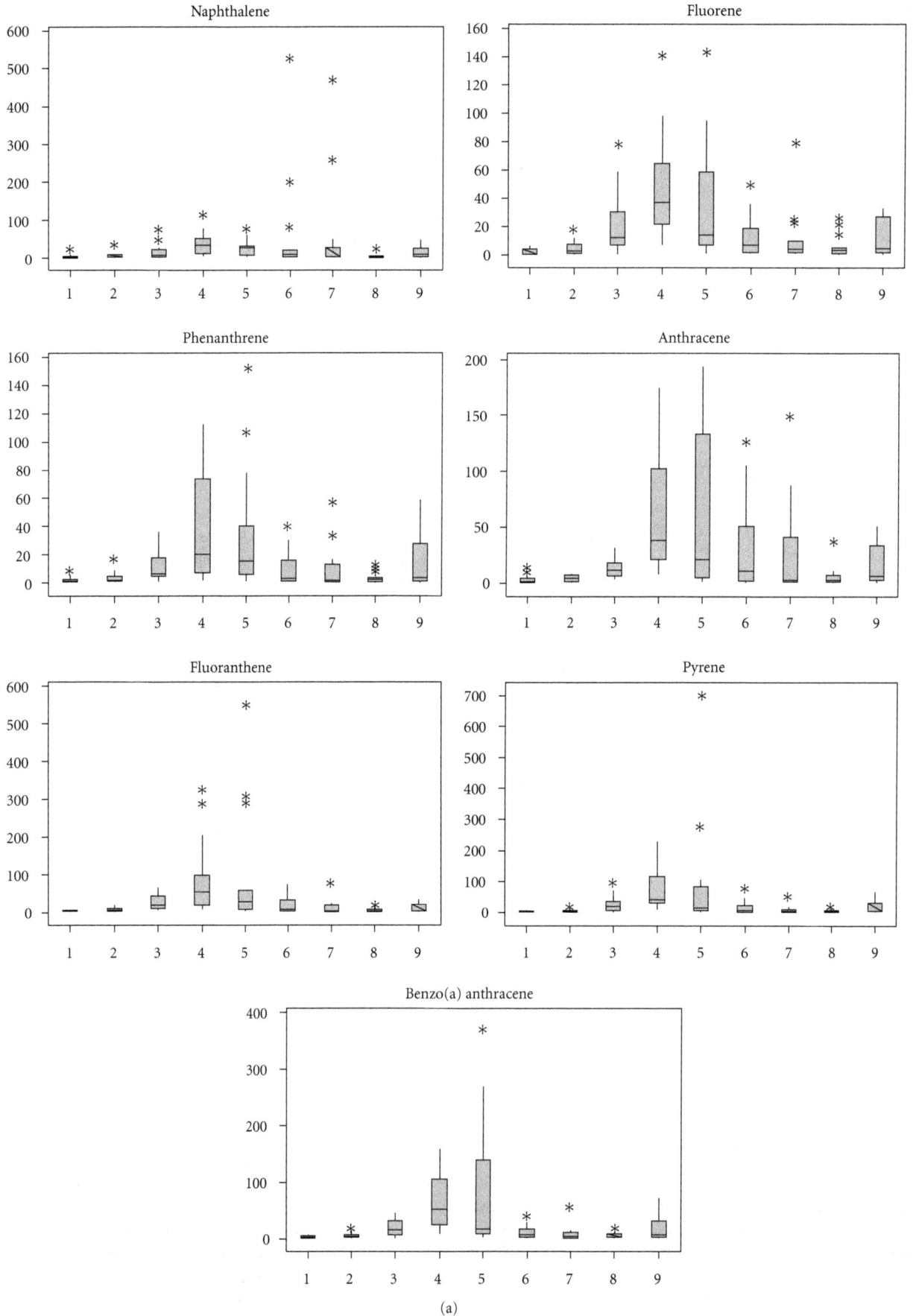

(a)

FIGURE 2: Continued.

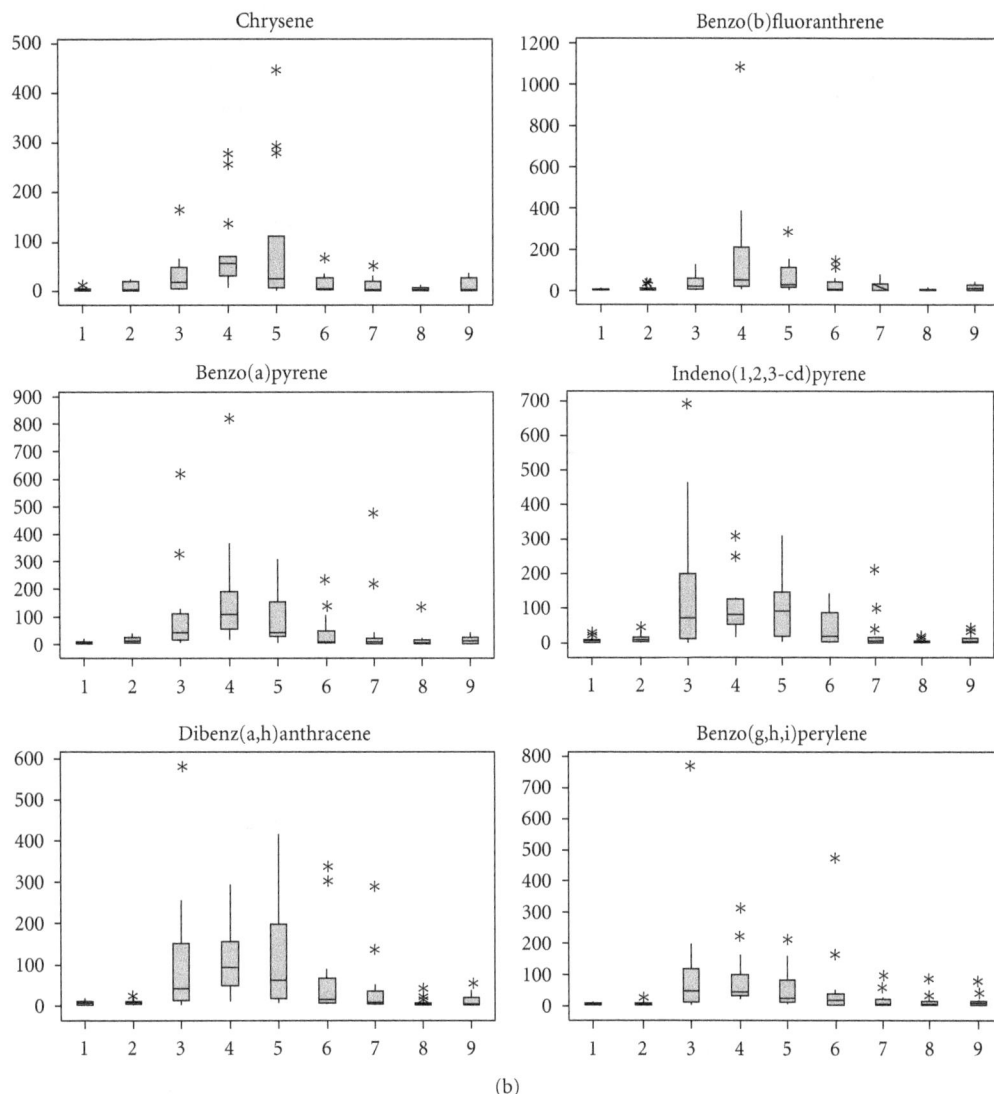

FIGURE 2: Observed mass distribution of analytes by sediment particle sizes. Notes: x-axis is particle size range (μm), where 1 = <45; 2 = 45–90; 3 = 90–180; 4 = 180–355; 5 = 355–710; 6 = 710–1400; 7 = 1400–2800; 8 = >2800 (w/o LOM*); 9 = >2800 (LOM). *with large organic material removed. y-axis weight in μg of analyte associated with particle fraction in 1 kg of total sediment.

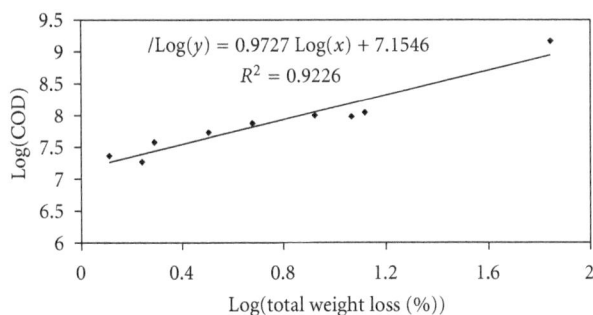

FIGURE 3: Log percentage of volatile content and log COD regression analysis.

similar to the distribution observed for PAH concentrations of the sediment fractions (Tables 1 and 4). Similarly, a strong,

statistically significant linear relation was observed between measured sediment fraction COD and log transformed percent volatile organic carbon content (Figure 3).

Regression analyses were also employed to examining the relationships between COD and PAH content of the sediment samples. For regressions at each sampling location relating COD by particle size and their PAH content, the slope term of the regression was significant ($P < 0.05$) for 193 out of a total of 351 cases (9 particle size ranges, 13 analytes, and 3 locations). A larger number of significant cases were found for sediment fractions with higher volatile content (larger and smaller sediment fractions compared to intermediate sizes). There was no particular trend seen based on the PAH's molecular weight or number of aromatic rings for the PAH, indicating there is no clear evidence of varied affinity of the studied PAHs based on sediment organic content, irrespective of their molecular weights.

TABLE 4: Thermal chromatography sediment volatile organic content and COD analysis results.

Size range (μm)	% Volatile organic content	SCOD (mg/kg dry sediment)
<45	1.1	74,000
45–90	1.45	54,000
90–180	0.49	23,000
180–355	0.45	18,000
355–710	0.8	38,000
710–1400	3.54	99,000
1400–2800	6.39	110,000
>2800 (w/o LOM*)	6.51	95,000
>2800 LOM	37.98	1,500,000

* With large organic material removed.

4. Conclusions

As expected for developed urban creeks that carry high runoff flows, the creek sediment samples were composed of fewer fines than would be expected in surface stormwater runoff samples. However, the concentrations of the contaminants in each particle size category are likely representative of the same particle size categories found in stormwater. These results can be used to characterize PAH characteristics by particle size in stormwater (in conjunction with a suitable particle size distribution), to indicate the likely transport and fate of discharged PAHs after discharge and to indicate the potential for stormwater treatment of particulate-bound PAHs. In this study, most of the sediment samples (by mass) were dominated by particles in the size range of 90–710 μm. Observed volatile organic content and SCOD and PAHs concentration of the size fractionated urban creek sediment particles showed a bimodal distribution: smaller and larger particles had relatively higher concentrations of organic content, PAHs, and SCOD than the intermediate particles. The presence of fragmentary plant material in the larger size fractions and the high clay/silt content along the high surface area in the smaller size fractions are believed to cause higher volatile organic content and hence the associated PAH concentrations in these size ranges. Among all the size fractions, highest PAH concentrations were observed in LOM fractions; however, LOM fractions represented only small portions of the sediment samples, by mass, and the amount of PAHs associated with these fractions also was small compared to other size fractions.

These results highlight the challenges associated with treating PAHs in urban runoff using traditional sedimentation methods. These small particle sizes are more favorably removed using treatment technologies such as media filtration, although large amounts can be removed by well-designed sedimentation practices. Sedimentation, therefore, should precede any media filter treatment method as a pretreatment unit process. The associations with LOM indicate that floatables control will be needed to remove the organic matter (leaves, grass clippings) and associated PAHs. Overall, a treatment train containing complimentary unit processes

such as floatable and grit control, sedimentation designed to remove small particles, and finally media treatment (as in the MCTT [21]) should result in excellent PAH removals from stormwater. This treatment approach can be expected to result in significant reductions of PAH contamination of urban stream sediments.

Acknowledgment

This material is based upon work supported by the National Science Foundation under Grant no. EPS-0447675. Any opinions, findings, and conclusions or recommendations expressed in this material are those of the author(s) and do not necessarily reflect the views of the National Science Foundation.

References

[1] V. P. C. Metre, B. J. Mahler, and E. T. Furlong, "Urban sprawl leaves its PAH signature," *Environmental Science & Technology*, vol. 34, no. 19, pp. 4064–4070, 2000.

[2] J. R. Kucklick, S. K. Sivertsen, M. Sanders, and G. I. Scott, "Factors influencing polycyclic aromatic hydrocarbon distributions in South Carolina estuarine sediments," *Journal of Experimental Marine Biology and Ecology*, vol. 213, no. 1, pp. 13–29, 1997.

[3] S. McCready, D. J. Slee, G. F. Birch, and S. E. Taylor, "The distribution of polycyclic aromatic hydrocarbons in surficial sediments of Sydney Harbour, Australia," *Marine Pollution Bulletin*, vol. 40, no. 11, pp. 999–1006, 2000.

[4] E. D. Stein, L. L. Tiefenthaler, and K. Schiff, "Watershed-based sources of polycyclic aromatic hydrocarbons in urban storm water," *Environmental Toxicology and Chemistry*, vol. 25, no. 2, pp. 373–385, 2006.

[5] R. Huston, Y. C. Chan, T. Gardner, G. Shaw, and H. Chapman, "Characterisation of atmospheric deposition as a source of contaminants in urban rainwater tanks," *Water Research*, vol. 43, no. 6, pp. 1630–1640, 2009.

[6] K. O'Reilly, J. Pietari, and P. Boehm, "PAHs review, Polycyclic aromatic hydrocarbons in stormwater and urban sediments," *Stormwater*, September 2010, http://www.stormh2o.com.

[7] R. Pitt, R. Field, M. Lalor, and M. Brown, "Urban stormwater toxic pollutants: assessment, sources, and treatability," *Water Environment Research*, vol. 67, no. 3, pp. 260–275, 1995.

[8] E. A. Guertal, D. J. Eckert, S. J. Traina, and T. J. Logan, "Parking lot sealcoat: an unrecognized source of urban polycyclic aromatic hydrocarbons," *Environmental Science & Technology*, vol. 39, no. 15, pp. 5560–5566, 2005.

[9] H. M. Hwang and G. D. Foster, "Characterization of polycyclic aromatic hydrocarbons in urban stormwater runoff flowing into the tidal Anacostia River, Washington, DC, USA," *Environmental Pollution*, vol. 140, no. 3, pp. 416–426, 2006.

[10] C. J. Diblasi, H. Li, A. P. Davis, and U. Ghosh, "Removal and fate of polycyclic aromatic hydrocarbon pollutants in an urban stormwater bioretention facility," *Environmental Science & Technology*, vol. 43, no. 2, pp. 494–502, 2009.

[11] J. L. Zhou, T. W. Fileman, S. Evans et al., "Fluoranthene and pyrene in the suspended particulate matter and surface sediments of the Humber estuary, UK," *Marine Pollution Bulletin*, vol. 36, no. 8, pp. 587–597, 1998.

[12] B. Rushton, Broadway Outfall Stormwater Retrofit Project, Monitoring CDS Unit and Constructed Pond. South Florida

Water Management District and City of Temple Terrace: W241, Brooksville, Fla, USA, 2006.

[13] H. Li, J. Chen, J. Wu, and X. Piao, "Distribution of polycyclic aromatic hydrocarbons in different size fractions of soil from a coke oven plant and its relationship to organic carbon content," *Journal of Hazardous Materials*, vol. 176, no. 1–3, pp. 729–734, 2010.

[14] G. Guggenberger, M. Pichler, R. Hartmann, and W. Zech, "Polycyclic aromatic hydrocarbons in different forest soils: mineral horizons," *Zeitschrift fur Pflanzenernahrung und Bodenkunde*, vol. 159, no. 6, pp. 565–573, 1996.

[15] S. Müller, W. Wilcke, N. Kanchanakool, and W. Zech, "Polycyclic aromatic hydrocarbons (PAHs) and polychlorinated biphenyls (PCBs) in particle-size separates of urban soils in Bangkok, Thailand," *Soil Science*, vol. 165, no. 5, pp. 412–419, 2000.

[16] A. K. Haritash and C. P. Kaushik, "Biodegradation aspects of Polycyclic Aromatic Hydrocarbons (PAHs): a review," *Journal of Hazardous Materials*, vol. 169, no. 1–3, pp. 1–15, 2009.

[17] Metropolitan Washington Council of Governments, Urban Runoff in the Washington Metropolitan Area. U.S. Environmental Protection Agency, Nationwide Urban Runoff Program, 1983.

[18] J. R. Bathi, *Associations of polycyclic aromatic hydrocarbons (PAHs) with urban creek sediments*, Ph.D. thesis, Department of Civil, Construction and Environmental Engineering, The University of Alabama, Tuscaloosa, Ala, USA, 2008.

[19] H. Ray, *Street dirt as a phosphorus source in urban stormwater*, MSCE, Department of Civil and Environmental Engineering, University of Alabama at Birmingham, Birmingham, Ala, USA, 1997.

[20] A. Krein and M. Schorer, "Road runoff pollution by polycyclic aromatic hydrocarbons and its contribution to river sediments," *Water Research*, vol. 34, no. 16, pp. 4110–4115, 2000.

[21] R. Pitt, B. Robertson, P. Barron, A. Ayyoubi, and S. Clark, Stormwater Treatment at Critical Areas: The Multi-Chambered Treatment Train (MCTT). U.S. Environmental Protection Agency, Wet Weather Flow Management Program, National Risk Management Research Laboratory, EPA/600/R-99/017, Cincinnati, Ohio, USA, 1999.

15

Nonlinear Seismic Response Analysis of Curved and Skewed Bridge System with Spherical Bearings

Junwon Seo,[1] Daniel G. Linzell,[2] and Jong Wan Hu[3]

[1] *Department of Civil, Construction, Environmental Engineering, Iowa State University, Ames, IA 50011, USA*
[2] *Department of Civil Engineering, The University of Nebraska, Lincoln, NE 68588, USA*
[3] *Department of Civil and Environmental Engineering, University of Incheon, Incheon 406-772, Republic of Korea*

Correspondence should be addressed to Jong Wan Hu; jongp24@incheon.ac.kr

Academic Editor: John Mander

A three-dimensional (3D) modeling approach to investigate nonlinear seismic response of a curved and skewed bridge system is proposed. The approach is applied to a three-span curved and skewed steel girder bridge in the United States. The superstructure is modeled using 3D frame elements for the girders, truss elements for the cross-frames, and equivalent frame elements to represent the deck. Spherical bearings are modeled with zero-length elements coupled with hysteretic material models. Nonlinear seismic responses of the bearings subjected to actual ground motions are examined in various directions. Findings indicate that the bearings experience moderate damage for most loading scenarios based on FEMA seismic performance criteria. Further, the bearing responses are different for the loading scenarios because of seismic effects caused by interactions between excitation direction and radius of curvature.

1. Introduction

Studies related to the design and analysis of curved and skewed steel bridges have focused on modeling and design for static and pseudo-static loads [1–4], and only a few investigations have looked at seismic behavior [5]. To design and assess curved steel bridges in high and moderate seismic zones, it is of interest to more extensively examine seismic analysis methods so that reliable 3D modeling approaches are developed. Studies have been undertaken that applied modeling approaches to predict the seismic response of straight steel girder bridges [6]. Similar simplified modeling approaches have been proposed for curved steel bridges, but the approaches were applied to static events [1]. These studies have shown that modeling using a 3D approach can provide improved accuracy relative to line girder analyses by incorporating member depths. For a curved and skewed bridge, where significant lateral displacements may be induced at the bearings under a seismic event, modeling structural component depths would be assumed to be important.

For these reasons, a 3D modeling approach is used herein to investigate seismic responses of a curved steel I-girder bridge system with skewed supports. The 3D approach is applied to a three-span continuous curved steel I-girder bridge system in the United States. Following the approach recommended by previous research [1], the bridge is modeled using elastic frame elements for the I-girders, truss elements for the cross-frames, and elastic frame elements for the deck. Preliminary seismic responses at the bearings are presented for the bridge under El Centro ground motions.

2. 3D Modeling Approach

All elements used for 3D model were generalized using OpenSees [8]. Curved bridge framing was represented using frame elements with lumped masses being placed at each node, with those masses calculated using tributary dimensions. Model construction initiated with calculation of superstructure and substructure section properties. Superstructure included girder, cross frame, concrete deck, and rigid link element, while substructure included pier column, cap, abutment, and footing. Spherical bearings were modeled in OpenSees using ZeroLength elements. All rotational degrees

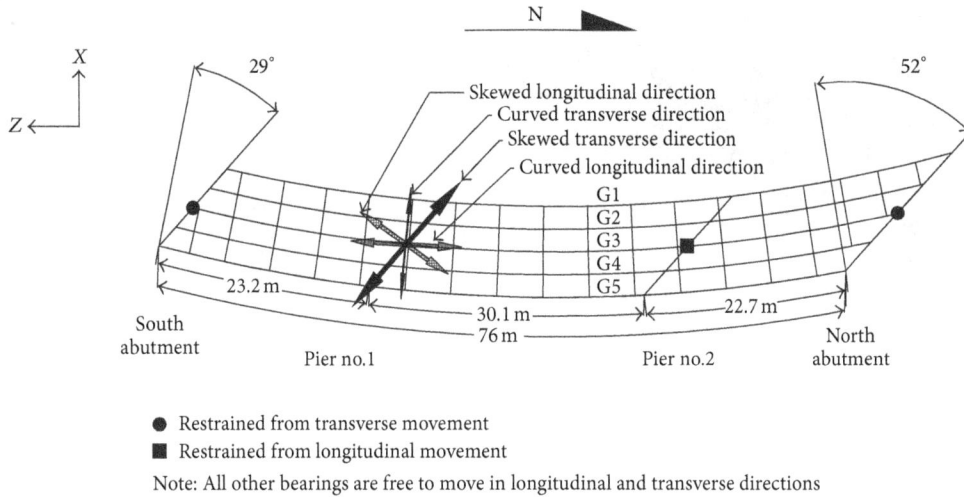

FIGURE 1: Studied bridge framing plan [5].

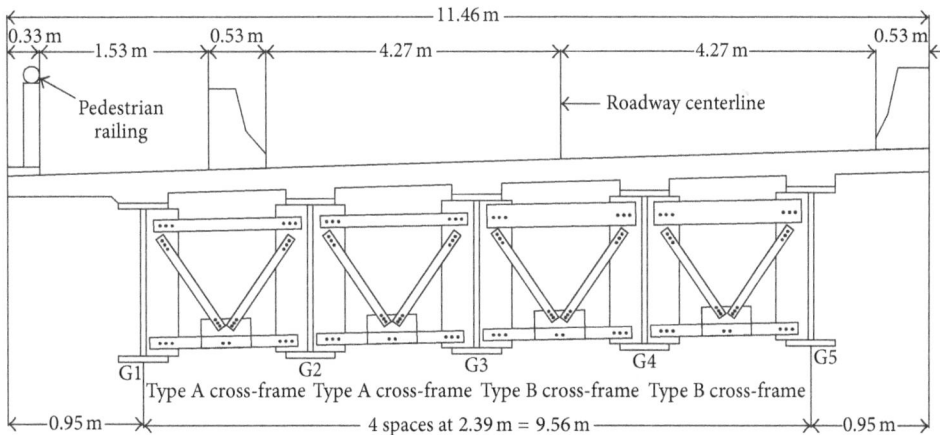

FIGURE 2: Typical bridge cross-section.

of freedom for the bearings, which accommodated rotations about various axes, were unrestrained. To simulate the bearings' moment-rotational behavior, a combination of different material models available in OpenSees was utilized. Appropriate nominal material properties were used for the steel and concrete.

3. Application to the Selected Bridge

3.1. Bridge Description. The bridge used for the study is a curved and skewed steel I-girder bridge located in Pennsylvania. The three-span continuous bridge has radius of curvature of 178.49 m and is composed of five ASTM A572 grade 50 steel plate girders with an abutment skew that varies between 29° and 52° (south to north) relative to the traffic direction as shown in Figure 1. Bridge support conditions are as shown in Figure 1. This figure shows that two bearings are restrained from transverse movement, one bearing is restrained from longitudinal movement, and all other bearings are free to move in both the longitudinal

and transverse directions. Girders are spaced 2.39 m center-to-center as shown in Figure 2. All girders have 1219 mm × 13 mm webs with 356 mm wide top and bottom flanges of varying thickness as shown in Table 1. Two different K-shaped cross-frame types are used in the bridge. Type A top and bottom chords are composed of 3.5 × 3.5 × 3/8 double angles. Type A diagonals are 3.5 × 3.5 × 3/8 angles. Type B top chords are WT14 × 49.5 s, and type B bottom chords are 3.5 × 3.5 × 3/8 double angles. Type B diagonals are composed of 3.5 × 3.5 × 3/8 angles. The superstructure is supported by multi-circular column piers with 914.4 mm wide by 1066.8 mm deep reinforced concrete pier caps [5]. Concrete pier columns on the foundation wall which is 11.9 m long, 3.4 m wide, and 0.7 m thick are spaced 4.0 m apart. The abutments are supported by the spread footings with a 1.6 m tall backwall. More detailed description of the substructure units can be found elsewhere [5].

3.2. Spherical Bearings. Spherical bearings have been utilized to support curved and skewed steel bridge superstructures

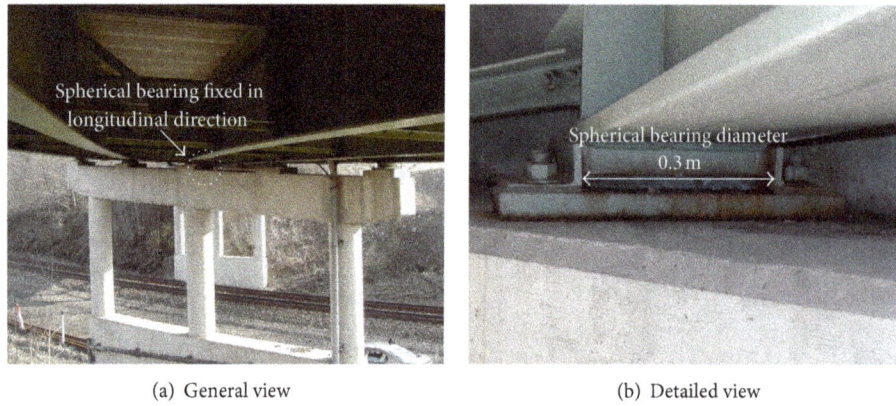

(a) General view

(b) Detailed view

FIGURE 3: Spherical bearing system.

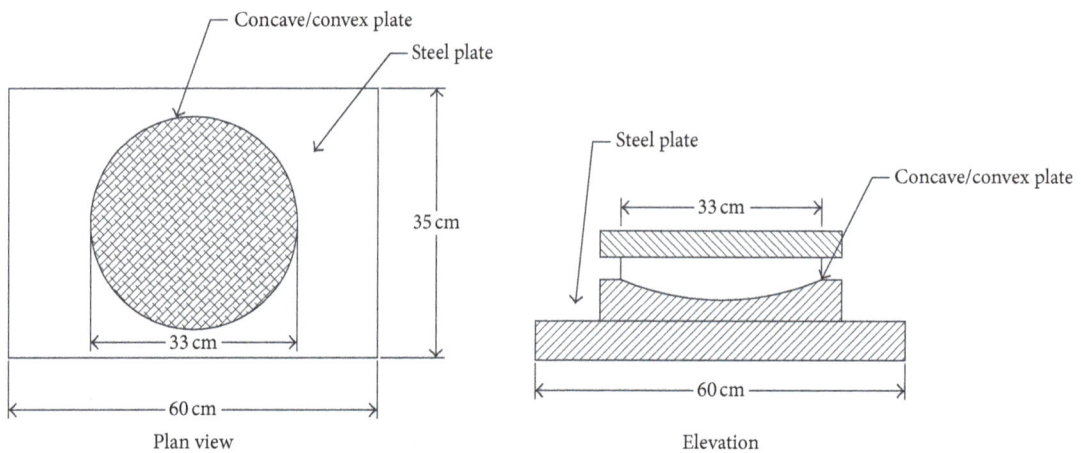

Plan view

Elevation

FIGURE 4: Spherical bearing details.

TABLE 1: Steel plate girder element dimensions (width × thickness).

Girder	Top flange (mm)	Web (mm)	Bottom flange (mm)
G1, G2	356 × 16	1219 × 13	356 × 25
G3, G4, G5	356 × 16	1219 × 13	356 × 32

to accommodate rotations that may occur about more than one axis. They are used in the curved bridge being examined herein. In general, bearings are divided into two main categories, fixed and expansion. A fixed bearing permits rotational movement and prevents translation in one or more directions, while an expansion bearing permits both rotation and translations. A spherical bearing fixed in the longitudinal direction exists for middle girder G3 at interior pier no. 2 (see Figure 1), and spherical bearings fixed in the transverse direction exist for G3 at the abutments. All other locations had spherical expansion bearings that are free to move in both the transverse and longitudinal directions. Figure 3 shows representative bearings used in the curved bridge, and Figure 4 illustrates spherical bearing details.

3.3. 3D Model. The superstructure of the curved steel bridge was idealized based upon the proposed 3D modeling

approach in OpenSees [8]. Figure 5 shows the 3D curved bridge model. The mesh consists of steel girders and the concrete deck modeled using elastic beam column elements. These elements were used because they were developed to simulate 3D beam behavior, including biaxial bending and torsion. Small straight sections were used to simulate the curvature of the girders and concrete deck. Nodes were placed at cross-frame locations. Longitudinal and transverse elements that represent the behavior of the slab were used. Member properties that reflect the slab dimensions were used along with appropriate steel girder and cross-frame properties in the model. Boundary conditions were implemented based on actual support conditions and attempted to account for spherical bearing moment-rotational behavior. This behavior was modeled in OpenSees using Steel01 and hysteretic material models in parallel as shown in Figure 6(a). The spherical bearings used in the bridge were made of A36 steel and the Steel01 material reflecting a bearing having an initial stiffness, K_e, of 200 GPa and a strain-hardening ratio, b, of 0.014. To approximate nonlinear hysteretic behavior, the hysteretic material model used four linear stiffness functions, including an initial stiffness, K_1, of 312.5 GPa, a second stiffness, K_2, of 3 GPa, a third stiffness, K_3, of 1.25 GPa, and a final stiffness, K_4, of −312.5 GPa. All stiffness values were determined via a

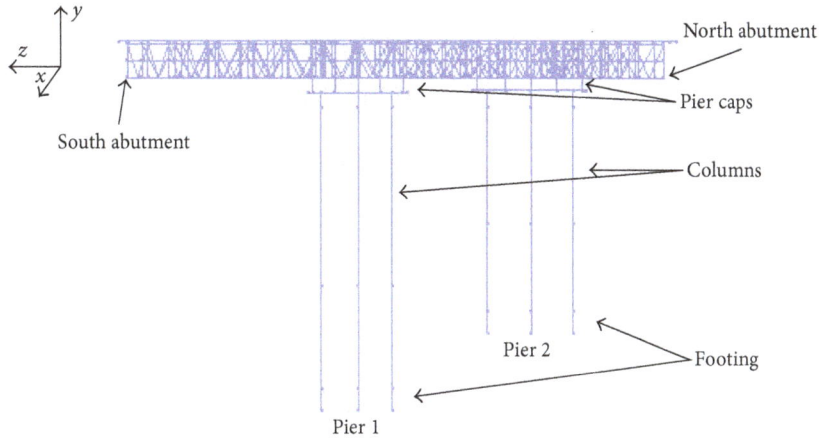

FIGURE 5: 3D bridge model.

trial and error procedure that compared model predictions to actual data from Roeder et al. [7] that examined spherical bearing under cyclic loads. Figure 6(b) shows a comparison between experimental and analytical moment-rotational hysteresis loops. In this figure, the analytical model provides reasonable approximation of real bearing behavior at 10,000 cycles. Substructure units, including the pier columns and caps, abutment, and footings, were idealized in the 3D OpenSees model following recommendations by Nielson [6]. Included in the substructure models were the pier columns and caps, abutments, and footings. The detailed modeling description for the substructure can be found elsewhere [5].

4. Seismic Response of Curved and Skewed Bridge

Preliminary results from seismic analyses of the curved and skewed bridge using the 3D model are presented. These results focus on seismic bearing response as a result of longitudinal and transverse earthquake loadings.

4.1. Curved and Skewed Longitudinal Earthquake Loadings. Since the structure being examined is horizontally curved and rests on skewed supports, directions both parallel to and perpendicular to the skewed supports were identified as those for the ground motions. Applying ground motions in this fashion has been shown to be preferred for inducing critical skewed bridge response [9, 10]. To apply these motions, two "longitudinal directions" were defined. The direction tangential to the chord of each curved girder at the abutment and/or pier was referred to as the "curved longitudinal" direction as shown in Figure 1. The direction perpendicular to substructure units at each bearing was referred to as the "skewed longitudinal" direction as shown in Figure 1. To capture critical superstructure response, El Centro ground motions, which had a peak horizontal ground acceleration of 0.313 g, were applied to the bridge in the curved longitudinal direction initially and then the skewed longitudinal direction. Bearing rotations were examined at all supports while the

earthquake loading was applied to the bridge. Existing literature indicates that these rotations provide key information in relation to assessing bridge susceptibility to earthquake damage [11].

To explore seismic behavior in the curved longitudinal direction, bearing rotations were monitored in the global x-axis direction as shown in Figure 1. Figure 7(a) shows the seismic response of a representative spherical bearing when acted on by the curved longitudinal earthquake loading. The response of this fixed spherical bearing, located underneath G3 at the south abutment, depicts rotations about the x-axis direction exceeding −0.02 radians. Seismic responses for the spherical bearing subjected to the skewed longitudinal earthquake loading are shown in Figure 7(b). As expected, rotations about the x-axis are different when the bridge is acted on by skewed and curved longitudinal earthquake loadings because of different seismic bending-torsion coupled effects being enacted based on relationships between the excitation direction and the girder radius of curvature. The hysteresis loop shown in Figure 7(b) depicts rotation about the x-axis exceeding −0.03 radians and moments beyond 30 kN-m, values that would classify this bearing as being slightly damaged based on existing research (greater than ±0.02 radians) if those rotations exceeded any existing clearance in the bearing [7]. In addition, it has been stated that a spherical bearing having rotations exceeding −0.03 radians may moderately damage an abutment or pier [11].

4.2. Curved and Skewed Transverse Earthquake Loadings. The transverse direction perpendicular to the chord of each curved girder at the abutment and/or pier was referred to as the "curved transverse" direction as shown in Figure 1. The direction parallel to substructure units was referred to as the "skewed transverse" direction as shown in Figure 1. Similar to the curved and skewed longitudinal earthquake loading cases, El Centro ground motions were also applied to the bridge in the curved and skewed transverse directions. Figure 8(a) shows the seismic response of the spherical bearing due to the curved transverse El Centro ground motions, again presented as moment-rotation hysteresis curves. The

Steel01 material

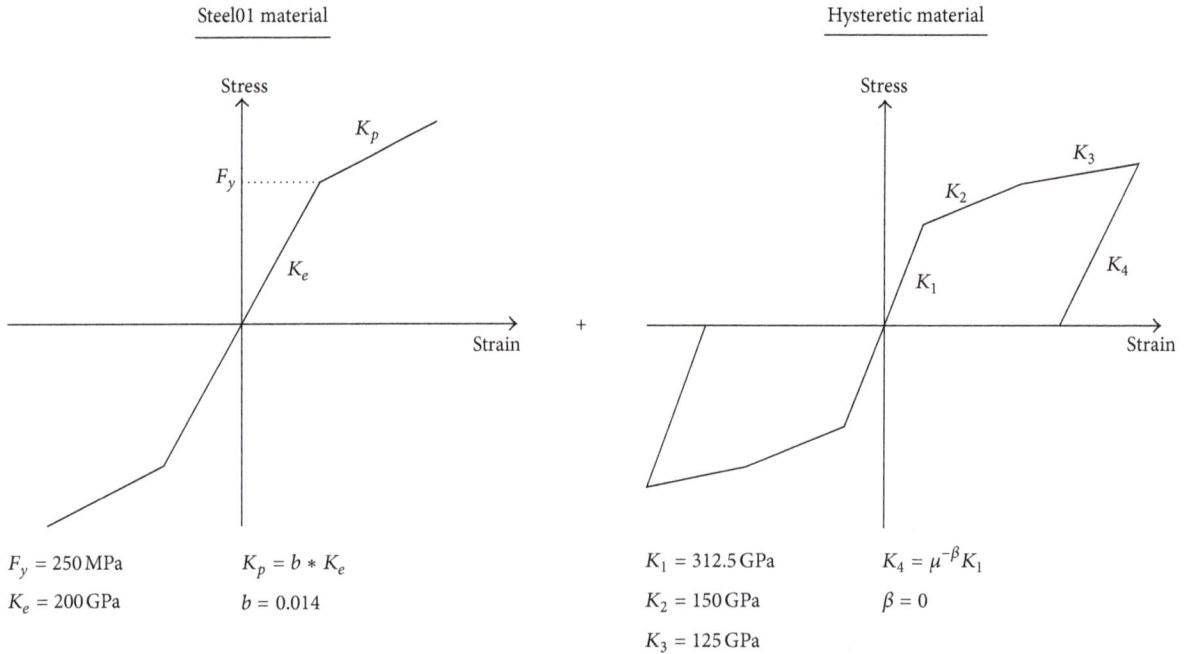

$F_y = 250\,\text{MPa}$ $K_p = b * K_e$

$K_e = 200\,\text{GPa}$ $b = 0.014$

Hysteretic material

$K_1 = 312.5\,\text{GPa}$ $K_4 = \mu^{-\beta} K_1$

$K_2 = 150\,\text{GPa}$ $\beta = 0$

$K_3 = 125\,\text{GPa}$

(a) Spherical bearing OpenSees model components modeled in parallel

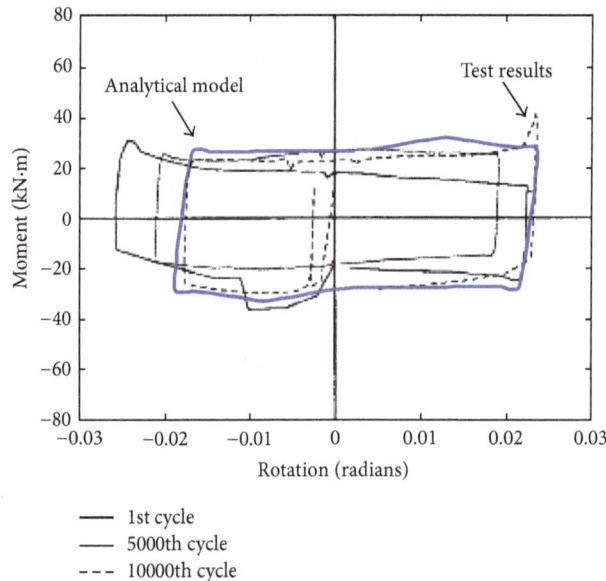

—— 1st cycle
—— 5000th cycle
--- 10000th cycle

(b) Moment-rotation behavior [5, 7]

FIGURE 6: Spherical bearing model.

hysteresis loop shown in Figure 8(a) depicts rotations about the x-axis direction exceeding -0.03 radians and moment greater than 30 kN-m. Again, this could be classified as moderately damaged according to FEMA [11]. The hysteresis loop shown in Figure 8(b), which examines bearing response when the bridge is subjected to the skewed transverse ground motions, depicts rotation about the x-axis direction reaching around -0.01 radian and moment of approximately 30 kN-m, values that would be indicative of slight damage according to FEMA [11]. In similar fashion to the longitudinal earthquake loading cases discussed earlier, moment-rotation

relationships for the spherical bearing were different between the skewed and curved transverse earthquake loading cases, due to the different behavior being attributed to relationships between the loading direction and radius of curvature.

5. Conclusions

Limited consideration has been given to seismic design and detailing of a curved steel bridge having skewed supports. The primary goal of this study was to investigate the seismic

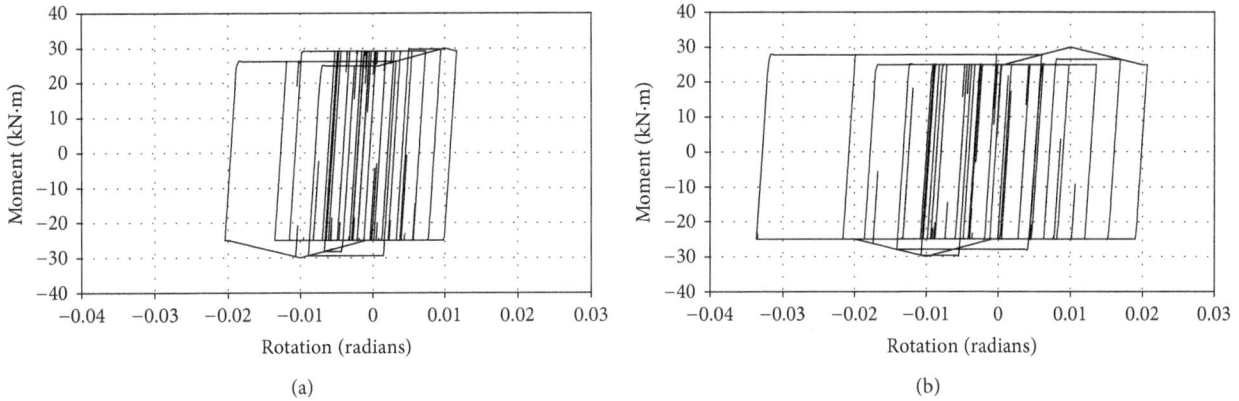

FIGURE 7: G3 spherical bearing moment-rotation response at the south abutment for (a) curved longitudinal direction and (b) skewed longitudinal direction.

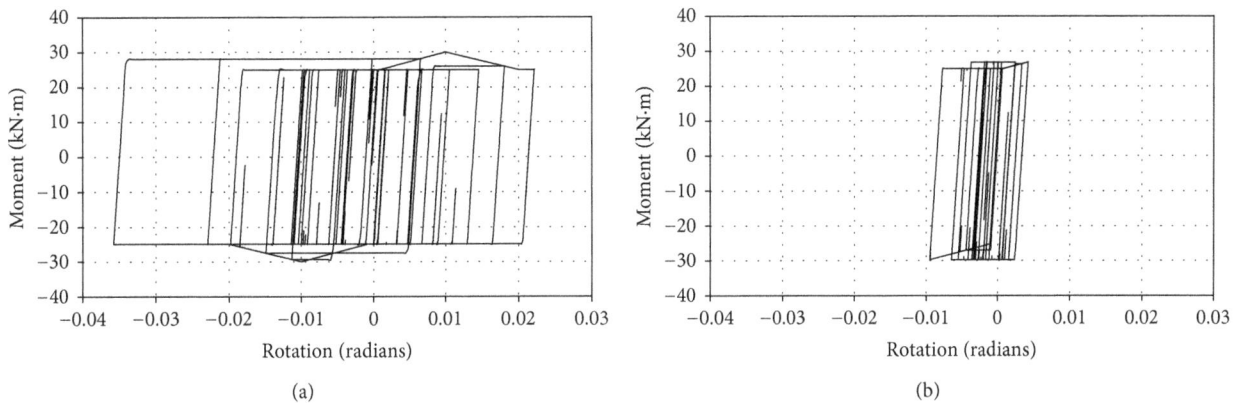

FIGURE 8: Spherical bearing response at the south abutment for (a) curved transverse direction and (b) skewed transverse direction.

bearing response of a curved and skewed steel I-girder bridge system using the 3D modeling approach being proposed in this study. Specifically, moment-rotation response relationships for representative spherical bearings were examined when a selected horizontally curved and skewed bridge was subjected to El Centro ground motions. Rotations of these spherical bearings appeared to exceed −0.03 radians when the bridge was subjected to the skewed longitudinal and curved transverse ground motions. Therefore, these bearings may experience moderate damage under the imposed ground motions based upon FEMA criteria (2003). In addition, bearing response differed for the considered earthquake loading scenarios because of different seismic bending-torsion coupled effects being enacted based on relationships between the excitation direction and radius of curvature.

Acknowledgments

This research was supported by the Basic Science Research Program through the National Research Foundation of Korea (NRF) and funded by the Ministry of Education, Science and Technology (Grant no. 2012R1A1A1041521). The authors gratefully acknowledge this support.

References

[1] C.-J. Chang, D. W. White, F. Beshah, and W. Wright, "Design analysis of curved I-girder bridge systems—an assessment of modeling strategies," in *Proceedings of the Structural Stability Research Council. Annual Stability Conference*, pp. 349–369, April 2005.

[2] W. H. Huang, *Curved I-girder systems [Ph.D. thesis]*, Department of Civil and Environmental Engineering, University of Minnesota, Minneapolis, Minn, USA, 1996.

[3] D. Nevling, D. Linzell, and J. Laman, "Examination of level of analysis accuracy for curved I-girder bridges through comparisons to field data," *Journal of Bridge Engineering*, vol. 11, no. 2, pp. 160–168, 2006.

[4] D. W. White, A. H. Zureick, N. Phoawanich, and S. K. Jung, "Development of unified equations for design of curved and straight steel bridge I-girders," Final Report to AISI, PSI Inc. and FHWA, pp. 547, October 2001.

[5] J. Seo and D. Linzell, "Nonlinear seismic response and parametric examination of horizontally curved steel bridges using 3-D computational models," *Journal of Bridge Engineering ASCE*, vol. 18, no. 3, 2013.

[6] B. G. Nielson, *Analytical fragility curves for highway bridges in moderate seismic zones [Ph.D. thesis]*, School of Civil and

Environmental Engineering, Georgia Institute of Technology, Atlanta, Ga, USA, 2005.

[7] C. W. Roeder, J. F. Stanton, and T. I. Campbell, "Rotation of high load multirotational bridge bearings," *Journal of Structural Engineering*, vol. 121, no. 4, pp. 747–756, 1995.

[8] S. Mazzoni, F. McKenna, M. H. Scott, and G. L. Fenves, "Open system for earthquake engineering simulation (OpenSees)," Pacific Earthquake Engineering Research Center, Version 1.7.3. 2006.

[9] H. M. O. Al-Baljat, *Behavior of horizontally curved bridges under static load and dynamic load from earthquakes [Ph.D. thesis]*, Department of Civil Engineering, Illinois Institute of Technology, Chicago, Ill, USA, 1999.

[10] G. Watanabe and K. Kawashima, "Seismic response of a skewed bridge," Japan-UK Joint Seminar, 2004.

[11] FEMA, HAZUS-MH, "MR1: Technical Manual," Earthquake Model. Federal Emergency Management Agency, Washington, DC, USA, 2003.

16

Electrokinetic Treatment for Model Caissons with Increasing Dimensions

Eltayeb Mohamedelhassan, Kevin Curtain, Matt Fenos,
Kevin Girard, Anthony Provenzano, and Wesley Tabaczuk

Department of Civil Engineering, Lakehead University, 955 Oliver Road, Thunder Bay, ON, Canada P7B 5W2

Correspondence should be addressed to Eltayeb Mohamedelhassan, eltayeb@lakeheadu.ca

Academic Editor: Jean-Herve Prevost

Electrokinetic treatment has been known in geotechnical engineering for over six decades, yet, the technique is rarely used. This stems from the absence of design guidelines and specifications for electrokinetic treatment systems. An important issue that need to be investigated and understood in order to devise guidelines from experimental results is the effect of the foundation element size on the outcome of the treatment. Also important is determining the optimum distance between the electrodes and estimating the energy consumption prior to treatment. This experimental study is a preliminary step in understanding some of the issues critical for the guidelines and specifications. Four model caissons with surface areas between 16000 and 128000 mm^2 were embedded in soft clayey soil under water and treated for 168 hr with a dc voltage of 6 V. From the results, a distance between the anode (model caisson) and the cathode equal 0.25 times the outside diameter of the model caisson was identified as optimum. Relationships between the surface area and axial capacity of the model caisson and the surface area and energy consumption were presented. The equations can be used to preliminary estimate the load capacity and the energy consumption for full-scale applications.

1. Introduction

Soft soils and marine deposits are very common around the world. There are many infrastructure projects and coastal high-rise buildings whose foundations are often supported by such soils of low shear strength and high compressibility. Furthermore, exploration and development of oil and gas fields around the world and expansion of wind farms has resulted in the construction of many platforms and towers on offshore soils with low shear strength. The construction of these projects on soft soils can lead to very expensive foundation systems. Moreover, the installation of traditional foundation elements, particularly driven piles or caissons, can destroy any naturally existing cohesion or cementation between the soil particles and disturb the structure of the soil in the close vicinity of the foundation, causing excessive settlement and further reduction in the foundation's loading capacity.

Electrokinetic treatment is an effective soil improvement technique to increase shear strength and load capacity of foundation elements in soft soils. Electrokinetics improves the strength properties of soft soils by inducing electrokinetic consolidation (e.g., [1]), generating electrokinetic cementation (e.g., [2]) and reducing the water content (e.g., [3]). Major benefits of using electrokinetic treatment are the limited disturbance the treatment may cause to the existing soil structure and the ability to control the zone of treatment. Electrokinetic treatment has been known in geotechnical engineering for over six decades [4, 5] with successful laboratory scale investigations (e.g., [6, 7]) and field experimentations (e.g., [8–10]). Furthermore, case records are reported where electrokinetics was successfully used to improve the load capacity of a friction pile [11] and control the pore water during excavation [12]. However, the technique is in fact seldom used on a professionally recognized scale. The reluctance of the ground improvement

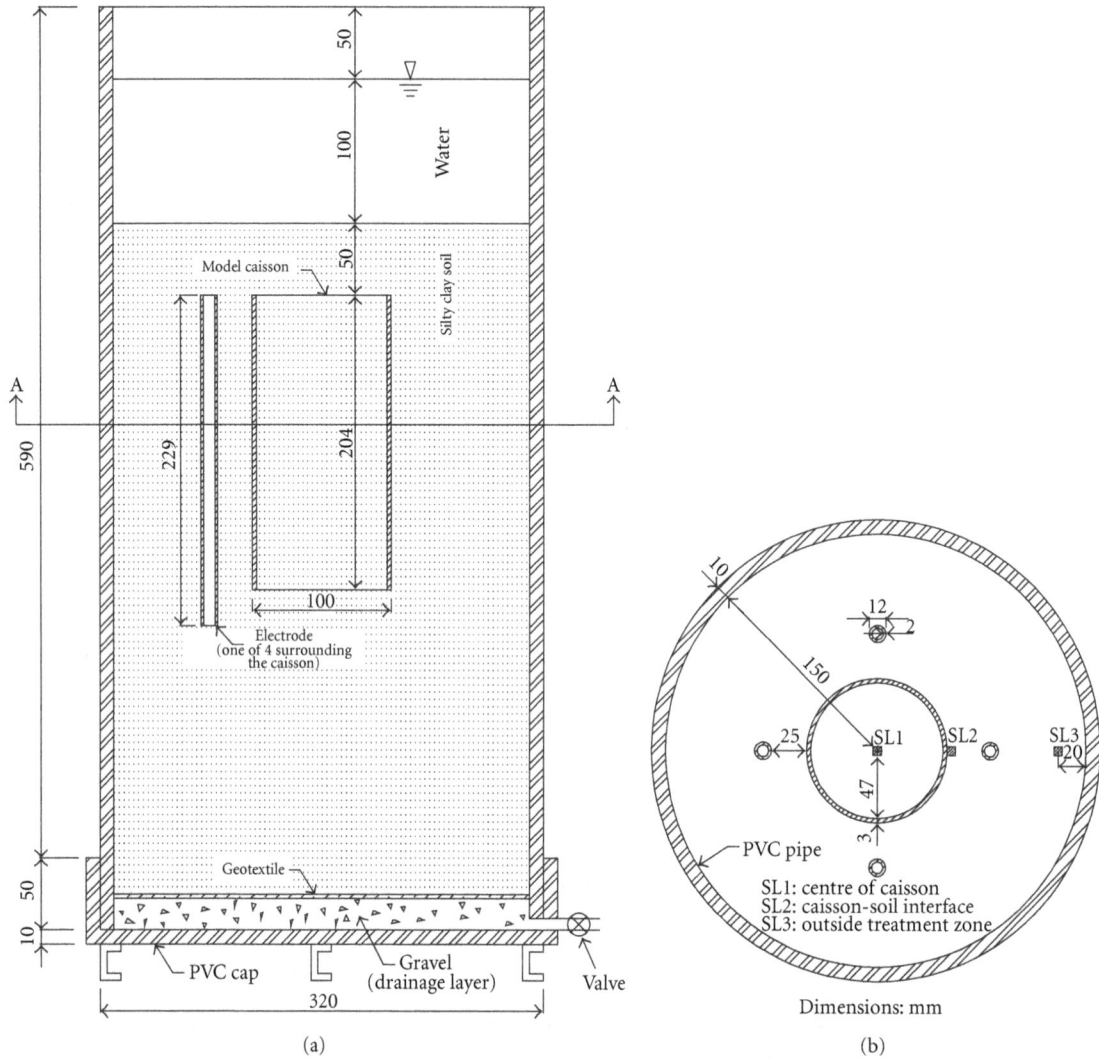

FIGURE 1: (a) Schematic of electrokinetic treatment cell; (b) Cross-section A-A (test 3 model caisson and configuration).

industry to embrace the technique is primarily due to the absence of design guidelines and specifications for electrokinetic treatment systems.

A critical issue that needs to be investigated and understood in order to devise design guidelines for electrokinetic treatment systems from experimental results is the effect of the foundation element size on the gained improvement. Equally important for the guideline is determining the optimum distance between the anode(s) and the cathode(s). Optimum electrode spacing focuses the electric field in the vicinity of the foundation element and controls the size of the treatment zone. Thus, the improvement in the strength properties of the soil occurs in foundation-soil interface with the lowest energy consumption. Finally, estimating the power consumption prior to an electrokinetic treatment is critical to evaluate the economic viability of the treatment. These are major issues that need to be investigated and understood in order to extrapolate, correlate, and/or model the results from bench-scale, laboratory-floor, and pilot tests for full-scale applications.

This experimental study is a preliminary step to address some of the issues important for devising guidelines for electrokinetic treatment. The study investigated the axial load capacity of model caissons with increasing dimensions embedded in soft soil under water after electrokinetic treatment. The study attempted to correlate the load capacity and energy consumption to the surface area of the model. The study proposed a formula for the distance between the electrodes.

2. Experimental Program

2.1. Material Properties. The soil used in this study was recovered from a construction site in Thunder Bay, ON. Grain size distribution analysis on the soil was performed in accordance with ASTM D422-63 [13] and showed that 15.5% of the soil is sand size and 84.5% is fines (silt and clay). The liquid and plastic limits of the soil, determined by ASTM D4318-10 [14], are 25 and 19, respectively. The Unified Soil

Classification System group symbol of the soil is CL-ML and the group name is silty clay with sand. The natural water content of the soil was 36% and the specific gravity is 2.72.

2.2. Experimental Setup and Procedure. Three identical electrokinetic treatment cells were designed and manufactured for the study. The cell, shown in Figure 1, was made from polyvinyl chloride (PVC) pipe 320 mm in outside diameter, 300 mm in inside diameter, 650 mm in length and with a volume capacity of 45 litres. One side of the pipe was covered with a PVC cap that served as a base for the cell. A drainage valve was installed at the base of the cell to facilitate saturation.

Four model caissons with 3 mm wall thickness and increasing diameters and lengths were manufactured from steel and used in the experiments. The surface area was doubled each time from the smallest to the largest model caisson. The outside diameter and length of the model caissons were 50 mm and 102 mm in test 1, 75 mm and 136 mm in test 2, 100 mm and 204 mm in test 3, and 150 mm and 272 in test 4. The corresponding surface areas (SAs) were 16000, 32000, 64000, and 128000 mm² respectively. The tests are summarized in Table 1.

A mass of the silty clay soil was placed in a concrete mixture drum. The volume of water required to increase the water content of the soil to 50% was measured and added to the drum. The soil and water were thoroughly mixed in the drum in order to produce a homogenous soft soil. The water content of the mixture was selected twice the liquid limit in order to produce a soil specimen with properties of reconstituted clay as described by Burland [15] and with virtually no shear strength.

Approximately 25 mm layer of clean gravel, 5–7 mm grain size diameter, was placed at the bottom of the cell as a drainage layer, which was overlain by a geotextile filter (Figure 1(a)). The soft clayey soil was then poured into the cell. The soft soil was allowed to settle and consolidate over its own weight for 48 hr. After settlement and consolidation, the soil specimen was approximately 465 mm high (150 mm shorter than the cell) and was overlain by a layer of water. The electrical conductivity of the soil, σ, was measured using ASTM G57-6 [16] and founded to be 0.1 S/m. The model caisson was then inserted into the cell with the centre of the caisson coinciding with that of the cell. The upper of the model caisson was 50 mm below the soil specimen as shown in Figure 1(a). The model caisson served as the anode during the treatment. Four electrodes serving as the cathode were inserted around the model caisson at equal distance from each other. The electrode was made of perforated steel pipe, 12 mm outside diameter, 8 mm inside diameter and was 25 mm longer than the model caisson. The perforation holes were 3 mm in diameter at spacing of 13 mm centre-to-centre. The tops of the electrode and the model caisson were at the same level while the tip of the electrode was 25 mm below the base of the caisson (Figure 1(a)).

The electric field was simulated for various distances between the anode (model caisson) and the cathode (four

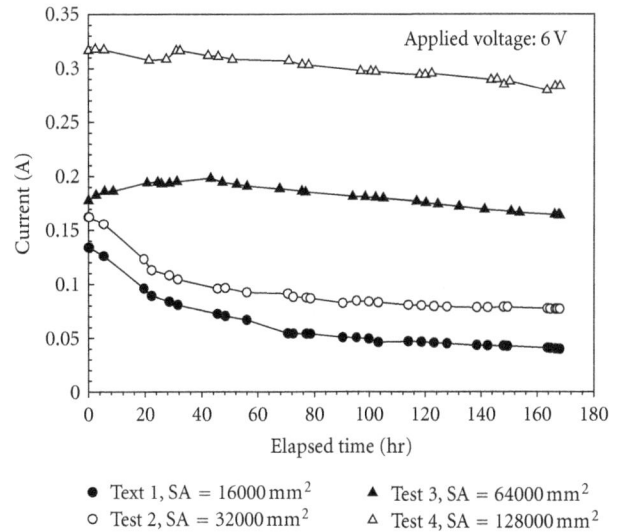

FIGURE 2: Electric current versus elapsed time of the test.

electrodes) by using QuickField [17], a field simulation software. The simulation aimed to provide the maximum electric field, and subsequently the maximum improvement in the strength properties of the soil, in the caisson-soil interface. A distance equal 0.25 times the outside diameter of the model caisson was identified as optimum for electric field between the caisson and cathode. Thus, 13, 19, 25, and 38 mm were the distances between electrode and model caisson in tests 1 to 4, respectively. The plan view of the electrodes layout in test 3 is shown in Figure 1(b). After installing the model caisson and the electrodes, the water above the soil specimen was raised to 100 mm and kept throughout the duration of the treatment and the shear strength and axial load tests.

A direct current (dc) voltage of 6 V was applied to the cell with the model caisson serving as the anode and four electrodes (A1 to A4) serving as the cathode. The electrokinetic treatment lasted for 168 hr with current intermittence intervals of 2 min on and 2 min off executed by a programmable timer. Current intermittence, the application of a pulse voltage at predetermined on/off intervals instead of a continuous dc voltage, was selected for its superior outcome in electrokinetic treatment as well as its effectiveness in reducing corrosion of the electrodes [7, 18, 19]. The electric current was monitored and reported during the treatment. For each electrokinetic treatment test in this study, a control test with identical soil and configuration but without electric field was carried out to provide baseline data for comparison.

3. Results and Discussion

3.1. Electric Field and Energy Consumption. Figure 2 shows the electric current across the tank versus the elapsed time of the test. The figure shows that for the same applied voltage, the electric current increases with the increase in the surface area of the model caisson. This is due to the

(a)

(b)

(c)

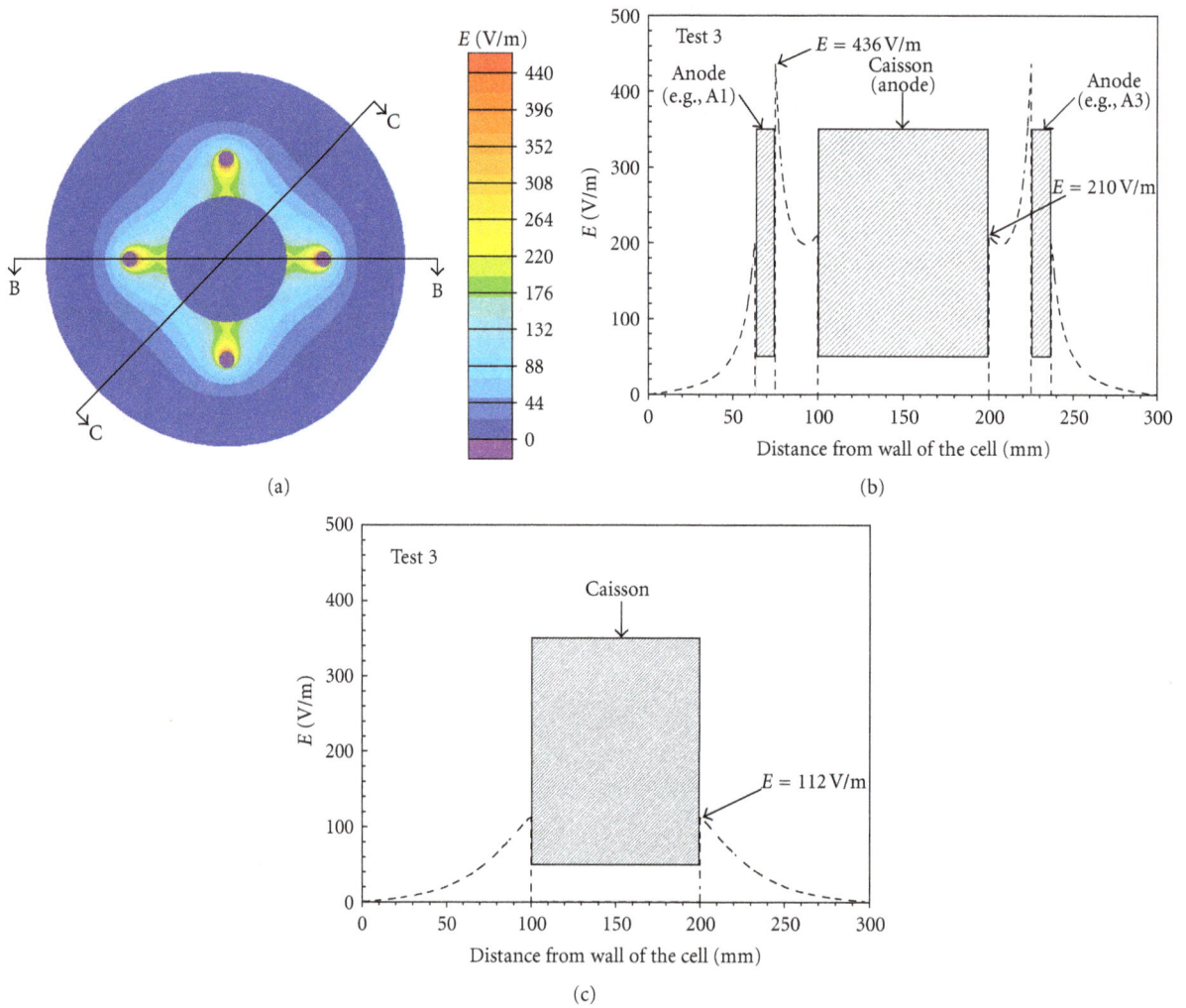

FIGURE 3: (a) Plan view of electric field intensity, E (V/m), distribution in the cell; (b) E along cross-section B-B of the cell; (c) E along cross-section C-C of the cell.

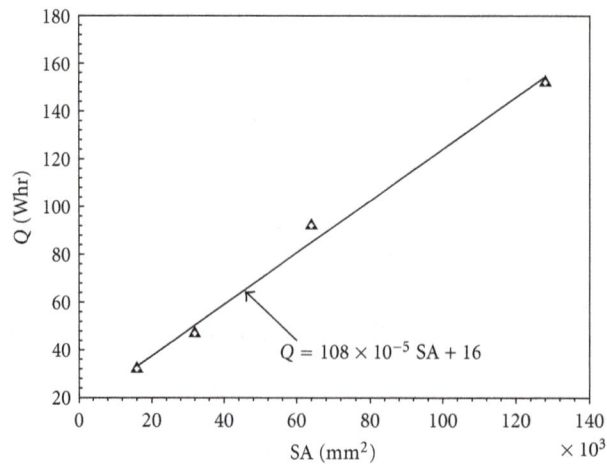

FIGURE 4: Energy consumption, Q, versus surface area, SA, of the model caisson.

TABLE 1: Summary of tests and results.

| Test | Model caisson | | | Distance between electrode & caisson mm | Energy consumption | | P_f | | Displacement at failure | |
	Dia. mm	Length mm	SA mm^2		Control	EK Whr	Control N	EK N	Control mm	EK mm
Test 1	50	102	16000	13	—	32	12	126	0.4	1.6
Test 2	75	136	32000	19	—	47	45	205	1.0	1.2
Test 3	100	204	64000	25	—	92	84	327	0.4	1.1
Test 4	150	272	128000	38	—	152	170	521	0.6	0.8

EK: electrokinetic.

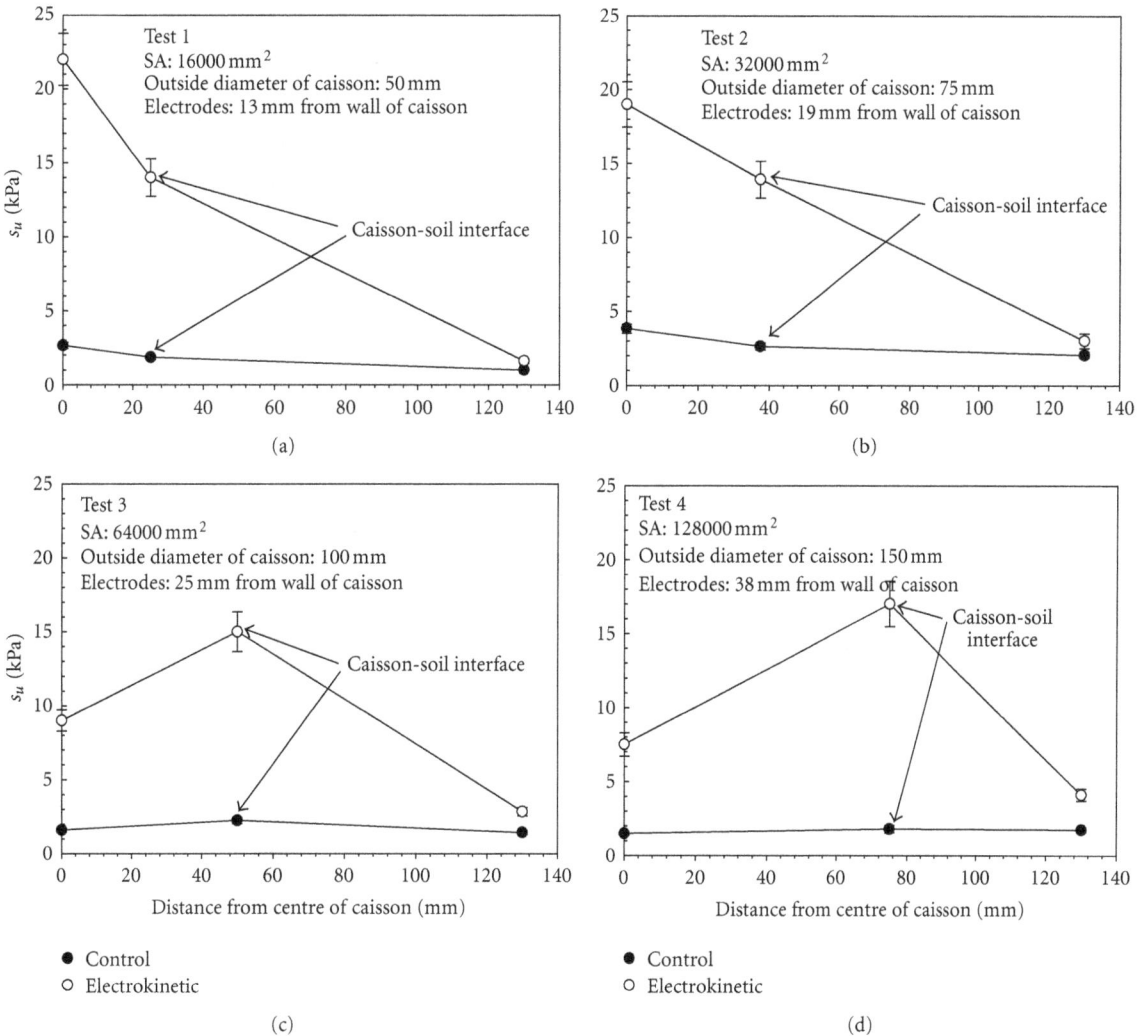

FIGURE 5: Undrained shear strength, s_u, versus distance from the centre of model caisson.

increase in the soil area subject to the electric current as the caisson increases. The electric current values shown in Figure 2 represent the integration of the current density over the soil area subjected to the current. As the surface area of the caisson increases, the area in the soil subjected to the current increases and so does the current.

Figure 2 shows electric current in tests 1 and 2 decreased throughout the test with the sharpest decrease occurring during the first 20 hr. In test 3, the current increased with time during the first 40 hr of the test. After, the current decreased slightly with time up to the end of the test. In test 4, the current slightly deceased during the first 20 hr of the test and then increased. The current then decreased with small rate until the end of the test. The change in electric current with time as shown in Figure 2 resulted from the change in electrical conductivity of the soil during the test. The change in the conductivity of soil during an electrokinetic process is a result of two opposing mechanisms. In general, as the pore fluid drained out of the soil mass (pore fluid dominates the bulk conductivity of the soil) by electroosmosis, the bulk electrical conductivity of the soil decreases. However, for water still remaining inside the soil pores, the electrical conductivity increases with the treatment time as a result of electrolytic reactions associated with the electrokinetic process [20, 21]. Therefore, the increase in the electrical conductivity of the pore fluid by the electrolytic reactions can sometimes become more dominant than the decrease in conductivity of the soil resulting from the draining of water. Thus the bulk conductivity of the soil, and thereby the electric current, may start to increase sometime after the start of the electrokinetic treatment as observed in tests 3 and 4. However, Figure 2 suggests that for all tests, the change in current and thereby the change in electrical conductivity was very small during most of the testing time.

Figure 3(a) shows distribution of the electric field intensity, E, during test 3 simulated using QuickField. The E distribution shown in Figure 3(a) was typical in all the tests. Figures 3(b) and 3(c) show plan views of E across the centre of the cell and two electrodes (section B-B) and across the centre of the cell and midway between electrodes (section C-C) for test 3. As shown in the figures, the highest E, and subsequently the highest current density (current density = $E\sigma$), occurred in the vicinity of the caisson. In test 3, E varied between 112 and 210 V/m (i.e., current density between 11.2 and 21 A/m^2) in the model caisson-soil interface compared to E and current density of zero in the soil near the wall of the cell.

The energy consumption, Q, during electrokinetic treatment was calculated for each test and is shown in Figure 4 versus the surface area (SA) of the model caisson. As shown in the figures, Q increased linearly with SA as

$$Q \text{ (Whr)} = 108 \times 10^{-5} \text{SA (mm}^2) + 16. \tag{1}$$

Thus, for soil with electrical conductivity of 0.1 S/m, applied voltage of 6 V, and electrodes layout similar to the configuration in this study, the energy consumptions per week of treatment can be estimated by (1).

3.2. Undrained Shear Strength. After the completion of the electrokinetic treatment, the undrained shear strength, s_u, was measured in three locations shown in Figure 1(b) (SL1, SL2, and SL3) using a shear vane. At each location s_u was measured at the top, mid, and bottom levels of the model caisson. An average value for s_u was determined from the three measurements for each location and presented in Figure 5. After the treatment, the average s_u varied between 14 ± 1.3 kPa and 17 ± 1.5 kPa in the model caisson-soil interface (SL2) and between 1.6 ± 0.2 kPa and 4.1 ± 0.5 kPa at 130 mm from the centre of the caisson (SL2). The corresponding s_u in the control tests ranged from 1 ± 0.1 kPa to 2.6 ± 0.2 kPa. The relationship between the magnitude of the electric field (stronger in the vicinity of the model caisson and weaker away from the caisson) and gained strength in the soil are illustrated for test 3 by Figures 3(b) and 5. At SL2, 201 V/m $\geq E \geq 112$ V/m and $s_u = 15 \pm 1.3$ kPa whereas

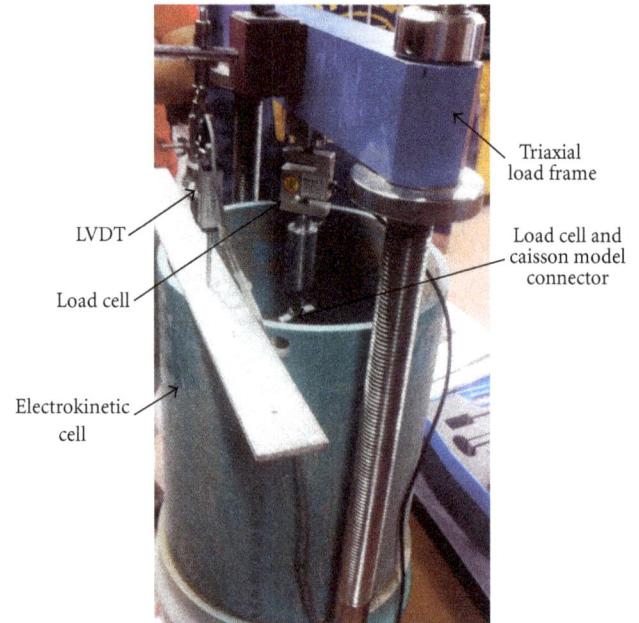

FIGURE 6: Electrokinetic cell during axial load testing for the model caisson.

at SL3, $E \leq 7$ V/m and $s_u \leq 2.8 \pm 0.3$ kPa. Thus focusing the electric field near the caisson significantly increased s_u in the caisson-soil interface as, a primary objective of the optimizing the distance between the electrodes while s_u away from the caisson remained approximately similar to that of the control.

As shown in Figure 5, the treatment also increased the strength of the soil inside the model caisson (SL1) with the highest shear strength reported in tests 1 and 2. The higher strength in tests 1 and 2 was likely due to the smaller size of the model caissons in the two tests. As the size of the model caisson deceased, more water was drained by electroomosis out of the enclosed soil since the caisson was serving as an anode. Increasing shear strength for soil inside a foundation element can generate a soil plug. For caisson foundations, a soil plug adds a toe bearing resistance component and thereby increases the axial load capacity.

3.3. Axial Load Capacity. After measuring s_u, the model caisson was axially loaded to failure by a triaxial load frame as shown in Figure 6. Figure 7 shows the axial load capacity, P, versus the vertical displacement of the model caisson after the electrokinetic treatment and for the control tests. The axial load capacity at failure, P_f, is marked on the figure. P_f was determined by the failure criterion proposed by Tani and Craig [22]. In this failure criterion, the failure was at the point of intersection of the load-displacement curve and the bisector line of the angle made by two tangents on both sides of the sharp bend of the load-displacement curve. As shown in Figure 7, P_f after electrokinetic treatment was 126 N in test 1 (SA = 16000 mm^2), 205 N in test 2 (SA = 32000 mm^2), 327 N in test 3 (SA = 64000 mm^2), and 521 N

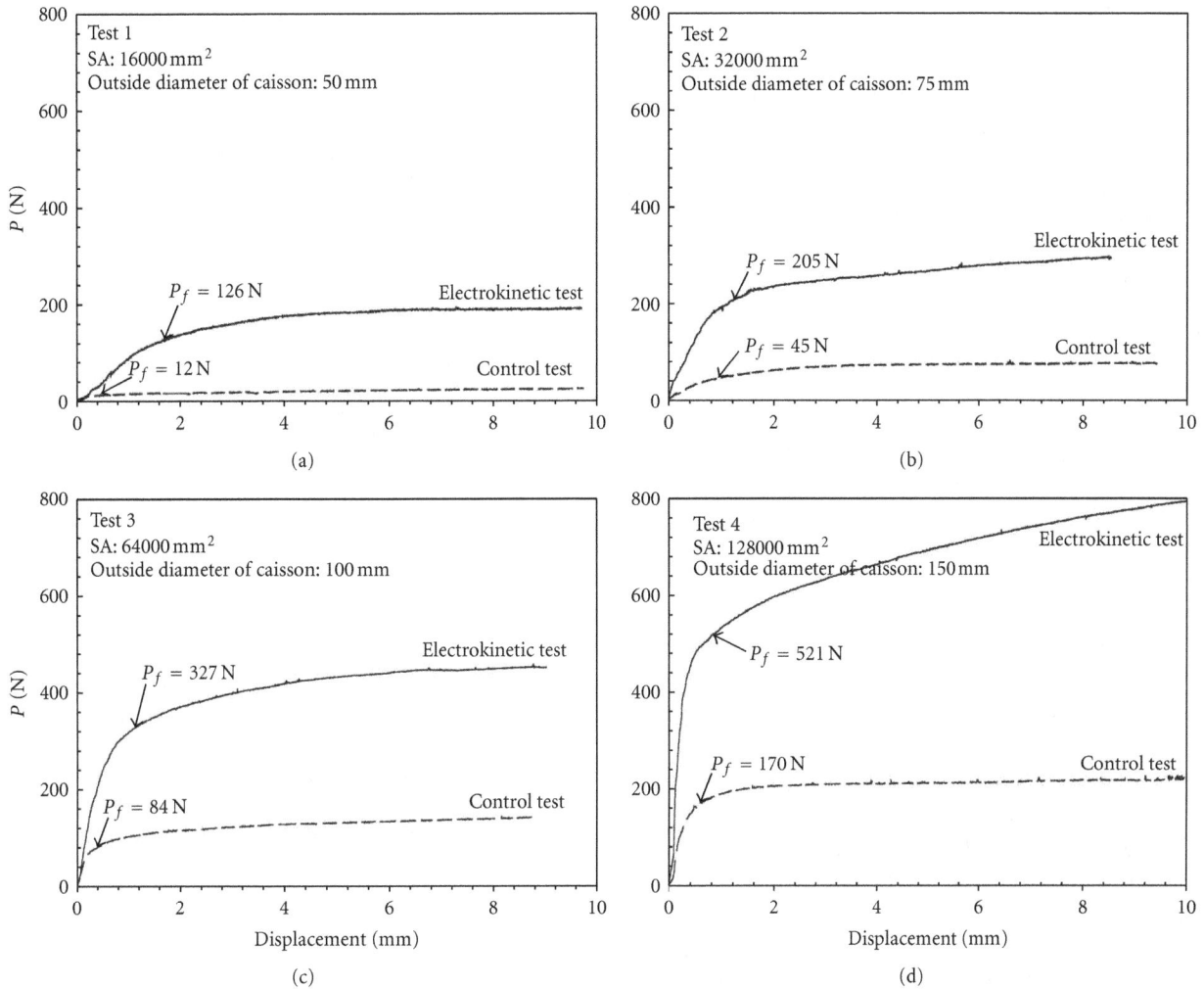

FIGURE 7: Axial load capacity, P, versus vertical displacement of the model caisson.

in test 4 (SA = 128000 mm^2). The corresponding P_f values in the control tests were 12, 45, 84, and 170 N, respectively. This represents an increase between 206 and 950% after electrokinetic treatment as compared to the control.

Figure 8 shows P_f (N) versus the surface area, SA (mm^2) of the model caisson for the four tests. As shown in the figure, P_f varied linearly with SA for the range covered in this study and the relationship is given by:

Control tests:

$$P_f \text{ (N)} = 1373 \times 10^{-6} \text{ SA } (\text{mm}^2) - 4. \qquad (2)$$

Electrokinetic treatment tests:

$$P_f \text{ (N)} = 3458 \times 10^{-6} \text{ SA } (\text{mm}^2) + 87. \qquad (3)$$

A linear relationship between P_f and SA is expected in the control tests as the commonly used formulas for the axial load capacity are linear function between the soil properties and the dimensions of the foundation elements. A linear relationship between P_f and SA after an electrokinetic

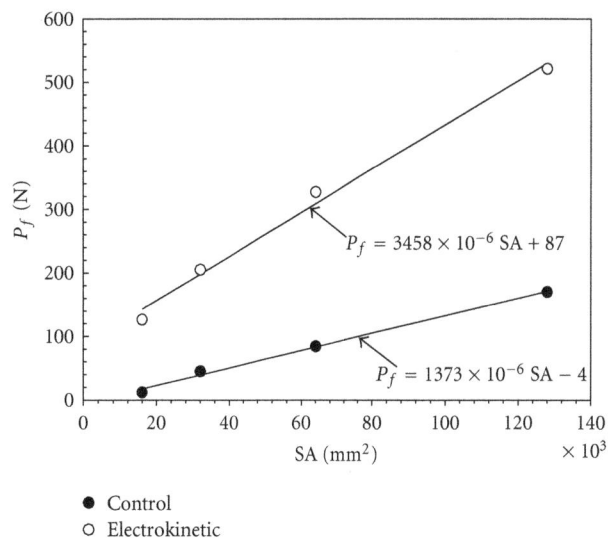

FIGURE 8: Axial load capacity at failure, P_f, versus surface area, SA, of the model caisson.

treatment means that the improvement in the strength properties of the soil after the treatment is independent caisson foundation dimensions. This is very important as it allows a preliminary estimation of the axial capacity for full-scale foundation element from laboratory testing of much smaller models. For example, P_f for a caisson foundation with a diameter of 1 m and a length of 8 m (SA = 25 m^2) inserted in soil with geotechnical properties similar to the silty clay used in this study and under water can be estimated from (2) as 34 kN. However, after an electrokinetic treatment with electric field configuration similar to the configuration in this study, P_f is estimated by (3) as 87 kN and the corresponding energy consumption from (1) is 27 kWhr.

4. Conclusions

This experimental study aimed to investigate some parameters that are critical for devising design guidelines for electrokinetic treatment. Four model caisson with surface areas between 16000 and 128000 mm^2 were embedded in soft clayey soil under water and treated with a dc voltage of 6 V for 168 hr. From the study we can conclude the following.

(i) The axial load capacity of the caissons after the electrokinetic treatment varied between 126 and 521 N compared to 12 to 170 N in the control tests.

(ii) A distance between the anode (model caisson) and the cathode equal 0.25 times the outside diameter of the caisson was identified as optimum.

(iii) A relationship between the surface area and the axial capacity of the caisson was presented. The equation can be used to preliminary estimate the load capacity of a full-scale foundation element after an electrokinetic treatment.

(iv) A correlation between the surface area of the caisson and the energy consumption was presented. The correlation can approximately predict the energy consumption for full-scale applications prior to treatment.

Acknowledgments

There is no financial gain for any of the authors of the paper for using Quickfield in this research. The authors wish to thank Mr. Conrad Hagstrom, Civil Engineering at Lakehead University and LH North General Contractor Limited for their valuable help during this research. The research is funded by Natural Science and Engineering Research Council of Canada (NSERC).

References

[1] M. I. Esrig, "Pore pressure, consolidation and electrokinetics," *Journal of the Soil Mechanics and Foundations Division*, vol. 94, no. 4, pp. 899–921, 1968.

[2] S. Micic, J. Q. Shang, and K. Y. Lo, "Electrocementation of a marine clay induced by electrokinetics," *International Journal of Offshore and Polar Engineering*, vol. 13, no. 4, pp. 308–315, 2003.

[3] J. Q. Shang, "Electrokinetic dewatering of clay slurries as engineered soil covers," *Canadian Geotechnical Journal*, vol. 34, no. 1, pp. 78–86, 1997.

[4] L. Casagrande, "Electroosmosis in soil," *Géotechnique*, vol. 1, no. 3, pp. 159–177, 1949.

[5] L. Casagrande, "Electro-osmosis stabilization of soils," *Journal of the Boston Society of Civil Engineers*, vol. 39, no. 1, pp. 51–83, 1983.

[6] J. Q. Shang and K. S. Ho, "Electro-osmotic consolidation behavior of two Ontario clays," *Geotechnical Engineering*, vol. 29, no. 2, pp. 181–194, 1998.

[7] S. Micic, J. Q. Shang, K. Y. Lo, Y. N. Lee, and S. W. Lee, "Electrokinetic strengthening of a marine sediment using intermittent current," *Canadian Geotechnical Journal*, vol. 38, no. 2, pp. 287–302, 2001.

[8] L. Bjerrum, J. Moun, and O. Eide, "Application of electroosmosis to a foundation problem in a Norwegian quick clay," *Géotechnique*, vol. 17, no. 3, pp. 214–235, 1967.

[9] K. Y. Lo, K. S. Ho, and I. I. Inculet, "Field test of electroosmotic strengthening of soft sensitive clay," *Canadian Geotechnical Journal*, vol. 28, no. 1, pp. 74–83, 1991.

[10] F. Burnotte, G. Lefebvre, and G. Grondin, "A case record of electroosmotic consolidation of soft clay with improved soil-electrode contact," *Canadian Geotechnical Journal*, vol. 41, no. 6, pp. 1038–1053, 2004.

[11] V. Milligan, "First application of electro-osmosis to improve friction pile capacity—three decades later," *Proceedings of the ICE: Geotechnical Engineering*, vol. 113, no. 2, pp. 112–116, 1995.

[12] W. Perry, "Electro-osmosis dewaters large foundation excavation," *Construction Methods and Equipment*, vol. 45, no. 9, pp. 116–119, 1963.

[13] ASTM, "Standard test method for particle-size Analysis of soils," Annual Book of ASTM Standards—Construction D422-63, ASTM, West Conshohocken, Pa, USA, 2007.

[14] ASTM, "Standard test methods for liquid limit, plastic limit, and plasticity index of soils," Annual Book of ASTM Standards—Construction D4318-10, ASTM, West Conshohocken, Pa, USA, 2007.

[15] J. B. Burland, "On the compressibility and shear strength of natural clays," *Géotechnique*, vol. 40, no. 3, pp. 329–378, 1990.

[16] ASTM, "Standard method for field measurement of soil resistivity using the wenner-four electrode method," Annual Book of ASTM Standards—Construction G57-06, ASTM, West Conshohocken, Pa, USA, 2006.

[17] Tera Analysis Ltd, "QuickField," Knasterhovvej 21, DK-5700, Svendborg, Denmark, 2011.

[18] R. H. Sprute and D. J. Kelsh, *Limited Field Tests in Electrokinetic Densification of Mill Tailings*, vol. 8034, U.S. Department of Interior, United States Bureau of Mines Report of Investigations, Washington, DC, USA, 1975.

[19] E. Mohamedelhassan and J. Q. Shang, "Effects of electrode materials and current intermittence in electro-osmosis," *Ground Improvement*, vol. 5, no. 1, pp. 3–11, 2001.

[20] B. Narasimhan and R. Sri Ranjan, "Electrokinetic barrier to prevent subsurface contaminant migration: theoretical model development and validation," *Journal of Contaminant Hydrology*, vol. 42, no. 1, pp. 1–17, 2000.

[21] E. Mohamedelhassan and J. Q. Shang, "Electrokinetics-generated pore fluid and ionic transport in an offshore

calcareous soil," *Canadian Geotechnical Journal*, vol. 40, no. 6, pp. 1185–1199, 2003.

[22] K. Tani and W. H. Craig, "Bearing capacity of circular foundations on soft clay of strength increasing with depth," *Soils and Foundations*, vol. 35, no. 4, pp. 21–35, 1995.

17

Implementation of a Probabilistic Structural Health Monitoring Method on a Highway Bridge

Adam Scianna, Zhaoshuo Jiang, Richard Christenson, and John DeWolf

Department of Civil and Environmental Engineering, University of Connecticut, Storrs, Connecticut 06269-2037, USA

Correspondence should be addressed to Richard Christenson, rchriste@engr.uconn.edu

Academic Editor: Piervincenzo Rizzo

This paper describes the application of a probabilistic structural health monitoring (SHM) method to detect global damage in a highway bridge in Connecticut. The proposed method accounts for the variability associated with environmental and operational conditions. The bridge is a curved three-span steel dual-box girder bridge located in Hartford, Connecticut. The bridge, monitored since Fall 2001, experienced a period of settling in the Winter of 2002-2003. While this change was not associated with structural damage, it was observed in a permanent rotation of the bridge superstructure. Three damage measures are identified in this study: the value of fundamental natural frequency determined from peak picking of autospectral density functions of the bridge acceleration measurements; the magnitude of the peak acceleration measured during a truck crossing; the magnitude of the tilt measured at 10-minute intervals. These damage measures, including thermal effects, are shown to be random variables and associated P values are calculated to determine if the current probability distributions are the same as the distributions of the baseline bridge data from 2001. Historical data measured during the settling of the bridge is used to verify the performance of the bridge, and the field implementation of the proposed method is described.

1. Introduction

Structural health monitoring (SHM) is a general term used in many engineering disciplines that describes a process to determine the integrity of a structure. There have been numerous techniques developed to ascertain the integrity of a structure including methods of visualization, nondestructive evaluation (NDE) as well as indentifying changes in vibration characteristics. All of these methods have a common goal to detect, locate, and quantify varying degrees of deterioration and damage within a structure.

Visual inspection and NDE are common methods of health monitoring. Visual inspection, however, is neither objective nor reliable, containing great uncertainty in identifying the existence, location, and degree of damage [1–3]. Additionally, visual inspection supplies information only at the time of inspection. Realistic SHM should supply information on a continuous basis if it is to be of real use. New technologies for the NDE of civil structures include ultrasonic testing, penetrant testing, visual testing (different than visual inspec-

tion), magnetic particle testing, radiographic testing, acoustic emission, and eddy current testing [4, 5]. These technologies determine local damage and typically require the general region of damage on the structure to be first identified. For many structures, locating the damage prior to inspection is difficult.

Using vibration measurements is a less subjective method for structural health monitoring that can allow for global as well as local evaluation. As such, vibration-based techniques are receiving much recent attention [6–14]. Employing methods that rely on ambient vibration or operational loads are challenging as these excitations tend to excite the lower frequency global modes that can be insensitive to local damage [15]. Furthermore, environmental and operational variability of civil structures can affect the natural frequencies and mode shapes, rendering SHM methods that rely on changes in these parameters ineffective except in the presence of extreme damage [16]. Despite these challenges, using vibration measurements for SHM has continued to receive the attention of researchers and is the focus of this study.

Bridge health monitoring (BHM), a type of SHM applied specifically to bridges, has been applied in Connecticut since 1984, when the University of Connecticut and the Connecticut Department of Transportation began evaluating the structural conditions of bridges in the state [17–19]. Since that time there have been various temporary sensor installations that have helped determine in-service behavior and justify rehabilitation and repair plans [19]. The portable monitoring system applications have provided benefits in the form of fewer or no repairs totaling over $2.5 M and have identified necessary repairs increasing safety. A long-term continuous bridge monitoring program has been in place since 1997, when permanent monitoring systems were installed on a variety of bridge types across Connecticut. The objective of the continuous monitoring is to identify global changes in a bridge's behavior over multiple years [17]. The bridge examined in this study is part of the long-term continuous BHM program in Connecticut.

Prior work conducted at the University of Connecticut studied the effects of damage on the dynamic properties of a bridge and the effects of temperature variation on these measurements. This previous work indicates that monitoring the bridge's modal information as well as peak values can provide a global measure of the bridge's structural integrity and identify major changes in the structural integrity [20]. In addition, tiltmeters are employed to measure the general orientation of the bridge as a measure of its health in terms of structural changes. The effect of temperature on modal information and tilt measurements have been identified and quantified in Liu and DeWolf [21] and Olund [22].

While the tools for statistical analysis are numerous and well established, there has been limited research applying these methods to BHM [12]. A statistical pattern recognition paradigm was proposed by Rytter [23]. More recently, Zhang [24] showed that by selecting a damage feature, characterized as a random variable with a normal distribution, a probability of damage could be determined by comparing a feature of an unknown condition to that of a known condition. Olund and DeWolf [20] established the basis for BHM in Connecticut showing a statistical approach that can be very robust with respect to environmental and operational variability.

This paper extends the statistical BHM approach of Olund and DeWolf [20] to examine the vibration characteristics and tilt of an actual highway bridge in Connecticut verifying the proposed approach using historical data collected over a period that included a permanent rotation in the bridge deck. This paper goes on to describe the implementation of a fully automated BHM system on an in-service highway bridge in Connecticut.

2. Bridge and Monitoring System Description

The bridge selected for this study, part of the long-term continuous BHM program in Connecticut, is referred to as the Hartford Flyover Bridge. This bridge connects I-84 east to I-91 north in Hartford, CT, USA. Figure 1 shows an aerial view of the Hartford Flyover Bridge.

The structure is comprised of two steel box girders with a composite concrete deck. The bridge in its entirety is nine

FIGURE 1: Aerial view of the Hartford Flyover Bridge.

spans consisting of three sets of three continuous spans which are simply supported. Previous studies have identified the fundamental natural frequency of the continuous bridge span being monitored being equal to approximately 1.52 Hz [22]. The monitoring system's 22 sensors include 8 accelerometers, 8 temperature transducers and 6 tiltmeters. Figure 2 shows the location of the various sensors on the bridge. All sensors are located in or on the box girder of the middle continuous three span segment. This is the portion of the bridge with the longer columns. The sensors were distributed to capture the behavior of this segment of the bridge in order to determine the cause of cracking in the supporting columns observed during the biannual visual inspection. The tiltmeters and accelerometers are located along the length of the bridge while all the temperature transducers are located at the midspan cross-section of span 4.

The accelerometers are PCB Piezotronics Model 393C quartz accelerometers with a measurement range of $\pm 2.5\,g$ and a frequency range of 0.025–800 Hz. Six of the accelerometers measure vertical accelerations and two measure horizontal accelerations. These are labeled as AV# and AH#, respectively, in Figure 2. The acceleration measurements are dynamic measurements collected at a sampling rate of 90.91 Hz for a duration of 30 seconds when a truck travels over the bridge and triggers sensor AV2 to exceed the prescribed 0.0095 g threshold. The data acquisition system used to collect data during 2002-2003 period employs a 2-pole analog low-pass filter with a cut-off frequency of 2 Hz. As such, the monitoring system is able to capture the frequency content of only the fundamental mode without distortion. The current upgraded system provides a bandwidth up to 500 Hz. A comparison of the autospectral density function measured from AV1 measured with the original and upgraded systems in shown in Figure 3. Both measurements show the fundamental natural frequency at 1.52 Hz as well as higher frequencies above 2 Hz. It is observed that higher modes are excited by the crossing truck traffic and higher modes can be used in current studies.

Tiltmeters 1, 2, 4, 5, and 6, applied geomechanics model 801-S ± 3 degrees, are located at the piers while tiltmeter 3 is located at midspan of Span 4, collocated with the array of

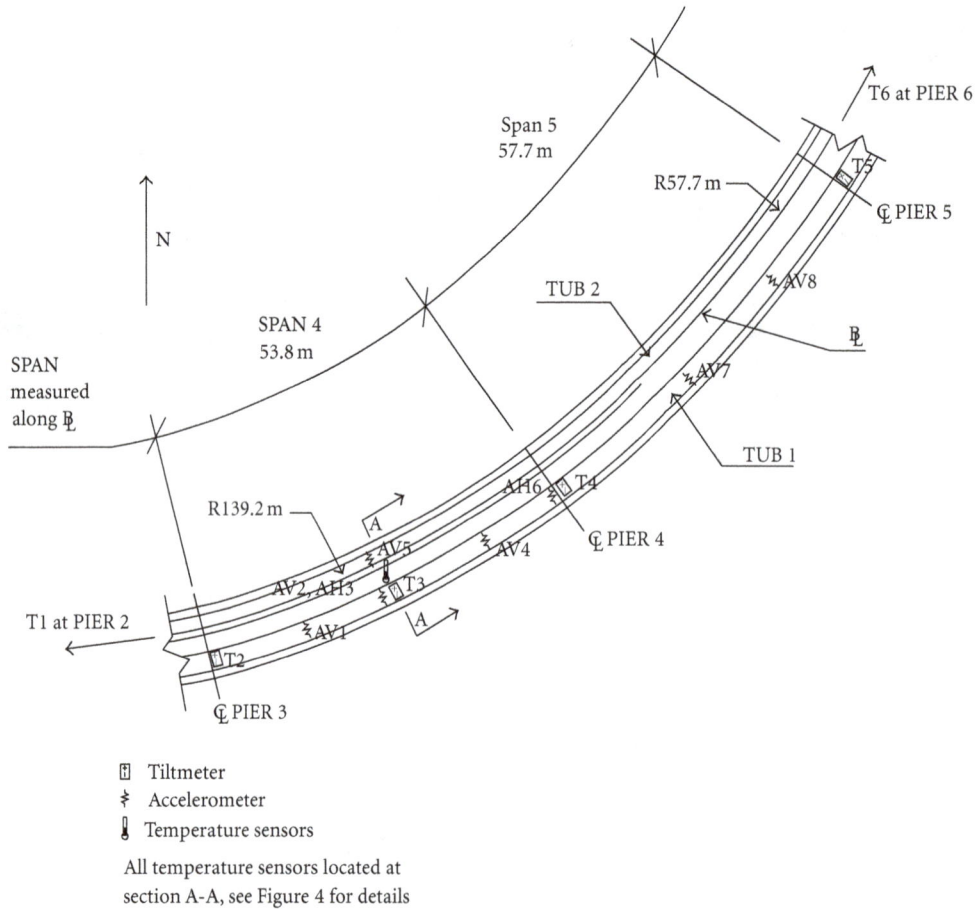

FIGURE 2: Sensor locations for the Hartford Flyover Bridge.

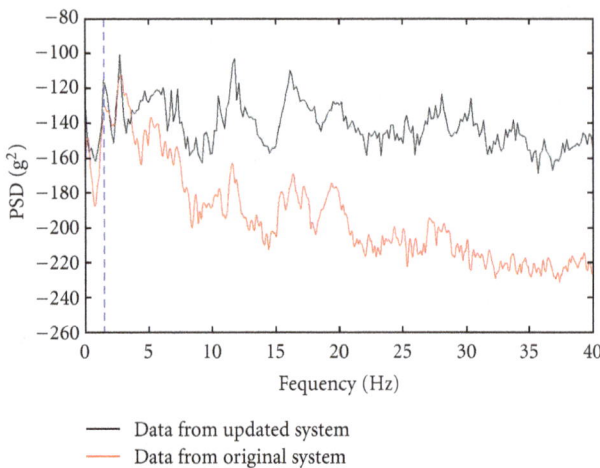

FIGURE 3: Autospectral density function of AV1 (vertical dashed line at fundamental natural frequency, 1.52 Hz).

temperature sensors. Tilt measurements are collected at regular ten-minute intervals.

The temperature transducers are RdF model 29258 RTD capsules for measuring ambient and inside deck temperatures. RdF surface flange mount model 22802 are used for measuring surface temperatures. The transducers provide

temperature measurements with a 0.041°C (0.074°F) resolution. Of the eight temperature transducers, two measure the concrete deck temperature, four measure the temperature in the steel tub corners, and the other two record ambient temperature both inside and outside the tub. These transducers were originally placed in this configuration to study the effects of a temperature differential on tilt in the cross-section of a steel girder bridge with a composite concrete deck [25]. The temperature detector locations are identified in Figure 4. Temperature measurements are also collected at regular ten minute intervals corresponding to the tilt measurements. The eight temperature measurements are closely correlated and as such only the measurement of the first temperature sensor, R1, is employed in this study.

3. Probabilistic Structural Health Monitoring Method

To address the challenges of environmental and operational variability, including unknown traffic excitation, inherent variability of the excitation, measurement error, and uncertainty in the structure itself, a probabilistic framework is adopted. In this probabilistic method damage measures (DMs) corresponding to the structural health of the bridge are determined from the structure's dynamic response and

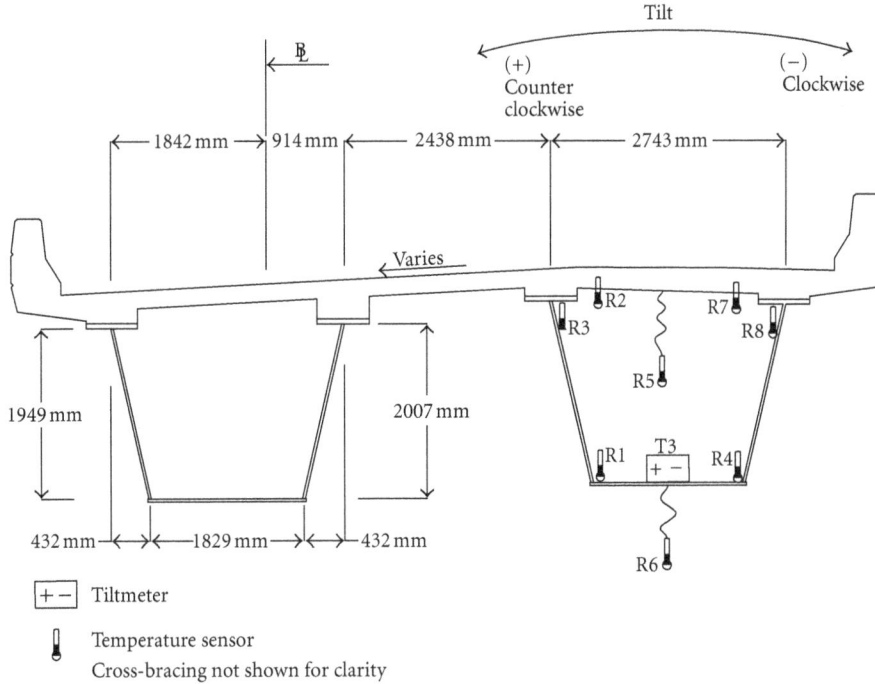

FIGURE 4: Cross-section of the Hartford Flyover Bridge with temperature sensor layout.

rotation of the bridge deck. The basis for this method is that a baseline damage measure (DM_B) can first be determined. At subsequent times, the current damage measure (DM_C) can be determined and compared to the baseline case. For this study the DMs used are: (i) the estimated fundamental natural frequency of the bridge, calculated as the first peak of the power spectral density function (PSD) of the acceleration; (ii) the peak acceleration resulting from a truck crossing over the bridge; (iii) the 10-minute tilt measurements of the bridge. Although almost any damage measure proposed in the literature can be used, these three benchmark parameters were chosen because of prior experience in BHM in Connecticut [18].

The DMs, calculated from measured responses for varying traffic loading, are observed from the approximate straight lines in the normal probability plots to be random variables with a Gaussian distribution. As random variables, not just one realization of the baseline DM is calculated, but a set of n baseline DMs are determined. The mean and variance of an original set of data for the baseline structure is determined so that the probability density function of DM_B is known. At each current time, n new DMs are calculated and the mean and variance are determined as before so that the probability density function of DM_C is now known. The basis for this method is then to compare the distribution of a current DM to the baseline DM to determine if there is a change in the underlying distribution, thus indicating a potential change in the structure and possible damage.

3.1. Statistical Test. In previous analytical work, a normal difference method was used to compare the distribution of DM_C to the DM_B [26]. This was appropriate because the

number of samples in each distribution was set equal to each other. For the actual implementation, the sample sizes of each distribution are likely to be different due to the actual varying truck volumes over specific time periods. It is also assumed that the variance of the two samples is unequal. With these considerations, Welch's t-test is used to compare distributions [27]. Welch's t-test defines the statistic t as:

$$t = \frac{\overline{X_H} - \overline{X_C}}{s_{\overline{X_H} - \overline{X_C}}}, \tag{1}$$

where

$$s_{\overline{X_H} - \overline{X_C}} = \sqrt{\frac{s_H^2}{n_H} + \frac{s_C^2}{n_C}} \tag{2}$$

$\overline{X_H}$ and $\overline{X_C}$, s_H^2 and s_C^2, and n_H and n_C are the mean, variance, and sample size, respectively, of the healthy and current distributions. The degrees of freedom v is approximated using the Welch-Satterthwaite equation:

$$v = \frac{\left((s_H^2/n_H) + (s_C^2/n_C)\right)^2}{(s_H^4/N_H^2(N_H - 1)) + (s_C^4/N_c^2(N_c - 1))}. \tag{3}$$

The t-distribution can be used to test the null hypothesis that the two DM means are equal at a certain significance level. If the current t-statistic is smaller in absolute value than the critical t-value for a particular level of significance, the null hypothesis is rejected; this indicates a statistically significant change in the mean and potential for damage in the structure. If the current t-statistic is greater in absolute value than the critical t-value for a particular level of significance, the null hypothesis cannot be rejected; this indicates no statistically significant change in the mean and that the structural integrity of the current scenario should be assumed the same as the structural integrity of the baseline scenario.

To provide further insight into the statistical test the P values instead of a general "pass" or "fail" result are observed. The P value is the probability of observing an event at least as extreme as the one actually observed, given that the null hypothesis is true. The P value for a two-tailed Welch's t-test can be calculated as:

$$P = 2[1 - \Phi(t)]. \tag{4}$$

Generally one rejects the null hypothesis when the P value is smaller than or equal to the significance level.

3.2. Thermal Effects. To account for thermal effects on the damage measures, each DM is paired with an associated temperature. Since tilt and temperature data are collected simultaneously, this data is already paired. The acceleration data is paired with the temperature record that has a time-stamp closest to that of the acceleration data. The paired data sets are distributed into temperature bins. For tilt data, the temperature bins have a range of $0.056°C$ $(0.1°F)$ and for the acceleration data the temperature bins have a range of $0.56°C$ $(1.0°F)$. A smaller bin is used for tilt data because it is more dependent on temperature than the acceleration-based DMs. When the probabilistic SHM method is applied the distribution of the DM from each temperature bin of the baseline year is compared to that of the same temperature bin for the current year. A temperature dependant P value, $P_{T_1 < T < T_2}$, is determined. The associated temperature dependant P values are averaged over all of the temperature bins such that:

$$\overline{P} = \frac{1}{N} \sum_{n=1}^{N} P_{\Delta T \cdot (n-1) < T < \Delta T \cdot n}, \tag{5}$$

where $\overline{P}$ is the average P value for all temperatures, N is the number of temperature bins, and ΔT is the size of the temperature bin. The $\overline{P}$ value is then compared to a prescribed significance level to determine if the null hypothesis, that the means are equal, can be rejected.

4. Verification of the Probabilistic SHM Method Using Historical Data

To verify the proposed method, actual bridge data is used. The baseline data from September 2001–August 2002, when monitoring of this bridge began, is compared to the most recent 10-month data set, from September 2003–August 2004. It should be noted that during each of these years data was only collected 10 months out of each year due to ongoing system maintenance. While bridge inspections over this three-year time period verify that there is no structural damage to the bridge, the bridge did experience a permanent rotation between September 2002 and August 2003. Additionally, an accelerometer sensor was observed to not function properly during this period. Both events provided the opportunity for verifying the ability of the SHM system to detect a change in

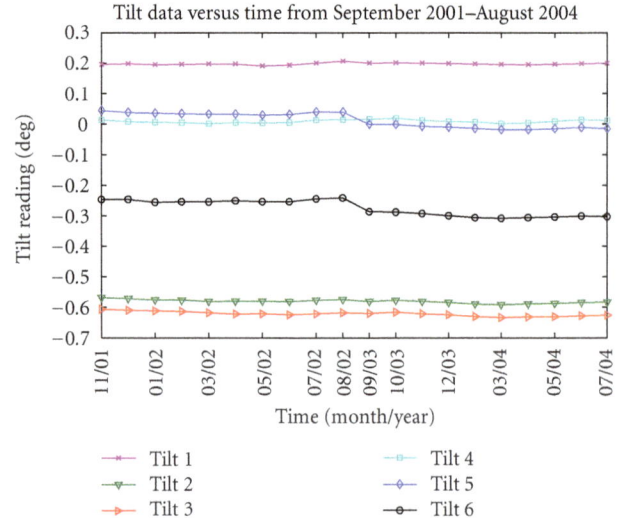

FIGURE 5: Tilt data versus time from September 2001 to August 2004 for Hartford Flyover Bridge.

the structural behavior of the system. This study considers the verification of the SHM method using actual measured data that happens to include a permanent tilt in the bridge structure. While further analytical studies including a high fidelity finite element model and a suite of likely damage scenarios would be necessary to verify the reliability and robustness of the proposed probabilistic approach for other damage conditions, the verification provided in this study is simply intended to illustrate, using actual data available on the bridge, the ability of the method to identify a major change in the structure.

In a previous study [22], the Hartford Flyover Bridge was observed to have undergone a structural change noted as a permanent rotation in two of the tiltmeters occurring between September 2002 and August 2003. This historical data was manually saved prior to and after this period. A plot of average monthly tilt versus time for all tilt sensors is shown in Figure 5. From this figure it can be seen that over the course of a year the average monthly tilt measurements vary with temperature between the seasons however between September 2002 and August 2003 a permanent rotation occurred in tiltmeters 5 and 6 that never recovered. Since there was no other observable permanent rotation in tiltmeters it is postulated that the movement in meters 5 and 6 is due to a movement in the structure and not drift in the sensor [22]. In this section the three DMs are identified and P values determined comparing the baseline and current DM distributions.

4.1. Natural Frequency Damage Measure. The premise of checking the natural frequency is that a change in the stiffness of the structure, presumably due to structural damage, will result in a change in the natural frequency. Checking the first, or fundamental, natural frequency will provide a global measure as to the structural health of the structure. To estimate the fundamental natural frequency of the bridge the power spectral density functions of the eight acceleration

TABLE 1: P values for comparison of natural frequency distributions.

	AV1	AV2	AH3	AV4	AV5	AH6	AV7	AV8
P values	0.4098	0.3904	0.3863	0.4019	0.3695	0.2531	0.3312	0.3262

TABLE 2: P values for comparison of peak acceleration distributions.

	AV1	AV2	AH3	AV4	AV5	AH6	AV7	AV8
P values	0.3734	0.3971	0.2392	0.3680	0.3916	0.0267	0.4328	0.4398

measurements are determined, where $G_{yy}(f)$ is the one-sided autospectra of the measured absolute acceleration [28]. The spectral density functions are defined as:

$$G_{yy}(f) = 2 \lim_{T \to \infty} \frac{1}{T} E\left[|Y_k(f,T)|^2 \right], \qquad (6)$$

where $Y_k(f,T)$ is calculated using the finite Fourier transform (FFT) of the measured output, f is frequency, T is the record length, and $E[\cdot]$ is the expected value operator over the ensemble index k in question. An FFT is performed on each record of collected acceleration data. Using these functions, the first natural frequency of the bridge can be estimated as the corresponding frequency, f, of the maximum value of the real-valued autospectral density function $G_{yy}(f)$ while searching over a range of $\pm 10\%$ of the measured baseline natural frequency.

The natural frequencies calculated during September 2001 to August 2002 represent the baseline scenario. The natural frequencies calculated during September 2003 to August 2004 represent the current scenario. Table 1 shows the P values as calculated from (5) for each accelerometer collecting data when Welch's t-test is applied at 5% significance with a null hypothesis that the mean natural frequency from each year is the same.

All P values exceed 5% (0.05) significance so the observation is consistent with the null hypothesis—that the means of the natural frequencies are the same for the current period as for the baseline period, and thus the structure is healthy.

4.2. Peak Acceleration Damage Measure.

The peak acceleration provides a local measure of the response and is expected to change in the presence of structural damage. This time domain measurement is able to provide meaningful results even in the presence of potential nonlinear behavior. The peak acceleration was calculated for each triggered truck crossing during September 2001 to August 2002 to represent the baseline scenario. The peak accelerations calculated during September 2003 to August 2004 represent the current scenario. Table 2 shows the P values as calculated from (5) for each accelerometer collecting data when Welch's t-test is apply at 5% significance with a null hypothesis that the mean peak acceleration from each year is the same.

Table 2 shows that with exception to AH6, the null hypothesis cannot be rejected and it is likely that the structure is still healthy. Sensor AH6 is observed during the current period to have a permanent offset. The sensor was removed and recalibrated to remove the offset. Thus the SHM method identified here a change in the sensor, not a change in the

TABLE 3: P values for comparison of tilt distributions.

	Tilt 1	Tilt 2	Tilt 3	Tilt 4	Tilt 5	Tilt 6
P values	0.1491	0.0084	0.0431	0.3096	0.0000	0.0000

structural system. This example illustrates how any SHM method should not be considered to replace regular bridge inspections but should supplement these inspections and indicate when there is a significant statistical event to warrant further investigation.

4.3. Tilt Damage Measure.

As Virkler [25] pointed out, the original purpose of using tiltmeters on the Hartford Flyover Bridge was to see if excessive tilting of the superstructure was the cause of the noticed column cracking. Thus, five of the tiltmeters were placed at piers and one at the midspan of Span 4. Olund [22] showed that during the winter of 2002-2003 there was a recorded permanent tilt of the bridge. The proposed probabilistic SHM method is applied to the tilt data. Applying Welch's t-test at 5% significance to the tilt distributions for the distributions of the baseline and current tilt DMs gives the following P values in Table 3.

Table 3 shows the P values for tiltmeters 5 and 6 are zero which verifies Olund's [22] conclusion. Additionally, the P values for tiltmeters 2 and 3 indicate there may have been more extensive, yet slight, tilt occurring over the length of the bridge whereby Span 4 in addition to the previously identified Span 6 may have undergone permanent drift. This result shows that using a probabilistic SHM method can show statistical significance in the difference of two values when a person physically observing the data might not be able to see this change.

5. Field Implementation of the SHM Method

The verification of the probabilistic SHM method was verified using actual measured data during a period where permanent tilt in the bridge was observed. The data used in this verification is from 2001–2004 was collected by a bridge monitoring system that automatically triggered and saved raw data and required the user to manually archive and process this data. A viable long term implementation of an SHM method on an actual highway bridge will require less manual effort. A fully automated system is being deployed on the Hartford Flyover Bridge that will trigger, process, save locally, and regularly archive data. The proposed SHM method is implemented at the bridge on a PC located in a cabinet

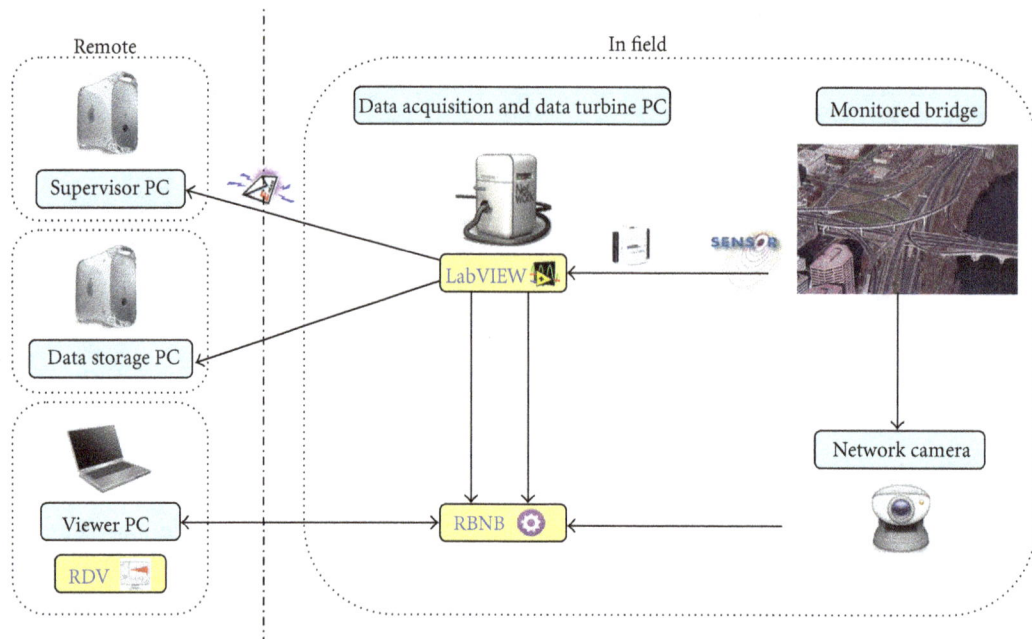

FIGURE 6: Automated bridge monitoring system architecture.

under the bridge with a Microsoft Windows Platform, using LabVIEW Software and National Instruments hardware for data acquisition. The system components and architecture are based on those proposed by Jiang [29]. A schematic of the system architecture can be seen in Figure 6.

Sensor measurements are collected by the National Instruments data acquisition system connected to the onsite PC. Once sampled and digitized, the measurements are processed in LabVIEW and data is collected and archived for later retrieval for SHM. In addition to automated processing and archiving to an FTP site, real-time viewing of streaming data and video is available.

Data acquisition refers to the sampling and analog to digital (A/D) conversion of the sensor measurements. The data acquisition system is a 32-channel analog input, 16-bit, 1.25 MS/s single-channel system, compatible with LabVIEW. Acceleration, tilt, and temperature measurements are acquired at different intervals and sampling rates. The tilt and temperature data are continuously collected at a sampling rate of 10 minutes (a single measurement is taken once every 10 minutes). Acceleration data is acquired by the system when triggered by an acceleration measurement of AV2 exceeding the predefined threshold of 0.0095 g. This trigger level corresponds to the crossing of a truck, which ensures sufficient excitation of the bridge. Once triggered, a record is saved with a sampling frequency of 100 Hz, starting 14 seconds prior to the triggering event, and lasting for a duration of 30 seconds.

The digitized data is processed in LabVIEW according to the probabilistic SHM method presented in this paper. If an anomaly in the data occurs during the SHM method, an email is generated stating the anomaly in the bridge behavior which is sent to another remote PC for a supervisor to investigate. Both raw and processed data are sent via FTP automatically to a remote PC at the Connecticut Department of Transportation for data storage.

Data is also streamed from the onsite PC to a data turbine program in real-time. Data Turbine is a high performance time-synchronized data streaming service [30]. Data Turbine allows data captured from a network video camera and the data acquisition system to be synchronized and stored in a buffer for viewing. This allows remote users to view the data and video in either real-time or in playback mode through the buffered data in Data Turbine. Data Viewing is accomplished through a Java-based software called the Real-Time Data Viewer (RDV) which was originally developed by The Network for Earthquake Engineering Cyberinfrastructure Center (NEESit) now NEEShub (http://nees.org/).

6. Conclusion

This paper uses an in-service highway bridge to verify a probabilistic SHM method. The bridge selected for this study is a curved steel box-girder with composite concrete deck located in Hartford, CT, USA. The bridge is being monitored as part of the University of Connecticut's long-term bridge monitoring project. The bridge is outfitted with accelerometers, temperature sensors, and tiltmeters. The damage measures studied are the fundamental natural frequency, peak accelerations, and tilt. The probabilistic SHM method uses Welch's t-test to compare damage measure distributions. The approach verifies the permanent rotation first observed by Olund [22] and that there is no structural damaged as observed by regular bridge inspections. It has also been shown through the error in accelerometer AH6 that this probabilistic BHM method is meant to complement regular bridge inspections and not to replace them. An automated bridge monitoring system is described with the probabilistic structural health monitoring method implemented on an actual highway bridge.

7. Acknowledgments

University of Connecticut, the Connecticut Transportation Institute, and the Connecticut Department of Transportation are acknowledged for their joint participation in the ongoing bridge monitoring project funded by the Connecticut Department of Transportation and the Federal Highway Administration. This paper, prepared in cooperation with the Connecticut Department of Transportation and the Federal Highway Administration, does not constitute a standard, specification, or regulation. The contents of this paper reflect the views of the authors who are responsible for the facts and the accuracy of the data presented herein. The contents do not necessarily reflect the views of the Connecticut Department of Transportation or the Federal Highway Administration.

References

[1] S. W. Doebling, C. R. Farrar, M. B. Prime, and D. W. Shevitz, "Damage identification and health monitoring of structural and mechanical systems from changes in their vibration characteristics: a literature review," Report LA-I3070-MS, Los Alamos National Laboratory, Los Alamos, NM, USA, 1996.

[2] S. W. Doebling, F. M. Hemez, L. D. Peterson, and C. Farhat, "Improved damage location accuracy using strain energy-based mode selection criteria," AIAA Journal, vol. 35, no. 4, pp. 693–699, 1997.

[3] S. W. Doebling, C. R. Farrar, M. B. Prime, and D. W. Shevitz, "A review of damage identification methods that examine changes in dynamic properties," Shock and Vibration Digest, vol. 30, no. 2, pp. 91–105, 1998.

[4] F. K. Chang, Structural Health Monitoring: Current Status and Perspectives, Technomic, Lancaster, Pa, USA, 1997.

[5] S. B. Chase and G. Washer, "Nondestructive evaluation for bridge management in the next century," Public Roads, vol. 61, no. 1, pp. 16–25, 1997.

[6] L. D. Olson and C. C. Wright, "Nondestructive testing for repair and rehabilitation," Concrete International, vol. 12, no. 3, pp. 58–64, 1990.

[7] C. R. Farrar, W. B. Baker, T. M. Bell et al., "Dynamic characterization and damage detection in the I-40 bridge over the Rio Grande," Report LA-12767-MS UC-906, Los Alamos National Laboratory, Los Alamos, NM, USA, June 1994.

[8] A. Chakraborty and D. Okaya, "Frequency-time decomposition of seismic data using wavelet-based methods," Geophysics, vol. 60, no. 6, pp. 1906–1916, 1995.

[9] O. S. Salawu and C. Williams, "Bridge assessment using forced-vibration testing," Journal of Structural Engineering, vol. 121, no. 2, pp. 161–173, 1995.

[10] C. R. Farrar and S. W. Doebling, "Lessons learned from applications of vibration based damage identification methods to large bridge structure," in Structural Health Monitoring: Current Status and Perspectives, F. K. Chang, Ed., pp. 351–370, Technomic Publishing, Lancaster, Pa, USA, 1997.

[11] O. S. Salawu, "Detection of structural damage through changes in frequency: a review," Engineering Structures, vol. 19, no. 9, pp. 718–723, 1997.

[12] C. R. Farrar, S. W. Doebling, and D. A. Nix, "Vibration-based structural damage identification," Philosophical Transactions of the Royal Society A, vol. 359, no. 1778, pp. 131–149, 2001.

[13] J. Caicedo, S. J. Dyke, and E. A. Johnson, "Health monitoring based on component transfer functions," in Advances in Structural Dynamics, J. M. Ko and Y. L. Xu, Eds., vol. 2, pp. 997–1004, 2000.

[14] F. K. Chang, Structural Health Monitoring 2000, Technomic, Lancaster, Pa, USA, 2000.

[15] H. Sohn, Charles R. Farrar, Michael L. Fugate, and Jerry J. Czarnecki, "Structural health monitoring of welded connections," in Proceedings of the 1st International Conference on Steel & Composite Structures, Pusan, Korea, June 2001.

[16] H. Sohn, M. Dzwonczyk, E. G. Straser, A. S. Kiremidjian, K. H. Law, and T. Meng, "An experimental study of temperature effect on modal parameters of the Alamosa Canyon Bridge," Earthquake Engineering and Structural Dynamics, vol. 28, no. 7-8, pp. 879–897, 1999.

[17] J. T. Dewolf, M. P. Culmo, and R. G. Lauzon, "Connecticut's bridge infrastructure monitoring program for assessment," Journal of Infrastructure Systems, vol. 4, no. 2, pp. 86–90, 1998.

[18] DeWolf and T. John, "Structural health monitoring of three bridges in connecticut," in Proceedings of the 88th Annual Meeting Transportation Research Board Annual Meeting (TRB '09), 2009.

[19] R. G. Lauzon and J. T. DeWolf, "Connecticut's bridge monitoring program: making important connections last," TR News, no. 224, p. 46, 2003.

[20] J. Olund and J. DeWolf, "Passive structural health monitoring of Connecticut's bridge infrastructure," Journal of Infrastructure Systems, vol. 13, no. 4, pp. 330–339, 2007.

[21] C. Liu and J. T. DeWolf, "Effect of temperature on modal variability of a curved concrete bridge under ambient loads," Journal of Structural Engineering, vol. 133, no. 12, pp. 1742–1751, 2007.

[22] J. Olund, Long term structural health monitoring of Connecticut's bridge infrastructure with a focus on a composite steel tub-girder bridge, M.S. thesis, The University of Connecticut, Storrs, Conn, USA, 2006.

[23] A. Rytter, Vibration based inspection of civil engineering structures, Ph.D. thesis, Aalborg University, Aalborg, Denmark, 1993.

[24] Q. W. Zhang, "Statistical damage identification for bridges using ambient vibration data," Computers and Structures, vol. 85, no. 7-8, pp. 476–485, 2007.

[25] C. Virkler, Continuous structural monitoring coupled with finite element modeling for a composite steel box girder building, M.S. thesis, The University of Connecticut, Storrs, Conn, USA, 2005.

[26] A. M. Scianna and R. Christenson, "Probabilistic structural health monitoring method applied to the bridge health monitoring benchmark problem," Transportation Research Record, no. 2131, pp. 92–97, 2009.

[27] A. H.-S. Ang and W. H. Tang, Probability Concepts in Engineering Planning and Design, Basic Principles, vol. 1, John Wiley & Sons, New York, NY, USA, 1975.

[28] J. S. Bendat and A. G. Piersol, Random Data Analysis and Measurement Procedures, Probability and Statistics, John Wiley & Sons, New York, NY, USA, 3rd edition, 2000.

[29] Z. Jiang, *The use of NEES cyberinfrastructure in structural health monitoring and earthquake education*, M.S. thesis, The University of Connecticut, Storrs, Conn, USA, 2008.

[30] DataTurbine: http://www.dataturbine.org/.

18

Strain-Based Evaluation of a Steel Through-Girder Railroad Bridge

Andrew N. Daumueller and David V. Jáuregui

Department of Civil Engineering, New Mexico State University, Hernandez Hall Box 30001, Las Cruces, NM 88003, USA

Correspondence should be addressed to Andrew N. Daumueller, adaumuel@nmsu.edu

Academic Editor: Husam Najm

In the state of New Mexico (USA), passenger rail began in 2008 between Belen and Santa Fe on the Rail Runner, following the acquisition of about 100 miles of existing rail and related infrastructure. Many of the bridges on this route are over 100 years old and contain fatigue prone details. This study focuses on a steel through-girder bridge along this corridor. To accurately evaluate these structures for load carrying capacity and fatigue, an accurate analytical model is required. Accordingly, four models were developed to study the sensitivity of a bridge in New Mexico to floor-system connection fixity and the ballast. A diagnostic load test was also performed to evaluate the accuracy of the finite-element models at locations of maximum moments. Comparisons between the simulated and measured bridge response were made based on strain profiles, peak strains, and Palmgren-Miner's sums. It was found that the models including the ballast were most accurate. In most cases, the pinned ended models were closer to the measured strains. The floor beams and girders were relatively insensitive to the ballast and end conditions of the floor-system members, whereas the stringers were sensitive to the modeling of the ballast.

1. Introduction

In the state of New Mexico, the Rail Runner commuter rail service was established in 2008. The state purchased about 100 miles of rail between Belen and Lamy, NM, for this purpose, which transferred responsibility of all related infrastructure to the state. One major responsibility that the state acquired was the periodic inspection and load carrying capacity evaluation of each of the bridges. The goal of this study is to develop analytical models that accurately describe the behavior of a steel through-girder bridge along the route purchased by the state. It was found in the literature that the effects of the end fixity of the floor-system members [1] and the ballast, ties, and rails [2, 3] may be significant. This model will subsequently be used in a fatigue evaluation. This study will also add to the area of evaluating railroad bridges, particularly those under commuter traffic.

1.1. Fatigue. Fatigue is generally evaluated in one of two ways, fracture mechanics or SN methods. Fracture mechanics methods often involve the equation developed by Paris

et al. [4], which relates the rate of crack growth (with respect to the number of stress cycles) to various constants that describe the material and the geometry/type of detail. When rearranged, the Paris equation can yield the number of stress cycles until failure at a given constant amplitude stress range magnitude. This approach is not commonly used for highway and railroad bridges in the U.S. since the corresponding bridge design specifications [5, 6] emphasize the SN method. This method requires curves that relate the stress range magnitude, S, to the number of constant amplitude fatigue cycles, N, that can be withstood by the detail before failure (i.e. SN curves). In the AASHTO [5] and AREMA [6] specifications, various curves are given that describe the fatigue limits for a range of common details. The SN curves provided in the specifications are based on full-sized fatigue tests using a uniform probability of failure.

One issue that must be addressed in using SN curves for bridges is that the curves are developed under constant amplitude stress ranges, whereas real bridge live loads are of variable amplitude since vehicles of varying weight traverse the structure. This issue is commonly addressed in one of two ways. The first is to determine an "equivalent" constant amplitude stress range, and the corresponding number of cycles. This is done by estimating the stress histories caused by historical truck or train traffic that has crossed the structure. Obtaining accurate stress histories is often one of the most difficult aspects of a fatigue evaluation since a representative structural model is required, along with reliable historical traffic data. Once the stress histories are obtained, further analysis is required to obtain the stress ranges which may be achieved by applying a rainflow algorithm. This analysis is simplified somewhat by the fact that the fatigue model used is typically a linear model, meaning that repeating vehicles only need to be evaluated once, and the result multiplied by the number of stress cycles caused by the vehicle based on the number of crossings. The effective stress range can then be determined as follows, using the root mean cube of stress ranges:

$$S_{Re} = \left(\sum_i \left(\frac{n_i}{N_{total}} S_i^3 \right) \right)^{1/3}, \quad (1)$$

where S_{Re} is the effective stress range; n_i is the number of stress cycles at the ith stress range magnitude; N_{total} is the total number of stress ranges considered; S_i is the ith stress range magnitude. The effective stress range with its corresponding N (N_{total} from (1)) can then be plotted and the suitability for fatigue can be determined based on whether that point lies above or below the corresponding SN curve.

The other common method for performing a fatigue evaluation based on SN curves is based on the equation developed by Palmgren [7] and popularized by Miner [8], known as the Palmgren-Miner's rule. This is a linear cumulative fatigue estimate as given by the following equation:

$$D = \sum_i \frac{n_i}{N_i}, \quad (2)$$

where D is the Palmgren-Miner's sum; n_i is the number of stress ranges at the ith stress range magnitude; N_i is the number of stress cycles that can be withstood by the detail at the ith stress range magnitude according to the corresponding SN curve. One advantage of this method is that the remaining life of the structure can be estimated. To do so, the Palmgren-Miner's sum to date is first determined and the effect of future traffic is then approximated by measuring the strains under live load for a representative period of time to create a "loading block." The same traffic pattern can then be assumed to repeat indefinitely or increase in volume and/or weight at some assumed rate (e.g., 5% per year). If the traffic is assumed to remain the same, the Palmgren-Miner's sum of all traffic in the "loading block" can be determined, and the remaining life may be calculated as:

$$T = \frac{D_f - D_h}{D_{LB}}, \quad (3)$$

where T is the time until fatigue failure; D_f is the Palmgren-Miner's sum at failure; D_h is the Palmgren-Miner's sum to date; D_{LB} is the Palmgren-Miner's sum accumulated under

Figure 1: Photograph of Rail Bridge at MP 880.37 (Algodones, NM).

the "loading block." The Palmgren-Miner's sum at failure is theoretically 1.0; however, some studies have shown that a lower value is appropriate, particularly when the magnitude of the stress ranges increases over time [9].

1.2. Bridge Location and Description. The state of New Mexico's purchase of the railroad track between Belen and Lamy, NM, to carry the New Mexico Rail Runner required the state to inspect and evaluate all the bridges on the line in accordance with federal regulations [10]. Since many are steel structures over 100 years old, fatigue may be an issue due to the large number of trains that have crossed these structures. Another reason for concern is that the first fatigue provisions entered the AREMA specifications in 1910 after some of these structures were designed [6], including the bridge evaluated in this paper.

The bridge investigated herein is located at milepost (MP) 880.37 near Algodones, NM and carries the railway over an arroyo (see Figure 1). This structure is a ballast-deck, steel through-girder bridge built in 1898. Ballasted decks have a deck that supports crushed stone, which in turn supports the ties and rails on which the trains travel. Through-girder structures consist of two plate girders located on either side of the railway; floor beams are placed transversely and support smaller longitudinal members called stringers, which in turn carry the deck. Currently, this Bridge only carries Rail Runner and Amtrak trains, but historically freight trains and other passenger trains crossed this structure on a regular basis.

The ballast depth of Bridge 880.37 is approximately 25.4 cm from the top of the timber deck to the bottom of the ties based on measurements taken during the last inspection in March 2011. The treated timber ties are spaced 49.5 cm center-to-center and are 16.5 cm high, 21.6 cm wide, and 2.59 m long. The rails are spaced at 1.44 m from inside to inside (i.e., gauge distance) and weigh 67.5 kg/m.

The floor system of bridge 880.37 consists of two 26.67 m long riveted plate girders that are built up using a 2.53 m × 0.953 cm web plate and two 15.2 cm × 15.2 cm × 2.22 cm angles for the top and bottom flanges. Cover plates that are 40.6 cm × 1.59 cm are also used; at midspan, four cover plates are used on both the top and bottom flanges. The floor beams are also built-up steel members that consist of a 61.0 cm × 1.43 cm web plate and 15.2 cm × 15.2 cm × 2.06 cm double angles. The stringers are rolled steel I 12X35 shapes that are

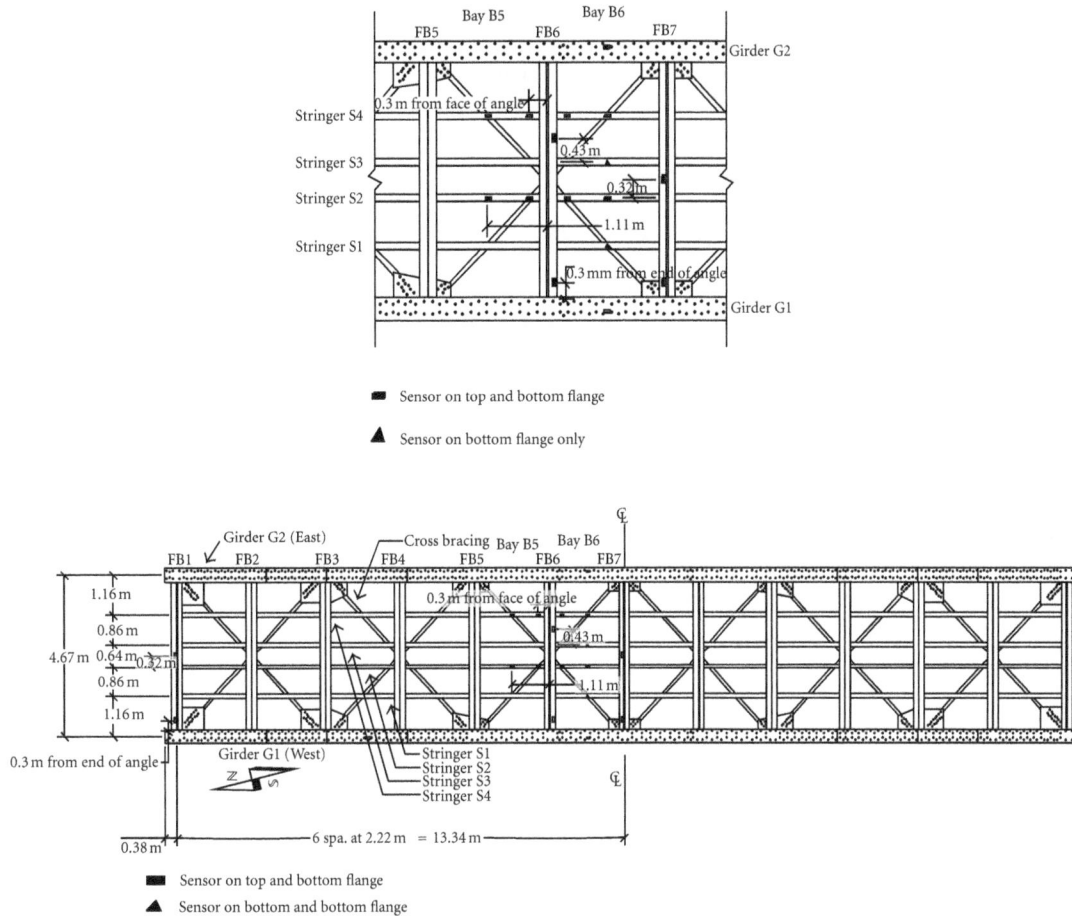

FIGURE 2: Framing plan and sensor locations for Bridge 880.37.

rivet connected to the floor beams using 15.2 cm × 15.2 cm × 1.27 cm double angles. The girders (labeled G1 and G2), floor beams (labeled FB1 through FB13), and stringers (labeled S1 through S4) are shown in Figure 2. All connections are made using rivets with a shank diameter of 2.22 cm.

The deck between the floor beams consists of 14.0 cm × 19.1 cm × 4.27 m treated timbers placed perpendicular to the stringers and resting directly on the top flanges. Over the floor beams, 1.27 cm thick steel apron plates bridge over the top flange so that all loads are distributed to the floor beams via the stringers. Timber ballast curbs are placed along the edges of the deck to ensure that the ballast remains in place.

2. Finite Element Models

Four static linear elastic finite-element models were developed. Models 1 and 2 only included the primary structural members (i.e., girders, floor beams, and stringers). Furthermore, model 1 used pinned connections between floor system elements, while model 2 used fixed connections. Models 3 and 4 were then created to more accurately model the live load distribution by adding the ballast and track structure to models 1 and 2.

2.1. Basic Models. In the basic models (models 1 and 2), the primary structural members of the floor system were modeled with frame elements. All of the frame elements were placed in the same vertical plane to simplify the floor system connections. Nodes were placed at locations of connectivity and approximately 0.152 m spacing for the floor beams and girders. For the stringers, nodes were spaced at 0.305 m intervals and at the connection locations. Moment influence lines were obtained using SAP2000 [11] by using a series of load cases, each with a unit load evenly distributed to the four stringers and representing a 0.305 m movement of the load.

Section properties were calculated for the builtup structural members. Since the girders and floor beams were riveted, the holes needed to be considered. Rivet holes in tension areas were deducted as per AREMA [6], whereas rivets in compression areas were assumed effective. As a result, the gross cross-sectional properties were used to calculate compressive stresses from simulated moments, and the net cross-sectional properties were used to calculate tensile stresses. Table 1 shows the net cross-sectional properties for girders and floor beams, and the gross properties for the stringers used in the basic models.

The moment influence lines for models 1 and 2 at select locations of the floor system members are shown in Figures

TABLE 1: Section properties of the floor system members.

	Area	Torsional constant	I_x	I_y	Major axis shear area	Minor axis shear area	S_x	S_y
	cm^2	cm^4	cm^4 ($*10^3$)	cm^4	cm^2	cm^2	cm^3 ($*10^3$)	cm^3
Girder X1[+]	490.8	196.2	5059	540.7	241.4	174.2	3837	1108
Girder X2[+]	513.7	238.7	5384	29611	241.4	200.0	3924	1457
Girder X3[+]	560.0	292.9	6215	29611	241.4	264.5	4882	1457
Girder X4[+]	674.2	1051	8122	47371	241.4	393.5	6302	2331
Girder X5[+]	787.7	3111	10076	65128	241.4	522.6	7724	3205
Girder X6[+]	901.9	7122	12078	82888	241.4	651.6	9148	4079
Floor beam	294.7	148.3	170.4	5927	86.64	125.8	5.590	291.7
Stringer	65.81	21.13	9.448	416.2	32.93	32.88	0.6194	20.48

[+]X1 through X6 denotes the different girder cross-sections based on the number of cover plates.

TABLE 2: Section properties of rails and ties.

	Area	Torsional constant	I_x	I_y	Major axis shear area	Minor axis shear area	S_x	S_y
	cm^2	cm^4	cm^4	cm^4	cm^2	cm^2	cm^3	cm^3
Rail	86.00	259.7	3921	601.0	23.47	31.09	462.1	31.05
Ties	356.5	16986	8097	13846	356.5	356.5	980.8	505.0

3 through 5. Comparing moment influence lines for models 1 and 2 shows the sensitivity of the bridge response to the end fixity of the floor system members. It is clear that there was virtually no effect on the girders, and only a small effect on the stringers. The fixity of the floor system connections did not affect the girders because the end conditions of the girders were not changed and therefore acted as a simply supported beam in all cases. The end fixity had the greatest effect on the floor beams, particularly floor beam FB1, which was likely due to bearing restraint. The bearings were modeled as fixed in the transverse direction, and therefore the end condition of floor beam FB1 represented that of a fixed-ended beam. Away from the bearings, the torsional rigidity of the girder plays a significant role in the degree of end-fixity of the floor beams. The moment influence lines for the stringers indicate that sharp peaks should be expected in the simulated strain histories. However, it is anticipated that this model may actually overestimate measured stresses since significant longitudinal load distribution may occur due to the 0.254 m of ballast.

2.2. Refined Models. Two additional models (models 3 and 4) were created that simulate the effect of the track structure and ballast on the distribution of train loads. For models 3 and 4, the ballast, ties, and rails were added to models 1 and 2 (see Figure 6). Table 2 shows the section properties for the rails and ties which were modeled using frame elements.

The ballast was modeled using elastic springs with a stiffness calculated based on the ballast beneath the ties. The ballast under each of the rail ties was assumed to extend a distance of $h/2$ in the longitudinal and transverse directions, where h is the ballast depth (see Figure 7). Assuming linear elastic behavior, Hooke's Law was used to determine the spring constant, which amounted to 506 MN/m. The elastic modulus for the ballast was taken as 137.9 MPa, which is

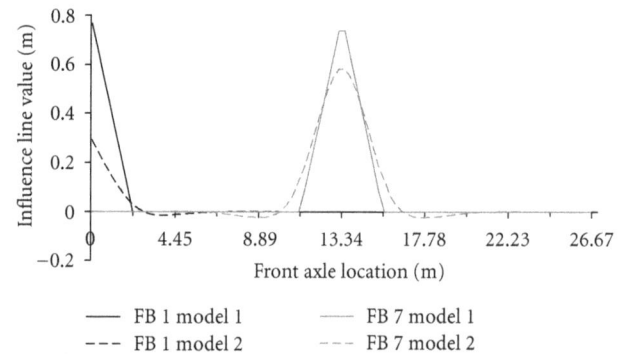

FIGURE 3: Moment influence lines for models 1 and 2 at midspan locations of floor beams FB1 and FB7.

approximately the average of the two values reported by Yang et al. [2] and Kuo and Huang [3] of 100 MPa and 170 MPa, respectively. In the models, the four springs beneath the tie and above each of the stringers were assumed to work in parallel, and therefore 1/4 of the total stiffness or 127 MPa/m was assigned to each spring element. Moment influence lines were created for Models 3 and 4 for comparison with models 1 and 2. Models 1 and 3 are compared in Figures 8 and 9 for the floor beams and stringers, respectively. It is important to note that similar to the connection fixity (see Figure 4), the ballast had essentially no effect on the girder response and therefore no results are shown. Since the girders are insensitive to both the fixity of the floor-system elements and the live-load spreading effect of the ballast, ties, and rails, measurements from the girders may be confidently used in a weigh-in-motion system since there are fewer parameters that affect these strain readings. As shown in Figure 8, the ballast slightly decreased the peak magnitude of the moment influence line of the interior floor beam, however, no change

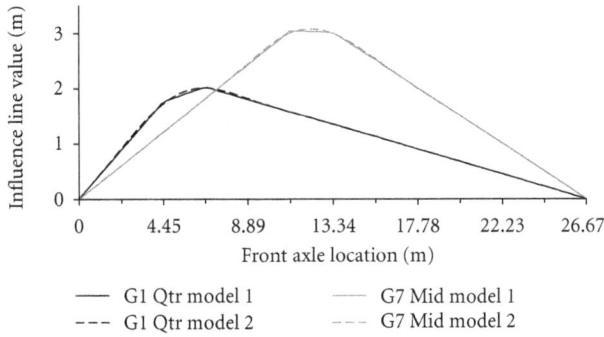

FIGURE 4: Moment influence lines for models 1 and 2 at Bay B6 midspan location and quarter point of girder G1.

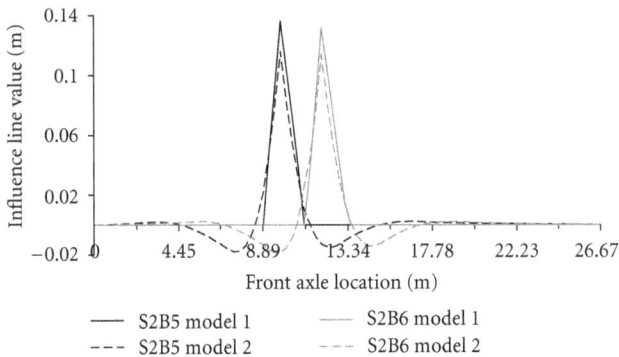

FIGURE 5: Moment influence lines for models 1 and 2 at Bays B5 and B6 midspan locations of stringer S2.

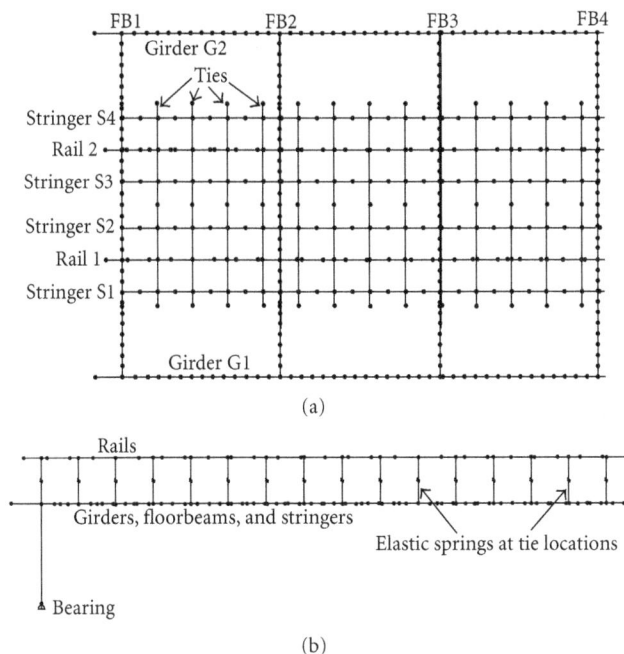

FIGURE 6: Models 3 and 4 of Bridge 880.37 near north support: (a) plan view and (b) profile view.

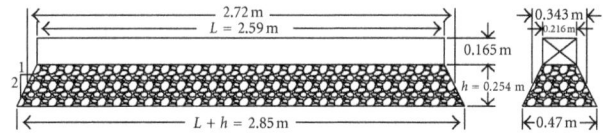

FIGURE 7: Ballast beneath ties used to determine spring constant.

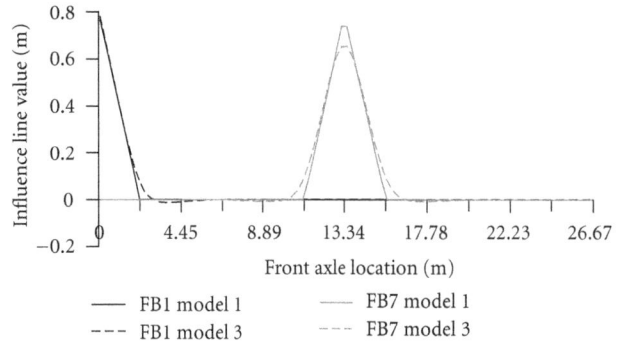

FIGURE 8: Moment influence lines for models 1 and 3 at midspan locations of floor beams FB1 and FB7.

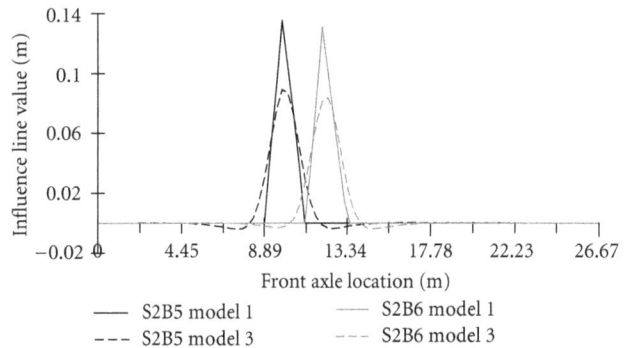

FIGURE 9: Moment influence lines for models 1 and 3 at Bay B5 and B6 midspan locations of stringer S2.

occurred at the end floor beam. The stringers were affected significantly by the modeling of the ballast, track, and rails. This is evident by the large reductions in the peak moment influence line magnitudes shown in Figure 9. Note that compared to the basic model (model 1) where the response is the same for all four stringers, the reduction was greater for the exterior stringers for model 3. The larger response of the interior stringers can be explained by the fact that the ties in the lateral direction act as a continuous beam over four elastic supports. As such, the interior supports support a greater portion of the load when the load is equidistant from the exterior and interior stringer.

2.3. Generation of Influence Profiles. The moment influence lines previously developed were used to simulate strain histories for comparison to measured strains from diagnostic load tests. This was achieved using the following matrix operation: $[M] = [IL] * [L]$, where $[M]$ is the moment history; $[IL]$ is a row vector containing the moment influence

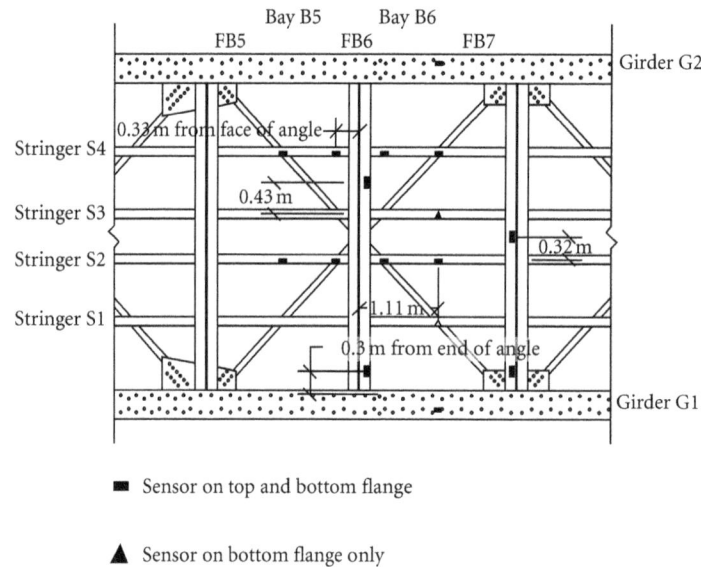

Figure 10: Sensor locations in Bays B5 and B6 of Bridge 880.37.

line ordinates; $[L]$ is a loading matrix that represents a moving train. The moment history can be easily converted to a strain history as follows: $[\varepsilon] = [M] / (S * E)$, where $[\varepsilon]$ is the strain history; S is the section modulus for the member; E is the elastic modulus for steel. The loading matrix must be developed carefully to ensure meaningful results. Each row of the $[L]$ matrix represents a longitudinal location which must correspond to each longitudinal location in the $[IL]$ vector. Each column of the $[L]$ matrix represents one load step (i.e., the first column represents the train front axle directly above floor beam FB1, and the 7th column is 1.38 m past floor beam FB1 using a 0.305 m increment). To simplify the matrix, the longitudinal spacing for the moment influence line ordinates should equal the distance the train travels for each load case. Consequently, only the first row of the loading matrix (which represents the train configuration) requires user input. The matrix can be completed by entering a zero in the first column of each row beneath the first, and the remaining cells taken as the value of the cell above and to the left, thus, representing the movement of the train. Further details regarding this process are given in [12].

3. Diagnostic Load Testing

To better understand the load distribution and determine the best-fit model of Bridge 880.37, a diagnostic load test was performed. The sensors used in this study were strain "intelliducers" (which self-identify to the data acquisition system) from Bridge Diagnostics, Inc., (BDI), and the data acquisition system was the Structure Testing System (STS) II equipment and software [13]. The gauges were attached to the bridge by bolting steel tabs to the sensor, which were subsequently attached to the bare steel of the bridge using a strong epoxy adhesive.

3.1. Instrumentation Plan and Setup. The primary purpose of the load test was to capture the live-load distribution of the structure. Strain transducers were placed near the ends of selected stringers and floor beams to evaluate the level of end fixity of the riveted connections. To assess the longitudinal and transverse load distribution, sensors were placed on all four stringers and both girders near the center of the bridge. Figures 2 and 10 show the sensor locations. The gauges were placed on the bottom of the girders and stringers, the top of the bottom flange for the floor beams, and at the bottom of the top flange for all members where a top sensor is used.

3.2. Train Loading and Testing Procedure. The strain transducers were attached to the structure and connected to junction boxes which were routed to the data acquisition box and laptop computer. Using the BDI testing software, the system was initiated and the sampling rate was set at the maximum of 100 Hz since the trains were traveling at full speed (about 127 kph).

To reduce the effect of temperature and random drift of the sensors, the gauges were rezeroed between trains. Readings initiated on sight of the train and stopped once the last axle of the train exited the structure. A total of six load tests were completed under six different trains as described in Table 3. Axle weights and spacings were extracted from Dick [14] except those for the Rail Runner locomotive which were found on the manufacturer's website [15]. Table 4 and Figure 11 describe the train configurations applied in the load tests.

3.3. Processing of Strain Measurements. The raw strain data was first cropped and then corrected for drift and adjusted for a time lag that was observed in 20 of the sensors due to a recording delay between the channels of the data acquisition system. To crop the data, the measurements taken at floor beam FB1 were used to determine when the train had entered the structure for southbound trains or when the train had travelled the length of the bridge for northbound

TABLE 3: Observed configurations of trains used in load tests.

Train Number	Train type	Time of crossing	Direction	Configuration
1	Rail runner	12:30 PM	Southbound	1 engine; 3 passenger cars
2	Amtrak	3:10 PM	Southbound	2 engines; 1 baggage car; 4 coach; 1 diner; 1 lounger; 3 sleepers
3	Rail runner	4:45 PM	Northbound	3 passenger cars; 1 engine
4	Rail runner	5:05 PM	Southbound	1 engine; 4 passenger cars
5	Rail runner	5:59 PM	Northbound	2 passenger cars; 1 engine
6	Rail runner	6:15 PM	Southbound	1 engine; 3 passenger cars

TABLE 4: Engine and rail car configurations.

	Weight per axle	LPFC*	AS*	TC*
	kN	m	m	m
Rail runner engine	322	20.7	2.74	13.1
Rail runner passenger car	122	25.9	2.90	19.5
Amtrak engine	296	21.0	2.75	13.2
Amtrak baggage car	133	25.9	1.68	18.1
Amtrak coach and lounger	178	25.9	2.59	18.1
Amtrak diner and sleeper	189	25.9	2.59	18.1

* Abbreviations defined in Figure 11.

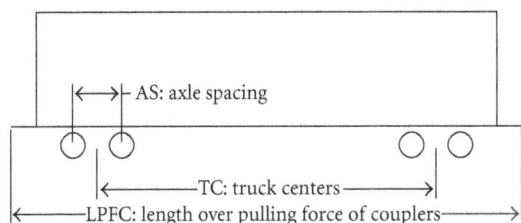

FIGURE 11: Definitions of abbreviations in Table 4.

trains. Sensor drift was then corrected using the following relationship:

$$\varepsilon_{n'.\text{adj}} = \varepsilon_{n'.\text{measured}} - \varepsilon_0 + \frac{n'}{N'} * (\varepsilon_{N'} - \varepsilon_0), \quad (4)$$

where $\varepsilon_{n'.\text{adj}}$ is the adjusted strain reading; $\varepsilon_{n'.\text{measured}}$ is the raw measured strain; ε_0 is the average strain before the train enters the structure; n' is the reading number; N' is the total number of readings between the train front axle entering and the last axle leaving the bridge; $\varepsilon_{N'}$ is the average strain after the last axle leaves the structure.

As mentioned previously, a time lag was observed in 20 of the 36 sensors and at one location, the strain readings on the top and bottom flanges were offset. Since the time between the first peaks in the readings represented the time lag, the lagging data was simply shifted to the left. Figure 12 shows an example strain data set for a stringer before and after processing.

3.4. General Behavior Based on Measured Strains. The measured strains for each of the six trains at all 19 instrumented locations were processed as discussed in Section 3.3. In general, the shapes of the strain histories were as expected

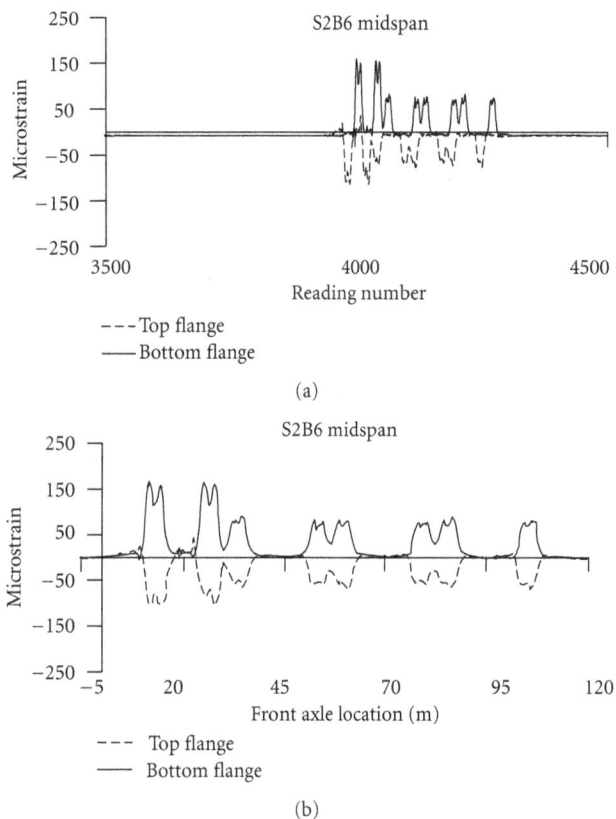

FIGURE 12: Typical response for stringer S2 at Bay B5 midspan: (a) raw strain data and (b) processed strain data corrected for drift, time lag, and axle location.

(e.g., a peak for each set of axles and peaks of similar magnitudes for each rail car). In Bay B6, the peak strain magnitudes were smallest in stringer S1 and largest in stringer S4, consistently increasing from west to east, as shown in Figure 13. This behavior was only observed under the locomotive, and possibly attributed to asymmetric weight distribution, or some dynamic effect of the diesel engine. Another observation was that tensile stresses developed in the stringers, particularly in Bay B5. At the instrumented ends of these stringers, pure tension was measured and at midspan, the tension flange strains were significantly larger than the compression flange strains. Similar behavior was observed in Bay B6, but to a smaller extent. This can be

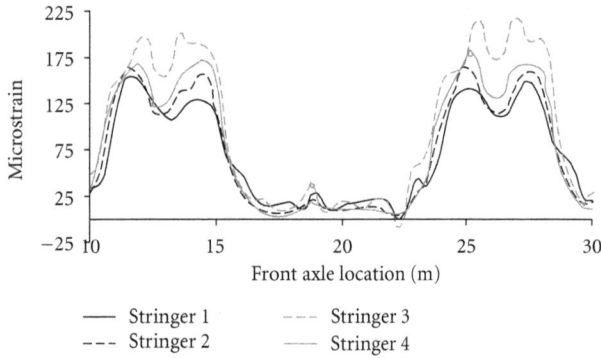

FIGURE 13: Illustration of asymmetric live-load distribution under the locomotive for Train 1.

TABLE 5: Neutral axis positions determined from drawings and strain measurements.

	From drawings	From strain data	Difference (%)
	cm	cm	%
Floorbeams	30.5	31.2	2.9%
Girder X5	132	123	−6.3%
Girder X6	133	131	−2.0%
Stringers	15.2	18.5	21.5%

Note: Neutral axis locations taken from bottom of member.

explained by the fact that the stringers behave in conjunction with the girders and resist some of the tensile flexural stresses that would otherwise be resisted by the girder bottom flanges.

Assuming linear strain distribution, the neutral axis positions were estimated at all locations where strains were measured at the top and bottom flanges. The average measured neutral axis positions are given in Table 5 compared to those calculated from the as-built plans. For the girders and floor beams, the average neutral axis locations based on strain measurements were close to the values obtained from the drawings. For the stringers, the average neutral axis location based on measured strains was significantly higher, which can be explained by the behavior discussed earlier, and partial composite action with the deck. It was also observed that minimal negative flexure developed at the ends of the stringers and floor beams with two exceptions: the stringers in Bay B5 and floor beam FB1. This may indicate negative flexure, since the instrumented location may have been near the points of inflection.

4. Evaluation of Best-Fit Model

The finite-element models were evaluated by comparing the shapes of the simulated and measured strain profiles, and also the peak strains from each passing train. It was found that the actual train speeds varied from 125 kph to 126 kph. The models were further evaluated based on Palmgren-Miner's sum for future use in a fatigue evaluation.

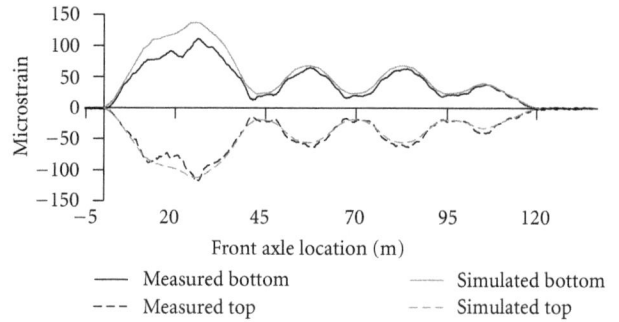

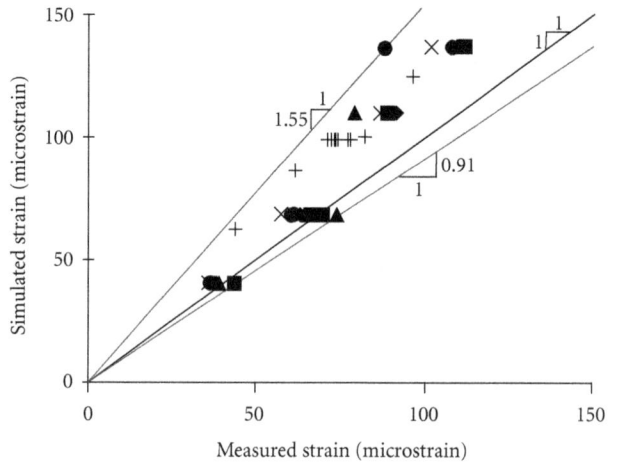

FIGURE 14: Simulated (Model 1) versus measured response near midspan of girder G2: (a) strain profiles for Train 1 and (b) peak strains for all trains.

4.1. Measured versus Simulated Girder Strains. Figure 14 shows the simulated versus measured strains near midspan of Girder G2 for model 1 since all four models behaved nearly the same. As shown in Figure 14(a), the strains match very well under Train 1, particularly for the rail cars, which are represented by the three smaller peaks in the data. For all the trains, it is also apparent that the locomotive weights may have been overestimated since the simulated strains consistently exceeded the measured strains under the locomotives as shown in Figure 14(b) (locomotive strains are represented by the first peak in the strain data). For the Rail Runner trains, the peak strain correlation was very close to the 1 : 1 line for the passenger cars, but as much as 1.5 : 1 for the locomotives. For the Amtrak train, both the locomotive and passenger cars showed higher simulated strains compared to the measured strains. The difference between the measured and simulated strains may be attributed to differences in the actual weights of the rail equipment when compared to the estimated weights taken from the manufacturer, which were used when developing the simulated strain histories.

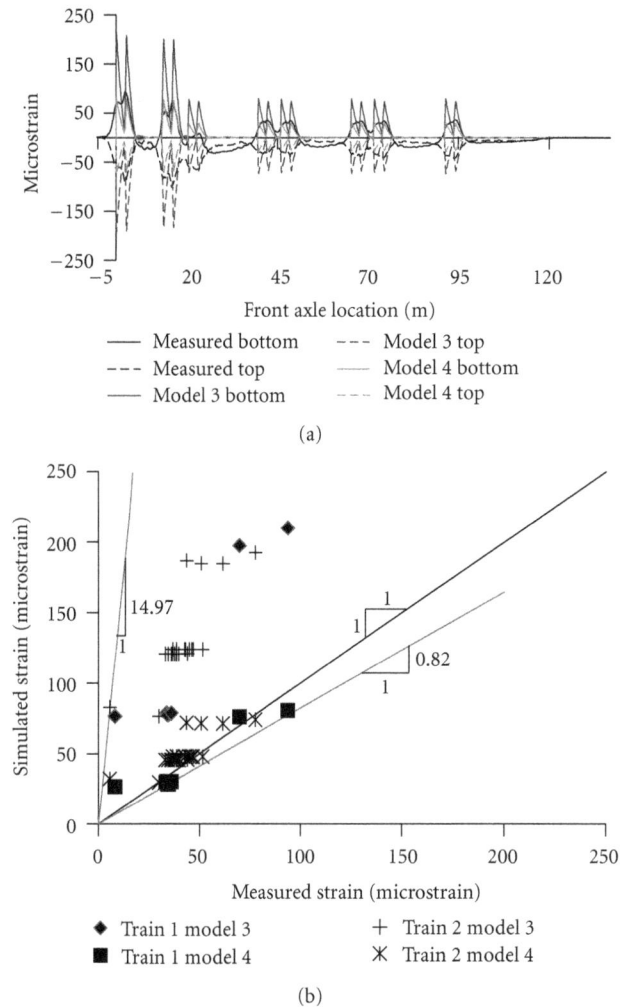

FIGURE 15: Simulated versus measured response at midspan of floor beam FB1: (a) strain profiles for Train 1 and (b) peak strains for Trains 1 and 2.

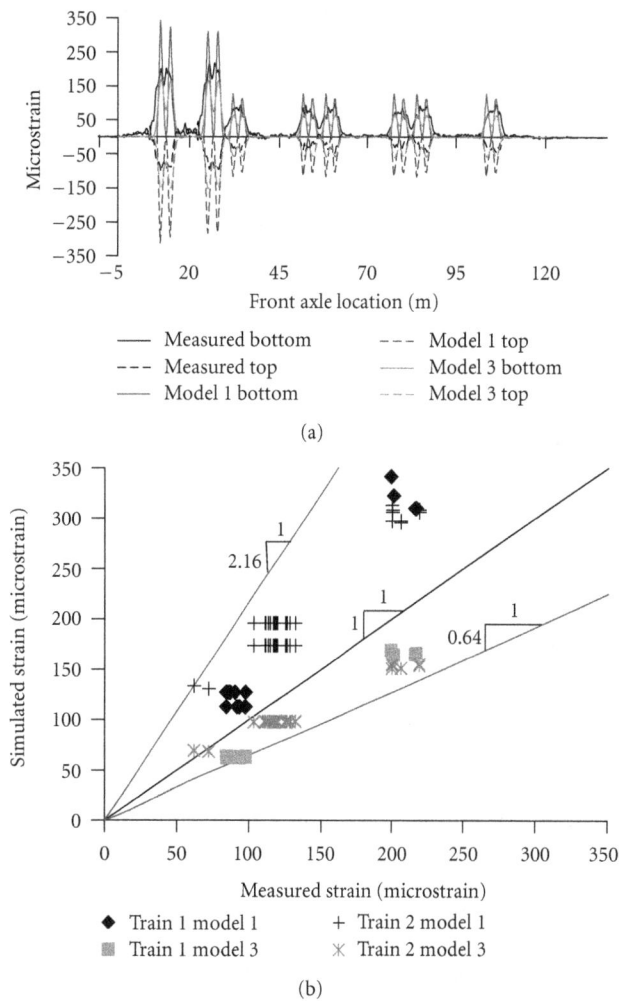

FIGURE 16: Simulated versus measured response at midspan of Stringer S4 in Bay B6: a) strain profiles for Train 1 and b) peak strains for Trains 1 and 2.

4.2. Measured versus Simulated Floor Beam Strains. Similar comparisons were made for the floor beams however, unlike the girders there were differences between the models. For the end floor beam FB1, the simulated response was significantly affected by the end fixity. As shown in Figure 15, the pinned connection model (i.e., model 3) significantly overestimated the measured strains whereas the fixed connection model (i.e., model 4) resulted in a very close agreement of peak strains, although the model does yield some results below the 1 : 1 line.

For the interior floor beams (FB6 and FB7), it was found that the influence of the end-fixity and ballast was minimal. The pronounced effect of the end-fixity on the end floor beam compared to the interior floor beams can be attributed to the bearings and torsional rigidity of the girders as mentioned earlier. None of the floor beams were affected significantly by the modeling of the ballast, ties, and rails. This is logical since, assuming the load is spread over a relatively small distance, the difference in the distribution to the floor beams is small. The results from this section

show potential in guiding a bridge inspector to locations that are more likely to experience fatigue related failure since it was shown that the end floor beam may develop negative moments at the ends, thus, causing stresses that were not likely considered in design.

4.3. Measured versus Simulated Stringer Strains. The strain comparisons are given in Figure 16 for Stringer S4 at the midspan of Bay B6. This location was chosen since the maximum measured stringer response occurred here. models 1 and 3 were compared since the stringers were more sensitive to load spreading through the ballast than end-fixity. Models 2 and 4 resulted in lower strains than models 1 and 3, respectively. The shapes of the curves also match much better when the pinned-end models are used. Based on the strain profile comparisons, model 3 appears to more accurately capture the shape of the measured strains and the peak magnitudes, although the results fell below the 1:1 line. For this model, the simulated strains for the interior stringers were greater than those for the exterior stringers. If the

Table 6: Palmgren-Miner's sums based on measured and simulated strains.

	Miner's sums					Difference from measured (%)			
	Measured $(*10^{-9})$	Model 1 $(*10^{-9})$	Model 2 $(*10^{-9})$	Model 3 $(*10^{-9})$	Model 4 $(*10^{-9})$	Model 1	Model 2	Model 3	Model 4
Train 1									
G1B3 midspan	1.27	2.26	2.32	2.26	2.31	78%	83%	78%	82%
G2B6 midspan	2.48	4.10	4.19	4.11	4.18	65%	69%	66%	68%
FB1 midspan	4.14	53.7	2.7	51.0	2.96	1196%	−34%	1131%	−29%
FB7 midspan	5.38	26.2	24	22.6	26.1	387%	348%	319%	385%
S2B6 midspan	1.30	17.0	8.51	3.11	1.99	1205%	555%	139%	53%
S4B6 midspan	2.72	17.0	7.93	1.43	0.85	524%	192%	−47%	−69%
Train 2									
G1B3 midspan	2.47	4.62	4.58	4.52	4.55	87%	86%	83%	84%
G2B6 midspan	4.09	7.95	7.92	7.81	7.90	94%	94%	91%	93%
FB1 midspan	10.4	152	8.1	150	8.86	1369%	−22%	1349%	−15%
FB7 midspan	12.3	80.1	66	64.2	67.6	551%	434%	422%	450%
S2B6 midspan	3.87	50.8	29.9	9.51	7.32	1211%	674%	146%	89%
S4B6 midspan	7.68	50.8	27.1	4.36	3.13	560%	253%	−43%	−59%

[*] G1B3 indicates girder G1 in Bay B3; G2B6 indicates girder G2 in Bay B6; S2B6 and S4B6 indicate stringers S2 and S4, respectively, in Bay B6.

maximum stringer response from model 3 (i.e., the response for an interior stringer) is compared to the measured strains at stringers S4, bay B6 midspan (as shown in Figure 16), the simulated strains more closely approximate the measured strains, and the slope of the lower-bound line is 0.81 rather than 0.64.

Modeling the stringers requires greater care than the modeling of the girders and floor beams since the stringers are sensitive to both the end-fixity and to the modeling of the ballast, ties, and rail. Correctly modeling the end-fixity is also more difficult for the stringers than for the floor beams since the level of fixity is affected by the rigidity of the double angle connections and the out-of-plane rigidity of the floor beam web. The potential contribution of the stringers in tension and the possibility for partial composite behavior with the deck also must be considered. Considering the many factors affecting the behavior of the stringers, the results from the finite-element models described the measured behavior quite well.

4.4. Comparisons Based on Palmgren-Miner's Sums. Apart from the comparisons of strain histories and peak strains, the Palmgren-Miner's sums were determined to further evaluate the models. For Bridge 880.37, the critical fatigue details for the girders and floor beams are classified as category D based on the AREMA specifications [6] and category A for the stringers away from the connections. To obtain the Palmgren-Miner's sums, the strain histories were first converted to stress histories (by multiplying by the modulus of elasticity) which were then processed using a rainflow counting algorithm [16] developed in MATLAB [17] to determine the stress range magnitudes and corresponding number of stress cycles. Subsequently, the

Palmgren-Miner's sums were determined using the following equation:

$$D = \sum_{1}^{k} \frac{n_i}{N_i} = \sum_{1}^{k} \frac{n_i}{n_o} * \left(\frac{\Delta \sigma_i}{\Delta \sigma_o} \right)^3, \qquad (5)$$

where D is the Palmgren-Miner's sum; n_i is the number of stress cycles at the ith stress range magnitude; $\Delta \sigma_i$ is the ith stress range magnitude; n_o is the number of cycles at the knee point of the SN curve; $\Delta \sigma_o$ is the stress range at the knee point; k is the number of different stress cycles. For the riveted details, n_o is 6 million cycles and $\Delta \sigma_o$ is 48.2 MPa as specified in AREMA [6] and for the stringers away from the connections, n_o is 2 million cycles and $\Delta \sigma_o$ is 165 MPa.

The Palmgren-Miner's sums based on the measured and simulated strains vary widely between models and between different locations for the same model (see Table 6).

Results are only shown for Trains 1 and 2; however, the trends and percentage of differences are consistent from train to train. The percentage of difference indicates the percentage of difference between the measured and simulated Palmgren-Miner's sum for a particular location. The pinned connection models (i.e., models 1 and 3) have higher Palmgren-Miner's sums than the fixed connection models (i.e., models 2 and 4) for the stringers and end floor beam locations. Models 3 and 4 have lower Palmgren-Miner's sums than Models 1 and 2 for the stringers because of the load spreading effect. Floor beam FB1 appears to be the least accurately depicted by the pinned connection models. At this floor beam, the Palmgren-Miner's Sum based on measured strains was much smaller. In addition, the Palmgren-Miner's sums based on simulated strains for floor beam FB1 were double the corresponding values for an interior floor beam. This behavior is mainly attributed to the larger degree of end-fixity developed at the end floor beam compared to the interior floor beams as discussed earlier. Thus, the fixed connection models match the measured data the best for the end floor beam.

At S4B6 midspan, model 3 underestimated the Palmgren-Miner's sum based on measured strains; however, if this model was used in an engineering analysis, the most critical stringer would be used, since the member with the highest stress ranges would be the critical member in the analysis. The critical stringer location using model 3 is S2B6 midspan and S4B6 midspan for the measured strains. If these Palmgren-Miner's sums are compared then model 3 is conservative by 14% for Train 1 and 24% for Train 2, which would yield very good results in a fatigue evaluation. The Palmgren-Miner's sums for the girders and interior floor beam were very similar from model to model, signifying insensitivity to the end-fixity of the floor-system members and the ballast modeling, and model 3 appeared to be the closest to the measured values. The models overestimated the Palmgren-Miner's sum by 65% to 94% for Girders G1 and G2, and by 319% to 551% for Floor beam FB7. More accurate locomotive weights would likely lead to greater convergence between the simulated and measured behavior.

5. Conclusions and Future Work

Based on the four models, it was found that those which consider the ballast, ties and rails perform better than the simpler models. For the girders and interior floor beams, the difference in response for the four models was small since these members were insensitive to the end-fixity of the floor-system members and to the modeling of the ballast, but model 3 was found to give the best results. This insensitivity to the parameters mentioned indicates that measurements taken at the girders may be useful in a weigh-in-motion system since only the axle weights will likely affect the measurements significantly. For the end floor beam, model 4 was closest to the actual response and model 2 also gave reasonably accurate results. There was a significant difference between the pinned and fixed ended models for the end floor beam due to the stiffness of the bearing in the transverse direction. Based on these results for the girders and floor beams, the level of analysis had a small impact on the accuracy of the simulated responses. Therefore, a relatively simple model is sufficient if only the girders and floor beams require evaluation; however, a slight improvement is achieved when the ballast, ties, and rails are included in the model.

The stringers were sensitive to both the end-fixity and the modeling of the ballast. It was found that if the critical stringer is evaluated, model 3 provides the most accurate results, particularly if a Palmgren-Miner's sum is desired for a fatigue evaluation. Although the stringers require a more complex model, it may not be required in many cases when evaluating fatigue. Based on the results of this study, the end-fixity of the stringers is minimal, and, therefore, the stress ranges at the ends are likely to be small, thus, the midspan of the member is critical for fatigue. Since the stringers are rolled steel shapes, they are relatively unsusceptible to

fatigue failure, and as such an analysis utilizing a simple, yet significantly conservative model may be sufficient to indicate that the stringers are unlikely to pose a fatigue-related risk. For the end floor beam, in spite of being closer to the actual response, the simulated values were at times underestimated. This can be remedied by applying a factor that accounts for the modeling error. Furthermore, if this model was used in a fatigue evaluation using the AREMA [6] specifications, factors would also be applied to account for vertical impact and rocking effects.

Work based on this study, including a fatigue evaluation of Bridge 880.37 to estimate the remaining fatigue life is published elsewhere [18]. To achieve this, representative trains were developed for various eras dating back to the construction of the bridge, which were used along with the known current train traffic. These loads were then applied to the most accurate models, as determined in this paper. The matrix multiplication discussed in Section 2.3 can be rearranged to obtain the moment influence line for a specific location based on strain measurements. These influence lines can then be used to observe changes in the live load distribution of the bridge that may indicate structural damage. The moment influence lines can also be used in lieu of analytical models for evaluating the bridge under historical loading [18].

References

[1] M. Al-Emrani and R. Kliger, "FE analysis of stringer-to-floor-beam connections in riveted railway bridges," *Journal of Constructional Steel Research*, vol. 59, no. 7, pp. 803–818, 2003.

[2] L. A. Yang, W. Powrie, and J. A. Priest, "Dynamic stress analysis of a ballasted railway track bed during train passage," *Journal of Geotechnical and Geoenvironmental Engineering*, vol. 135, no. 5, pp. 680–689, 2009.

[3] C. M. Kuo and C. H. Huang, "Two approaches of finite-element modeling of ballasted railway track," *Journal of Geotechnical and Geoenvironmental Engineering*, vol. 135, no. 3, pp. 455–458, 2009.

[4] P. C. Paris, M. P. Gomez, and M. E. Anderson, "A rational analytical theory of Fatigue," *Trends in Engineering*, vol. 13, pp. 9–14, 1961.

[5] AASHTO, *LRFD Bridge Design Specifications*, American Association of State Highway and Transportation Officials, Washington, DC, USA, 5th edition, 2010.

[6] AREMA, *Manual for Railway Engineering*, American Railway Engineering and Maintenance-of-Way Association, Lanham, Md, USA, 2007.

[7] A. Palmgren, "Die Lebensdauer von Kugellagern. (The life of roller bearings)," *Zeitschrift des Vereines Deutscher Ingenieure*, vol. 68, no. 14, pp. 339–341, 1924 (German).

[8] M. A. Miner, "Cumulative damage in fatigue," *ASME Journal of Applied Mechanics*, vol. 12, no. 3, pp. A159–A1964, 1945.

[9] H. Agerskov, "Fatigue in steel structures under random loading," *Journal of Constructional Steel Research*, vol. 53, no. 3, pp. 283–305, 2000.

[10] Federal Register, "Bridge safety standards," vol. 75, no. 135, 2010.

[11] *SAP2000 Version 14.2.2*, Computers and Structures, Berkely, Calif, USA, 2010.

[12] A. Daumueller and D. Jauregui, "Development of a structural health monitoring system for the assessment of critical transportation infrastructure," Tech. Rep. Sand2012-0886, Sandia National Laboratories, Department of Energy, February 2012.

[13] BDI, *Structural Testing System II Operation Manual*, Bridge Diagnostics, Boulder, Colo, USA, 2005.

[14] S. M. Dick, *Bending moment approximation for use in fatigue life evaluation of steel railway girder bridges [Ph.D. thesis]*, Department of Civil, Environmental and Architectural Engineering, University of Kansas, Lawrence, Kan, USA, 2002.

[15] MotivePower, "MPX Commuter Locomotive," Locomotives, 2010, http://www.motivepower-wabtec.com/locomotives/commuter/mpxpress.php.

[16] A. Niesłony, "Determination of fragments of multiaxial service loading strongly influencing the fatigue of machine components," *Mechanical Systems and Signal Processing*, vol. 23, no. 8, pp. 2712–2721, 2009.

[17] *MATLAB R2010b*, The MathWorks, Natick, Mass, USA, 2010.

[18] A. Daumueller, *Strain-based evaluation of a riveted steel railroad bridge [Ph.D. dissertation]*, New Mexico State University, Las Cruces, NM, USA, 2012.

19

Seismic Response of Torsionally Coupled System with Magnetorheological Dampers

Snehal V. Mevada and R. S. Jangid

Department of Civil Engineering, Indian Institute of Technology Bombay, Powai, Mumbai 400 076, India

Correspondence should be addressed to Snehal V. Mevada, snehalvm@iitb.ac.in

Academic Editor: Andreas Kappos

The seismic response of linearly elastic, idealized single-storey, one-way asymmetric building with semiactive magnetorheological (MR) dampers with clipped-optimal algorithm is investigated. The response is obtained by numerically solving the governing equations of motion. The effects of eccentricity ratio, uncoupled time period, and ratio of uncoupled torsional to lateral frequency are investigated on peak responses which include lateral, torsional and edge displacements and their acceleration counter parts, base shear, and control forces. To study the effectiveness of control system, the controlled response of asymmetric system is compared with the corresponding uncontrolled response. Further, controlled response of asymmetric system is compared with corresponding symmetric system to study the effects of torsional coupling. It is shown that the implementation of semiactive dampers reduces the deformations significantly. Also, the effects of torsional coupling on effectiveness of semiactive system are found to be more sensitive to the variation of eccentricity and torsional to lateral frequency ratio.

1. Introduction

Many times, most of the real structures are prone to the severe response and damage during a seismic event due to their asymmetric nature. This asymmetry in the buildings arises primarily due to uneven distribution of mass and/or stiffness of the structural components for elastic range. Due to the asymmetric nature of buildings causing the torsional deformations, they are more vulnerable to the earthquake induced damage. The prime focus of the structural engineer is to reduce the torsional response mainly by avoiding the eccentricity which is produced due to uneven mass and stiffness distribution. However, there are many limitations for avoiding the eccentricity between mass and stiffness due to stringent architectural and functional requirements and hence in such cases, implementation of supplemental energy dissipation devices proves to be an effective solution to minimize the lateral-torsional response of the buildings.

In past, many researchers have investigated the performance of seismic control aspects of lateral-torsional deformations using various techniques such as passive control

namely base isolation and supplemental dampers and active control. Hejal and Chopra [1] investigated the effects of lateral-torsional couplings and demonstrated that the building response significantly depends on structural eccentricity and frequency ratio. Jangid and Datta [2] investigated that the effectiveness of base isolation for an asymmetric system reduces for higher eccentricity. Jangid and Datta [3] found that the effectiveness of multiple tuned mass dampers is overestimated by ignoring the system asymmetry. Goel [4] investigated that by implementing proper supplemental damping, edge deformations in asymmetric-plan systems can be reduced than those in the corresponding symmetric systems. Date and Jangid [5] investigated the effectiveness of active control system for an asymmetric system in controlling torsional and corner displacements and shown that effectiveness is overestimated by ignoring the effects of torsional coupling. Lin and Chopra [6] studied the effects of plan-wise distribution of nonlinear viscous and visco-elastic dampers and found that the asymmetric distribution of damping reduces the response more effectively as compared to symmetric distribution. De La Llera et al. [7] proposed the weak torsional balance condition for system installed with friction dampers such as to minimize the correlation between translation and rotation. Petti and De Iuliis [8] proposed a method to optimally locate the viscous dampers for torsional response control in asymmetric plan systems by using modal analysis techniques.

The semiactive control systems combine the attractive features of passive and active control systems and hence drawn the attention of many researchers, in recent past. Chi et al. [9] evaluated an asymmetric building with base isolation augmented with semiactive magnetorheological (MR) damper and found improvement in torsional behavior. Yoshida et al. [10] and Yoshida and Dyke [11] investigated the torsional response of asymmetric building using semiactive MR damper and noticed an increase of torsional response and decrease of translational response due to asymmetry. Moreover, the base torque increases and base shear decreases with an increase in eccentricity for strongly coupled system. Shook et al. [12] investigated the effectiveness of semiactive MR damper and observed that fuzzy logic controller is effective in decoupling lateral and torsional response with reduction in displacement and acceleration responses. H. N. Li and X. L. Li [13] developed an MR damper based on semi-geometric model for asymmetric building and found a greater reduction in displacement and acceleration responses compared to passive control case. Although, the above studies reflect the effectiveness of some of the semiactive systems in controlling the lateral-torsional responses, however, no specific study has been carried out to investigate the effects of torsional coupling especially by considering important system parameters on edge deformations as well as on control forces using semiactive MR dampers.

In this paper, the seismic response of idealized single storey, one-way asymmetric building, is investigated under different real earthquake ground motions. The specific objectives of the study are summarized as (i) to study the effects of torsional coupling on the effectiveness of the semiactive MR damper control system, (ii) to investigate the effectiveness of semiactive MR dampers with clipped-optimal control algorithm in controlling lateral, torsional and specifically the edge deformations, and (iii) to investigate the influence of important parameters on the effectiveness of semiactive MR dampers for asymmetric systems. The important parameters considered are eccentricity ratio of superstructure and uncoupled time period and ratio of uncoupled torsional to lateral frequency.

2. Model of Magnetorheological Damper

Magnetorheological (MR) dampers are the semiactive devices which use MR fluids to provide control forces and they are quite promising for civil engineering applications. They offer highly reliable operation at a modest cost and can be considered as fail-safe in the case of hardware malfunction in that they become passive dampers. MR fluids typically consist of micron-sized magnetically polarizable particles dispersed in a carrier medium such as silicon oil. When a magnetic field is applied to the fluids, particle chains

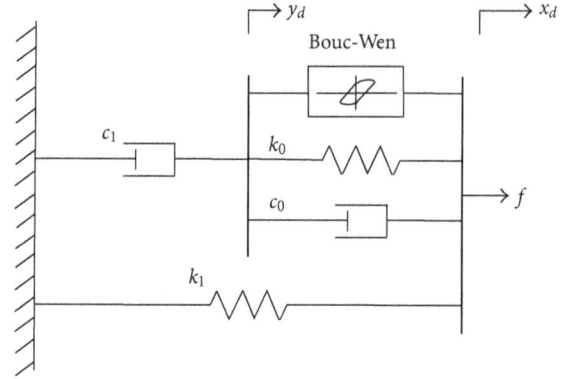

FIGURE 1: Mechanical model of the MR damper (Modified Bouc-Wen model) [14].

form, and the fluid becomes a semisolid, exhibiting plastic behavior. MR fluids have high yield strength, low viscosity, and stable hysteretic behavior over a wide temperature range. Transition to rheological equilibrium can be achieved in a few milliseconds, providing devices with high bandwidth. MR devices are capable of generating large forces [15].

In this study, the Modified Bouc-Wen model proposed by Spencer et al. [14] as shown in Figure 1 is used. The model has shown to accurately predict the behavior of the prototype MR damper over a broad range of inputs. The equations governing the force predicted by this model is [14]

$$
\begin{aligned}
f_i &= \alpha_i z_i + c_{0i}(\dot{x}_{di} - \dot{y}_{di}) + k_0(x_{di} - y_{di}) + k_1(x_{di} - x_0) \\
&= c_{1i}\dot{y}_{di} + k_1(x_{di} - x_0),
\end{aligned}
\tag{1}
$$

where the evolutionary variable z_i is governed by

$$
\begin{aligned}
\dot{z}_i &= -\gamma_m |\dot{x}_{di} - \dot{y}_{di}| z_i |z_i|^{n-1} - \beta_m (\dot{x}_{di} - \dot{y}_{di}) |z_i|^n \\
&\quad + A_m (\dot{x}_{di} - \dot{y}_{di}),
\end{aligned}
\tag{2}
$$

$$
\dot{y}_{di} = \frac{1}{(c_{0i} + c_{1i})} \{\alpha_i z_i + c_{0i}\dot{x}_{di} + k_0(x_{di} - y_{di})\}.
$$

In this model, the accumulator stiffness is represented by k_1 and the viscous damping observed at large velocities by c_{0i}. A dashpot, represented by c_{1i}, is included in the model to introduce the nonlinear roll-off in the force-velocity loops at low velocities, k_0 is present to control the stiffness at large velocities, and x_0 is the initial displacement of spring k_1 associated with the nominal damper force due to the accumulator.

To account for the dependence of the force on the voltage applied to the current driver and the resulting magnetic current, the suggested parameters are

$$
\alpha_i = \alpha_a + \alpha_b u_{di}, \qquad c_{0i} = c_{0a} + c_{0b}u_{di},
$$
$$
c_{1i} = c_{1a} + c_{1b}u_{di}.
\tag{3}
$$

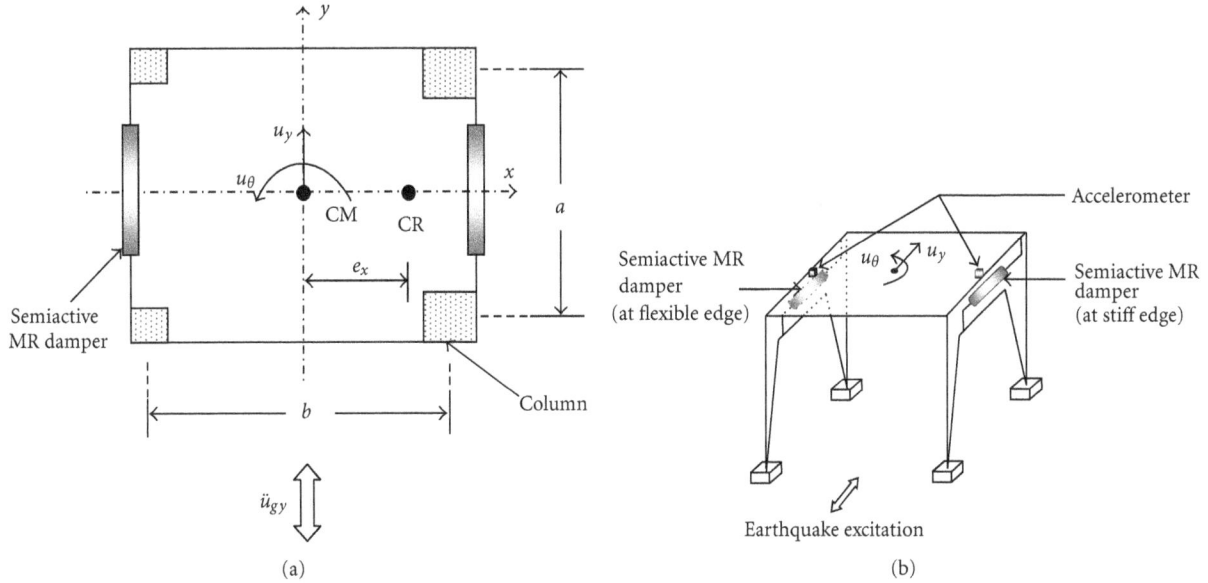

FIGURE 2: Plan and isometric view of one-way asymmetric system showing arrangement of dampers and accelerometers.

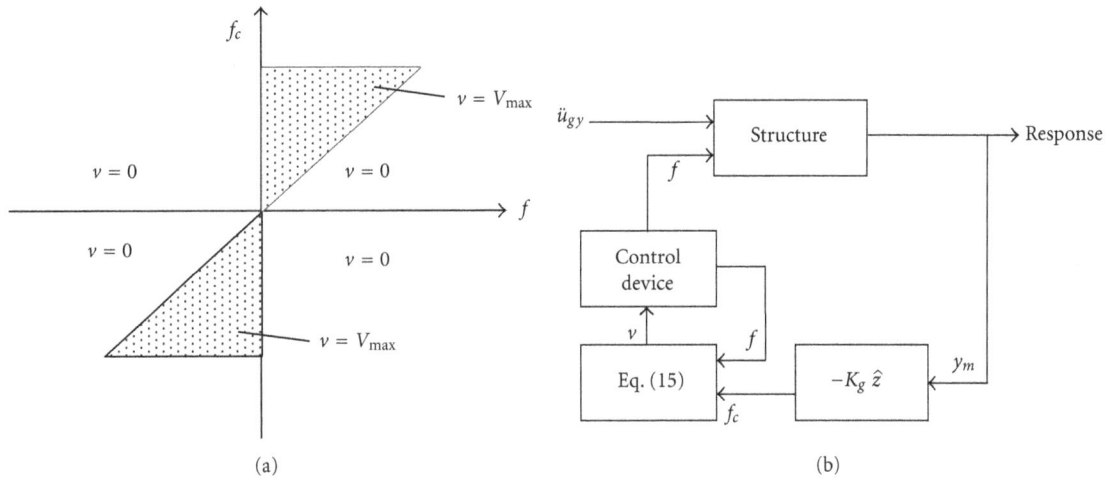

FIGURE 3: Graphical representation of clipped-optimal control algorithm [15] and semiactive control system.

In addition, the dynamics involved in the MR fluid reaching rheological equilibrium are accounted for through the first-order filter

$$\dot{u}_{di} = -\eta(u_{di} - v_i), \qquad (4)$$

where v_i is the commanded voltage sent to the current driver.

3. Structural Model and Solution of Equations of Motion

The system considered is an idealized one-storey building which consists of a rigid deck supported on columns as shown in Figure 2. Following assumptions are made for the structural system under consideration: (i) floor of the superstructure is assumed as rigid, (ii) force-deformation behaviour of the superstructure is considered as linear within elastic range, (iii) the structure is excited by uni-directional horizontal component of earthquake ground motion and the vertical component of earthquake motion is neglected, and (iv) the effect of time delay is neglected for control algorithm. The mass of deck is assumed to be uniformly distributed and hence centre of mass (CM) coincides with the geometrical centre of the deck. The columns are arranged in a way such that it produces the stiffness asymmetry with respect to the CM in one direction and hence, the centre of rigidity (CR) is located at an eccentric distance, e_x from CM in x-direction. The system is symmetric in x-direction and therefore, two degrees-of-freedom are considered for model, namely, the lateral displacement in y-direction, u_y and torsional displacement, u_θ as represented in Figure 2. The governing equations of motion of the building model with coupled lateral and torsional degrees-of-freedom are obtained by assuming that the control forces provided by

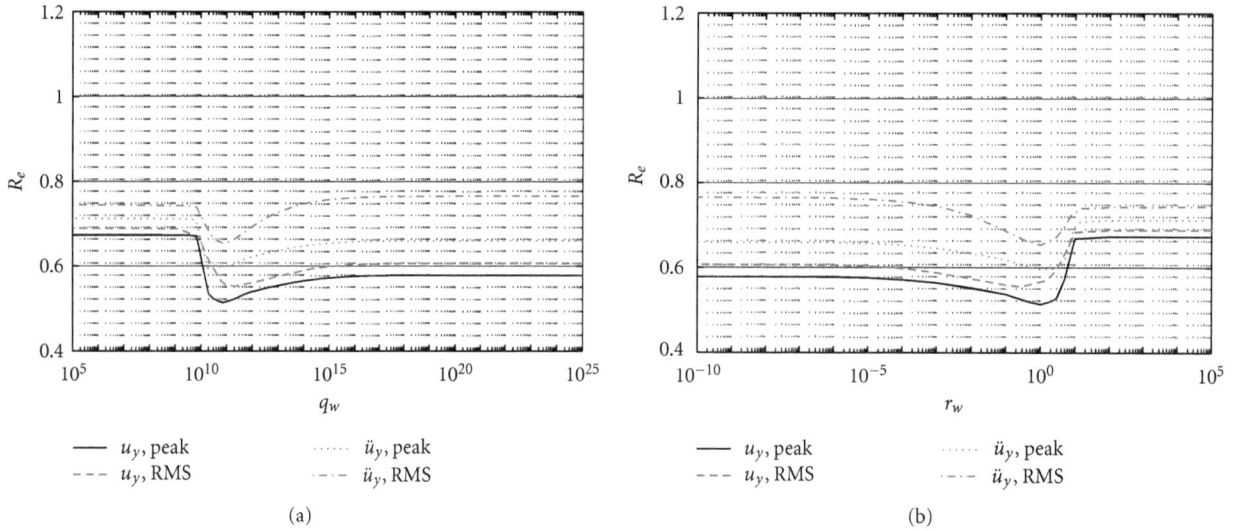

FIGURE 4: Effect of weighting parameters on response ratio, R_e for peak and RMS responses under Imperial Valley, 1940 earthquake ($T_y = 1$ s, $e_x/r = 0$).

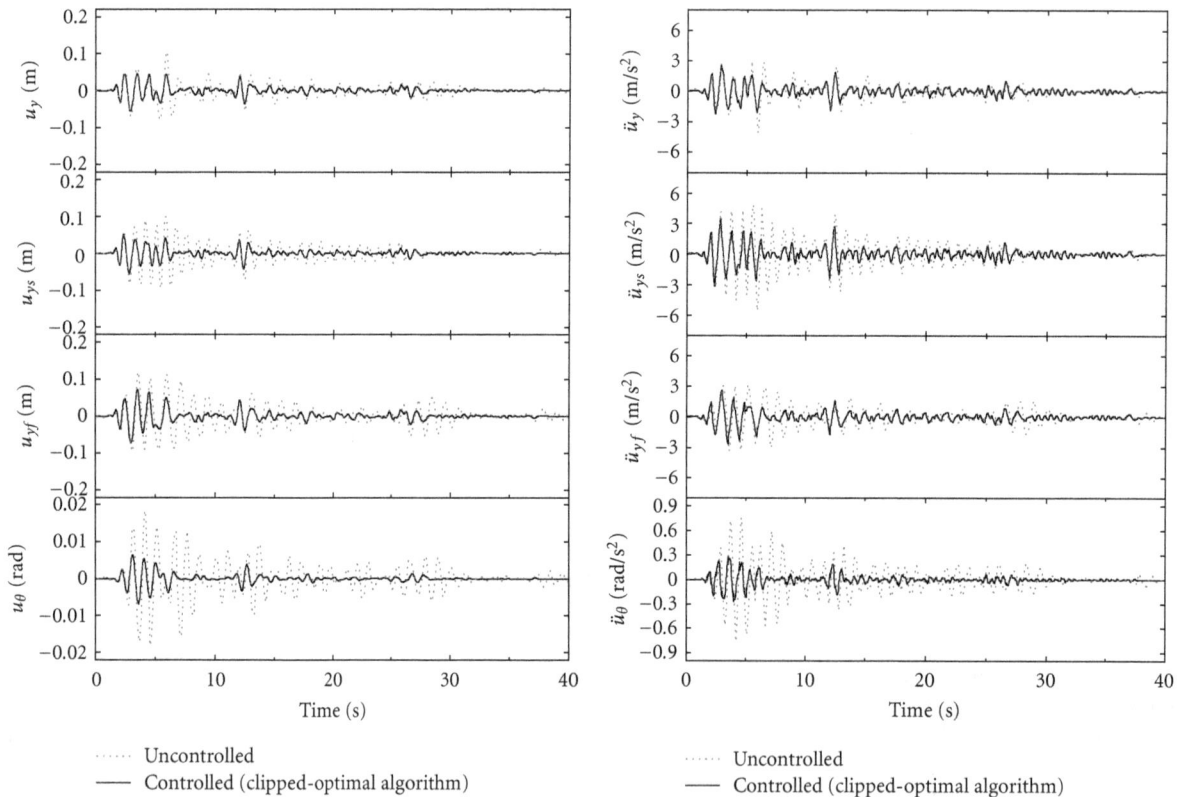

FIGURE 5: Time histories for various uncontrolled and controlled displacements and accelerations under Imperial Valley, 1940 earthquake ($T_y = 1$ s, $\Omega_\theta = 1$, $e_x/r = 0.3$).

the MR dampers are adequate to keep the response of the structure in the linear range. The equations of motion of the system in the matrix form are expressed as

$$\mathbf{M}\ddot{\mathbf{u}} + \mathbf{C}\dot{\mathbf{u}} + \mathbf{K}\mathbf{u} = -\mathbf{M}\Gamma\ddot{\mathbf{u}}_g + \Lambda\mathbf{f}, \qquad (5)$$

where $\mathbf{M}$, $\mathbf{C}$, and $\mathbf{K}$ are mass, damping, and stiffness matrices of the system, respectively; $\mathbf{u} = \{u_y \ u_\theta\}^{\mathrm{T}}$ is the displacement vector; Γ is the influence coefficient vector;

$\ddot{\mathbf{u}}_{\mathbf{g}} = \{\ddot{u}_{gy} \ 0\}^{\mathrm{T}}$ is the ground acceleration vector; $\ddot{u}_{gy}$ is the ground acceleration in y-direction, Λ is the matrix that defines the location of control devices, and $\mathbf{f}$ is the vector of control forces.

The mass matrix can be expressed as

$$\mathbf{M} = \begin{bmatrix} m & 0 \\ 0 & mr^2 \end{bmatrix}, \qquad (6)$$

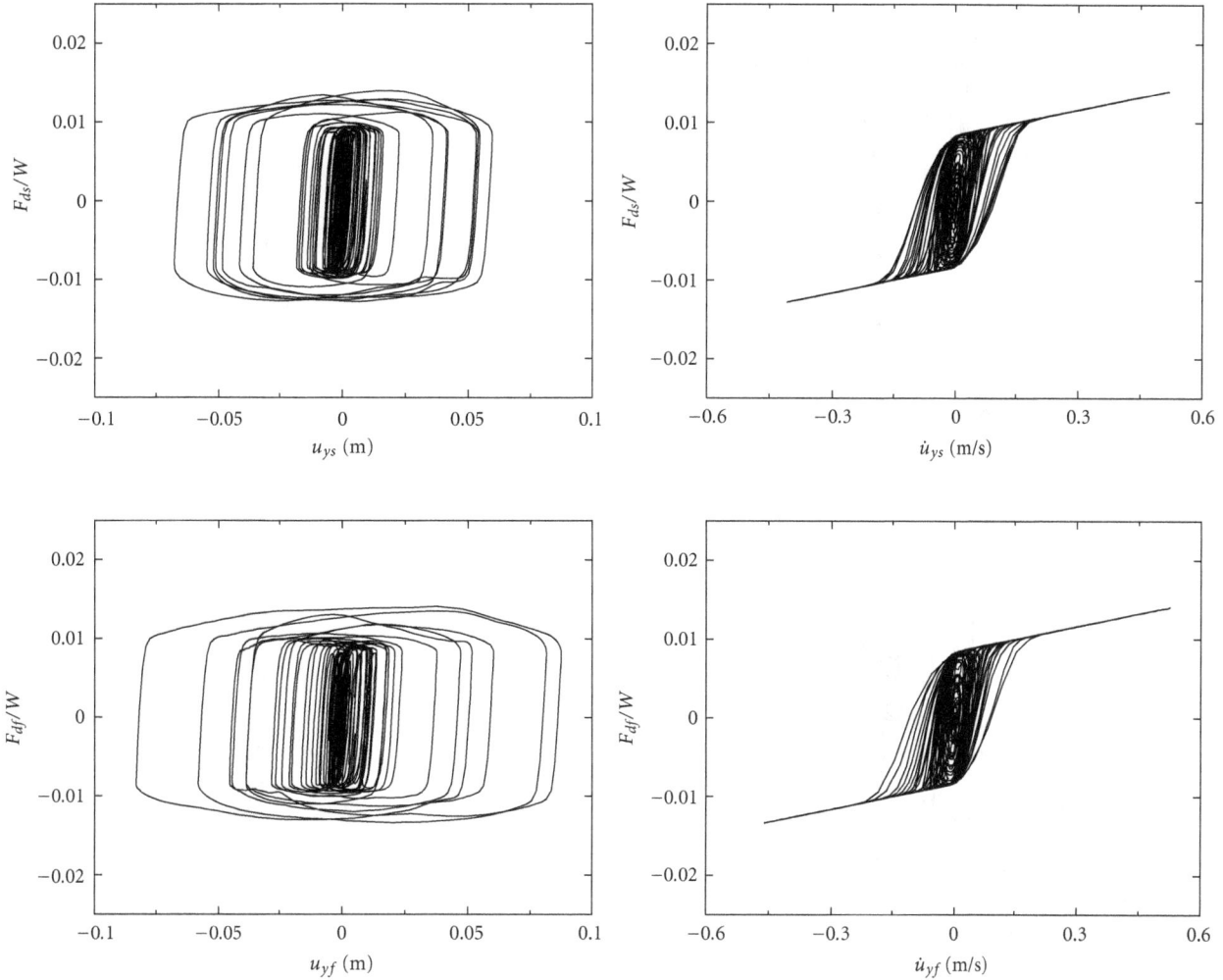

FIGURE 6: Normalized damper force-displacement and velocity loops for dampers located at stiff and flexible edges for passive-off algorithm under Imperial Valley, 1940 earthquake ($T_y = 1$ s, $\Omega_\theta = 1$, $e_x/r = 0.3$).

where m represents the lumped mass of the deck; r is the mass radius of gyration about the vertical axis through CM which is given by, $r = \sqrt{(a^2 + b^2)/12}$; where a and b are the plan dimensions of the building.

The stiffness matrix of the system is obtained as follows, after doing some algebraic manipulations [4]

$$\mathbf{K} = K_y \begin{bmatrix} 1 & e_x \\ e_x & e_x^2 + r^2\Omega_\theta^2 \end{bmatrix},$$

$$e_x = \frac{1}{K_y}\sum_i K_{yi}x_i, \qquad \Omega_\theta = \frac{\omega_\theta}{\omega_y},$$

$$\omega_\theta = \sqrt{\frac{K_{\theta r}}{mr^2}}, \qquad \omega_y = \sqrt{\frac{K_y}{m}},$$

$$K_{\theta r} = K_{\theta\theta} - e_x^2 K_y, \qquad K_{\theta\theta} = \sum_i K_{xi}y_i^2 + \sum_i K_{yi}x_i^2,$$

where K_y denotes the total lateral stiffness of the building system in y-direction; e_x is the structural eccentricity between CM and CR of the system; Ω_θ is the ratio of

uncoupled torsional to lateral frequency of the system; K_{yi} indicates the lateral stiffness of ith column in y-direction; x_i is the x-coordinate distance of ith element with respect to CM; ω_y is uncoupled lateral frequency of the system; ω_θ is uncoupled torsional frequency of the system; $K_{\theta r}$ is torsional stiffness of the system about a vertical axis at the CR; $K_{\theta\theta}$ is torsional stiffness of the system about a vertical axis at the CM; K_{xi} indicates the lateral stiffness of ith column in x-direction; y_i is the y-coordinate distance of ith element with respect to CM.

The damping matrix of the system is not known explicitly and it is constructed from the Rayleigh's damping considering mass and stiffness proportional as

$$\mathbf{C} = a_0\mathbf{M} + a_1\mathbf{K} \qquad (8)$$

in which a_0 and a_1 are the coefficients depends on damping ratio of two vibration modes. For the present study 5% damping is considered for both modes of vibration of system.

The governing equations of motion are solved using the state space method [16, 17] and rewritten as

$$\dot{\mathbf{z}} = \mathbf{A}\mathbf{z} + \mathbf{B}\mathbf{f} + \mathbf{E}\ddot{u}_g, \qquad (9)$$

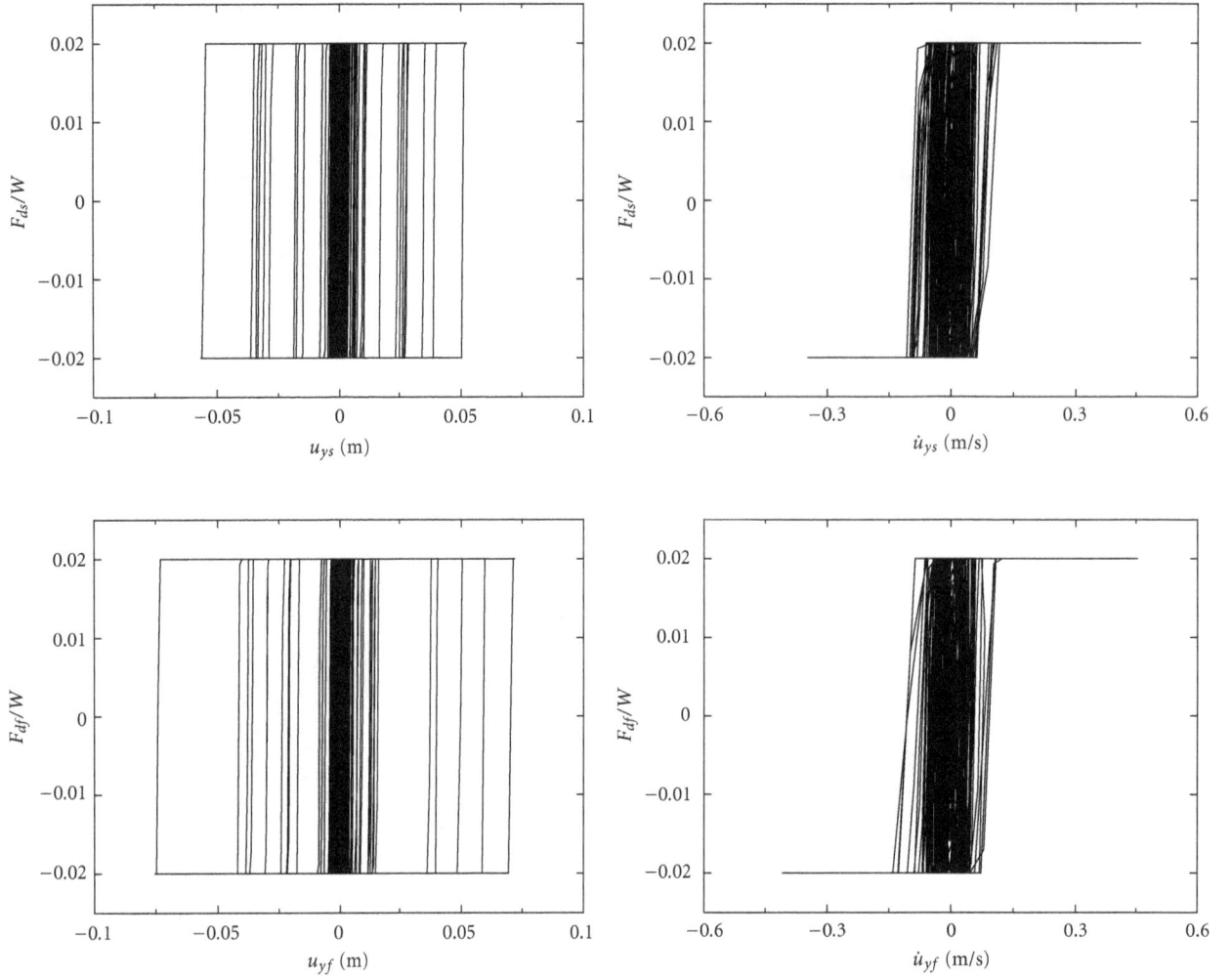

FIGURE 7: Normalized damper force-displacement and velocity loops for dampers located at stiff and flexible edges for passive-on algorithm under Imperial Valley, 1940 earthquake ($T_y = 1$ s, $\Omega_\theta = 1$, $e_x/r = 0.3$).

where $\mathbf{z} = \{\mathbf{u} \ \dot{\mathbf{u}}\}^T$ is a state vector; $\mathbf{A}$ is the system matrix; $\mathbf{B}$ is the distribution matrix of control forces; $\mathbf{E}$ is the distribution matrix of excitations. These matrices are expressed as,

$$\mathbf{A} = \begin{bmatrix} \mathbf{0} & \mathbf{I} \\ -\mathbf{M}^{-1}\mathbf{K} & -\mathbf{M}^{-1}\mathbf{C} \end{bmatrix}, \qquad \mathbf{B} = \begin{bmatrix} \mathbf{0} \\ \mathbf{M}^{-1}\mathbf{\Lambda} \end{bmatrix},$$

$$\mathbf{E} = -\begin{bmatrix} \mathbf{0} \\ \mathbf{\Gamma} \end{bmatrix}, \tag{10}$$

in which $\mathbf{I}$ is the identity matrix.

Equation (9) is discretized in time domain, and the excitation and control forces are assumed to be constant within any time interval, the solution may be written in an incremental form [16, 17]

$$\mathbf{z}[k + 1] = \mathbf{A_d}\mathbf{z}[k] + \mathbf{B_d}\mathbf{f}[k] + \mathbf{E_d}\ddot{\mathbf{u}}_{\mathbf{g}}[k], \tag{11}$$

where k denotes the time step; and $\mathbf{A_d} = e^{\mathbf{A}\Delta t}$ represents the discrete-time system matrix with Δt as the time interval. The

constant coefficient matrices $\mathbf{B_d}$ and $\mathbf{E_d}$ are the discrete-time counterparts of the matrices $\mathbf{B}$ and $\mathbf{E}$ and can be written as

$$\mathbf{B_d} = \mathbf{A}^{-1}(\mathbf{A_d} - \mathbf{I})\mathbf{B}, \qquad \mathbf{E_d} = \mathbf{A}^{-1}(\mathbf{A_d} - \mathbf{I})\mathbf{E}. \tag{12}$$

4. Semiactive Clipped-Optimal Control Algorithm

The clipped-optimal control algorithm based on acceleration feedback is used herein as shown in Figure 3(a). This algorithm has been found to be among the best performing of several non linear semiactive controllers for MR devices [15, 18].

The measurement equation is given by

$$\mathbf{y}_m = \mathbf{C}_m\mathbf{z} + \mathbf{D}_m\mathbf{f} + \mathbf{v}, \tag{13}$$

where $\mathbf{y}_m$ is the vector of measured outputs, and $\mathbf{v}$ is the measurement noise vector. In this study the measurements considered for control force determination include the accelerations at flexible and stiff edge of the structure.

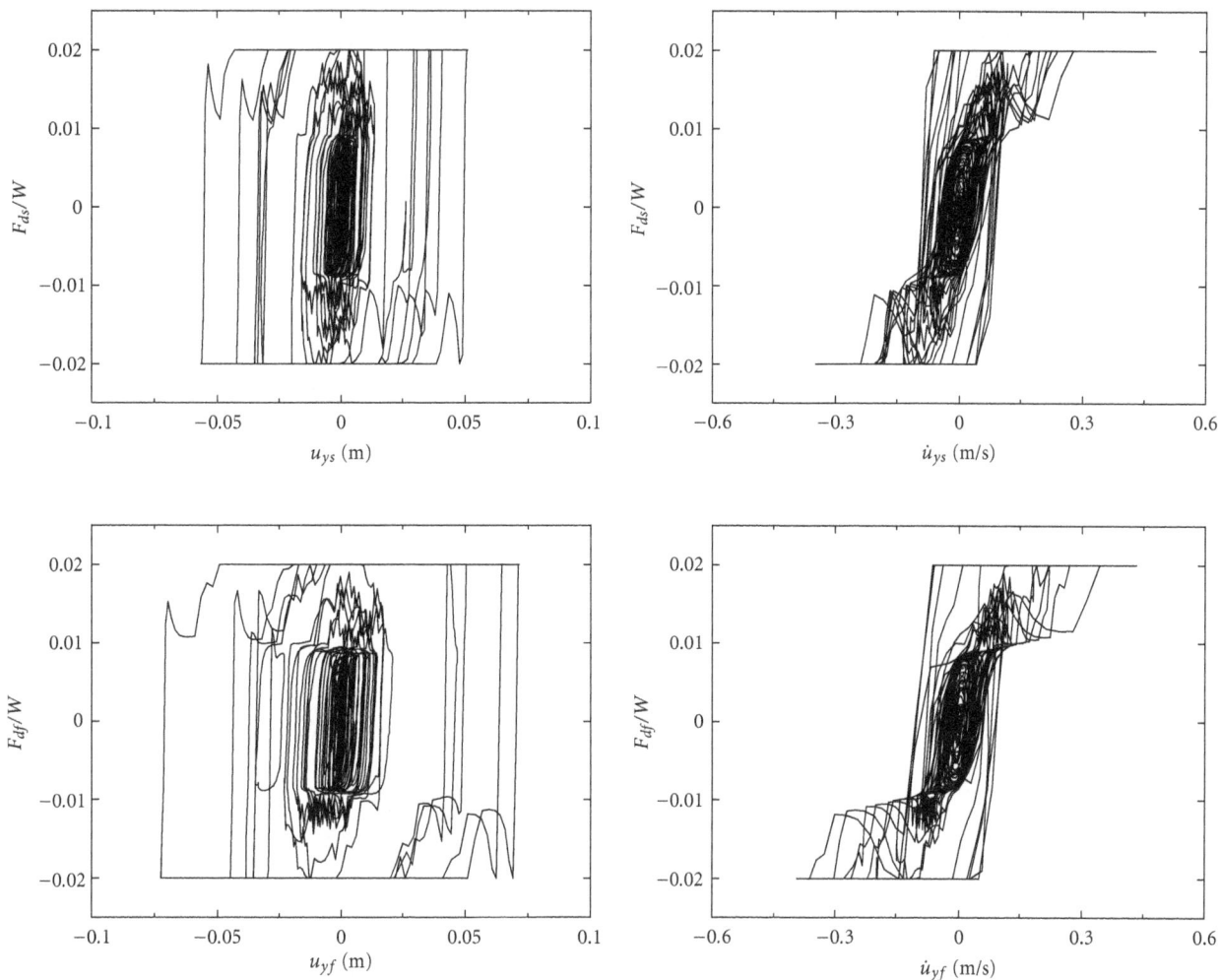

FIGURE 8: Normalized damper force-displacement and velocity loops for dampers located at stiff and flexible edges for semiactive clipped-optimal control algorithm under Imperial Valley, 1940 earthquake ($T_y = 1$ s, $\Omega_\theta = 1$, $e_x/r = 0.3$).

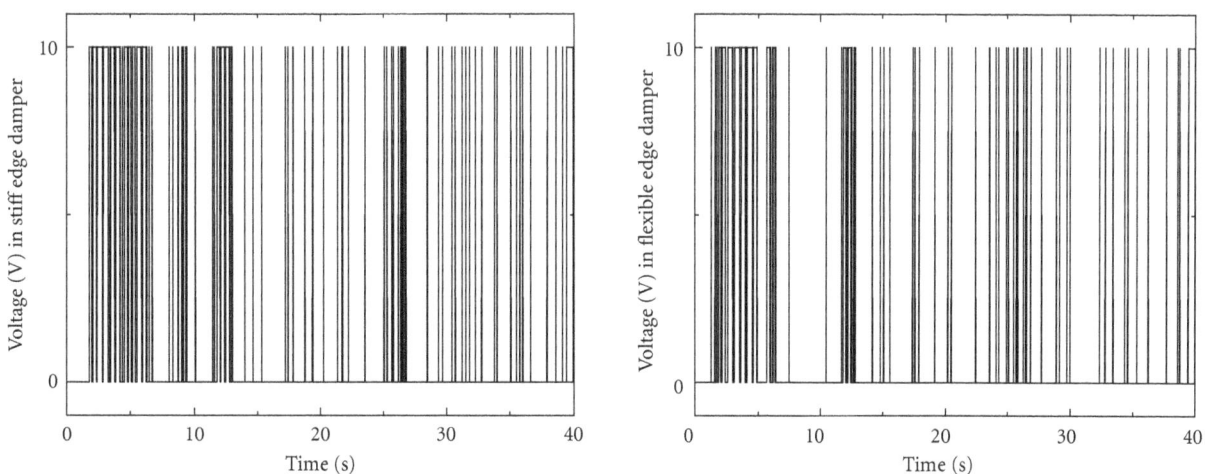

FIGURE 9: Time variation of applied voltage for dampers located at stiff and flexible edges for semiactive clipped-optimal control under Imperial Valley, 1940 earthquake ($T_y = 1$ s, $\Omega_\theta = 1$, $e_x/r = 0.3$).

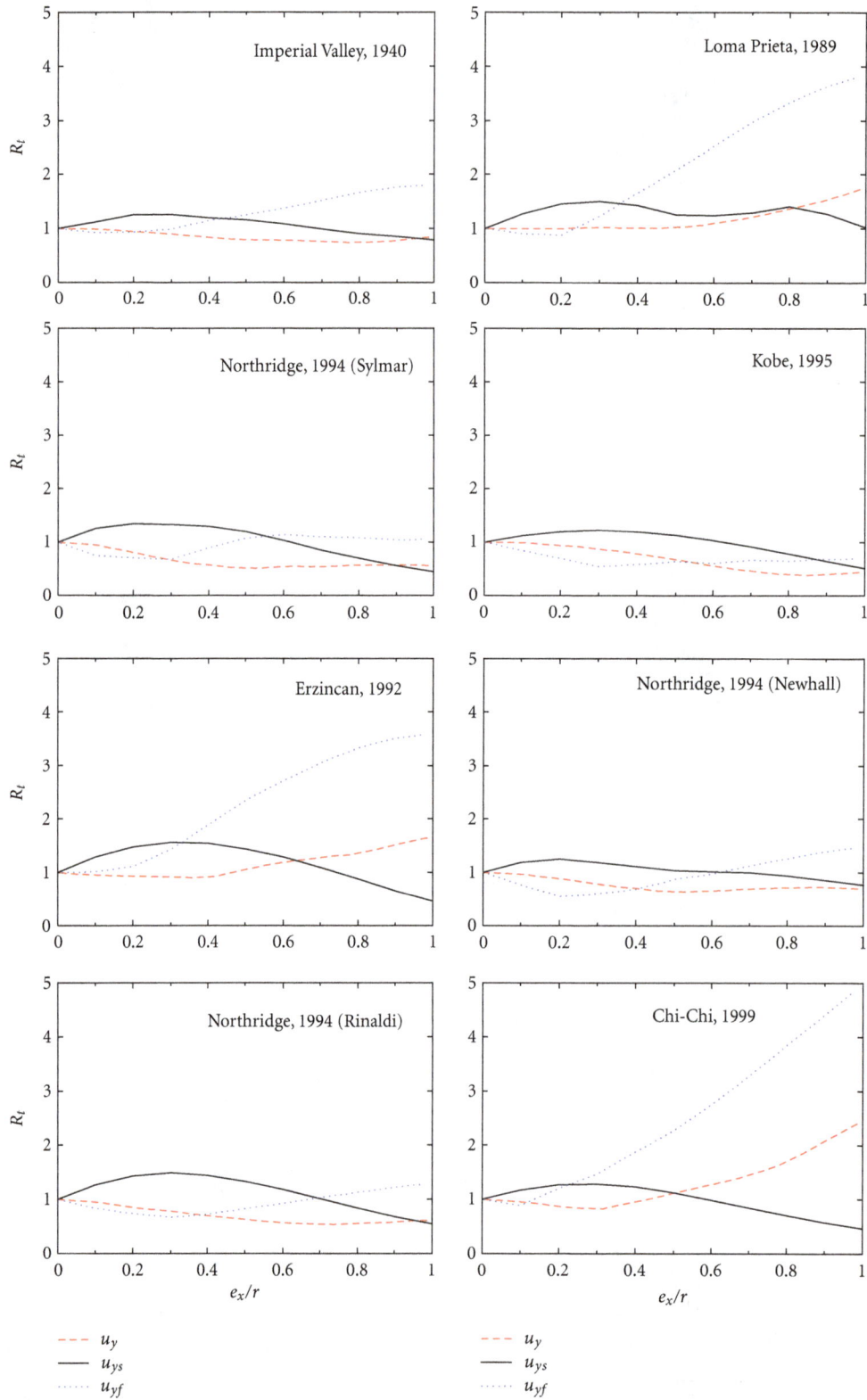

FIGURE 10: Effect of eccentricity on response ratio, R_t for various peak displacements ($T_y = 1$ s, $\Omega_\theta = 0.5$).

The matrices used in the measurement equation have the following form:

$$\mathbf{C}_m = \left[-\mathbf{P}_m\mathbf{M}^{-1}\mathbf{K} \quad -\mathbf{P}_m\mathbf{M}^{-1}\mathbf{C}\right],$$

$$\mathbf{D}_m = \left[\mathbf{P}_m\mathbf{M}^{-1}\mathbf{\Lambda}\right], \qquad \mathbf{P}_m = \begin{bmatrix} 1 & l_{a1} \\ 1 & l_{a2} \end{bmatrix}, \tag{14}$$

where l_{a1} and l_{a2} are the coordinates of the acceleration measurements of flexible and stiff edges, respectively. For the study carried out herein, the accelerometers are assumed to be fixed at the edges of building as shown in Figure 2(b).

The approach considered is to append a force feedback loop to induce the MR damper to produce approximately a desired control force f_{ci}. The force generated by the MR damper cannot be commanded; only the voltage v_i applied to the current driver for the MR damper can be directly changed as follows. When the MR damper is providing the desired optimal control force (i.e., $f_i = f_{ci}$), the voltage applied to the damper should remain at the present level. If the magnitude of the force produced by the damper is smaller than the magnitude of the desired optimal control force and the two forces have the same sign, the voltage applied to the current driver increases to the maximum level so as to increase the force produced by the damper to match the desired control force. Otherwise, the commanded voltage is set to zero. The algorithm for selecting the command signal is graphically represented in Figure 3(a) and can be stated as [15, 18]

$$v_i = V_{\max} H\{(f_{ci} - f_i)f_i\}, \tag{15}$$

where $V_{\max}$ is the voltage sent to the current driver, associated with saturation of the magnetic field in MR damper, and $H(\cdot)$ is the Heaviside step function. In this study, H_2/LQG (Linear Quadratic Gaussian) strategy is employed as a nominal controller to have desired control forces. Figure 3(b) represents the block diagram of the semiactive control system.

For the design of H_2/LQG controller, the absolute acceleration of the ground, $\ddot{u}_{gy}$, is taken to be a stationary white noise, and an infinite horizon performance index is chosen as

$$J = \lim_{\tau \to \infty} \frac{1}{\tau} E\left[\int_0^\tau \left\{\mathbf{y}_m^T\mathbf{Q}\mathbf{y}_m + \mathbf{f}^T\mathbf{R}\mathbf{f}\right\}dt\right], \tag{16}$$

where $\mathbf{Q}$ and $\mathbf{R}$ are weighting matrices for the vector of regulated/measured responses $\mathbf{y}_m = \{\ddot{u}_{mf}\ \ddot{u}_{ms}\}^T$ and of control forces $\mathbf{f} = \{F_{df}\ F_{ds}\}^T$, respectively. Where, $\ddot{u}_{mf}$ and $\ddot{u}_{ms}$ are the flexible and stiff edge accelerations measured by the accelerometers which are placed at the respective edges on the floor, as shown in Figure 2(b). For design purposes, the measurement noise vector $\mathbf{v}$ is assumed to contain identically distributed, statistically independent Gaussian white noise processes, with $S_{\ddot{u}_{gy}\ddot{u}_{gy}}/S_{v_iv_i} = \gamma_g = 50$.

The Nominal controller is represented as [10]

$$\hat{\mathbf{z}} = \left(\mathbf{A} - \mathbf{L}_g\mathbf{C}_m\right)\hat{\mathbf{z}} + \mathbf{L}_g\mathbf{y}_m + \left(\mathbf{B} - \mathbf{L}_g\mathbf{D}_m\right)\mathbf{f},$$

$$\mathbf{f}_c = -\mathbf{K}_g\hat{\mathbf{z}}, \tag{17}$$

where $\mathbf{L}_g$ is the gain matrix for the state estimator; $\hat{\mathbf{z}}$ is the estimated state vector; $\mathbf{K}_g$ is the gain matrix for the linear quadratic regulator obtained as follows:

$$\mathbf{K}_g = \mathbf{B}'\mathbf{R}^{-1}\mathbf{P}, \tag{18}$$

where $\mathbf{P}$ is the solution of the algebraic Ricatti equation given by

$$0 = \mathbf{P}\mathbf{A} + \mathbf{A}'\mathbf{P} - \mathbf{P}\mathbf{B}\mathbf{R}^{-1}\mathbf{B}'\mathbf{P} + \mathbf{C}'_m\mathbf{Q}\mathbf{C}_m,$$

$$\mathbf{L}_g = (\mathbf{C}_m\mathbf{S})', \tag{19}$$

where $\mathbf{S}$ is the solution of the algebraic Ricatti equation given by

$$0 = \mathbf{S}\mathbf{A}' + \mathbf{A}\mathbf{S} - \mathbf{S}\mathbf{C}'_m\mathbf{C}_m\mathbf{S} + \gamma_g\mathbf{E}\mathbf{E}'. \tag{20}$$

The response weighting matrix $\mathbf{Q}$ corresponding to the regulated output vector $\mathbf{y}_m$ is considered as follows:

$$\mathbf{Q} = q_w \cdot \begin{bmatrix} 1 & 0 \\ 0 & 1 \end{bmatrix}, \tag{21}$$

where $\mathbf{q}_w$ is the coefficient for weighting matrix $\mathbf{Q}$.

The weighting matrix $\mathbf{R}$ which is corresponding to the control force vector is considered as follows:

$$\mathbf{R} = \mathbf{r}_w \cdot \begin{bmatrix} 1 & 0 \\ 0 & 1 \end{bmatrix}, \tag{22}$$

where r_w is the coefficient for weighting matrix $\mathbf{R}$.

5. Numerical Study

The seismic response of linearly elastic, idealized single-storey, one-way asymmetric system installed with semiactive MR dampers with clipped-optimal control algorithm is investigated by numerical simulation study. The response quantities of interest are lateral and torsional displacements of the floor mass obtained at the CM (u_y and u_θ), displacements at stiff and flexible edges of the system (u_{ys} and $u_{yf} = u_y \pm bu_\theta/2$), lateral and torsional accelerations of the floor mass obtained at the CM ($\ddot{u}_y$ and $\ddot{u}_\theta$), accelerations at stiff and flexible edges of the system ($\ddot{u}_{ys}$ and $\ddot{u}_{yf} = \ddot{u}_y \pm b\ddot{u}_\theta/2$), base shear ($V_y$), as well as damper control forces at stiff edge (F_{ds}) and at flexible edge (F_{df}) of the building. The response of the system is investigated under following parametric variations: structural eccentricity ratio (e_x/r), uncoupled lateral time period of system ($T_y = 2\pi/\omega_y$) and ratio of uncoupled torsional to lateral frequency of the system ($\Omega_\theta = \omega_\theta/\omega_y$). The peak responses are obtained corresponding to the important parameters which are listed above and variations are plotted for the eight considered earthquake ground motions namely, Imperial Valley (1940), Loma Prieta (1989), Northridge (Sylmar, 1994), Kobe (1995), Erzincan (1992), Northridge (Newhall, 1994), Northridge (Rinaldi, 1994), and Chi-Chi (1999) for the present study with corresponding peak ground acceleration (PGA) values of

TABLE 1: Details of earthquake motions considered for the numerical study.

Earthquake	Recording Station	Component	Duration (sec)	PGA (g)
Imperial Valley, 19th May, 1940	El Centro (Array # 9)	ELC 180	40	0.31
Loma Prieta, 18th October, 1989	Los Gatos Presentation Center	LGP 000	25	0.96
Northridge, 17th January, 1994	Sylmar Converter Station	SCS 142	40	0.89
Kobe, 16th January, 1995	Japan Meteorological Agency	KJM 000	48	0.82
Erzincan, 13th March, 1992	Erzincan	ERZ-NS	20	0.51
Northridge, 17th January, 1994	Newhall Fire Station	NWH360	40	0.59
Northridge, 17th January, 1994	Rinaldi Receiving Station	RRS228	20	0.82
Chi-Chi, 20th September, 1999	TCU068	TCU068-N	90	0.57

TABLE 2: Parameters for the MR damper model (200 kN capacity) [19].

Parameter	Value	Parameter	Value
α_a	46.2 kN/m	k_1	0.0097 kN/m
α_b	41.2 kN/m/V	γ_m	164.0 m^{-2}
c_{0a}	110.0 kN s/m	β_m	164.0 m^{-2}
c_{0b}	114.3 kN s/m/V	A_m	1107.2
c_{1a}	8359.2 kN s/m	n	2
c_{1b}	7482.9 kN s/m/V	η	100 s^{-1}
k_0	0.002 kN/m	x_0	0.18 m

0.31 g, 0.96 g, 0.89 g, 0.82 g, 0.51 g, 0.59 g, 0.82 g, and 0.57 g as per the details summarized in Table 1. The input parameters considered for the present study are; Total weight of building, $W = 10,000$ kN and aspect ratio of plan dimension is kept as unity. For the numerical simulation study carried out herein, two MR dampers, each having 200 kN capacity [20]

are assumed to be installed at the edges of the building as shown in Figure 2. The parameters of mechanical model of a damper are presented in Table 2 [19].

In order to study the effectiveness of implemented semiactive control system, the response is expressed in terms of indices, R_e and R_t are defined as follows:

$$R_e = \frac{\text{Peak (or RMS) response of controlled asymmetric system}}{\text{Peak (or RMS) response of corresponding uncontrolled system}},$$

$$R_t = \frac{\text{Peak (or RMS) response of controlled asymmetric system}}{\text{Peak (or RMS) response of corresponding symmetric system}}.$$

(23)

The value of R_e less than one indicates that the implemented semiactive control system is effective in controlling the responses. On the other hand, the value of R_t reflects the effects of torsional coupling on the effectiveness of semiactive control system and seismic behavior of system. The value of R_t greater than one indicates that the response of controlled asymmetric system increases due to torsional coupling as compared to the corresponding symmetric system. The effects of various levels of excitations for particular earthquake motion are not studied by considering the fact that the main objective is to investigate the effects of torsional coupling on the effectiveness of semiactive device for asymmetric systems as compared to symmetric systems and the responses are obtained in terms of response ratios. Moreover, in past, some work have been reported with the use of semiactive MR damper with the assumption that the building behaves in linear range and

also without considering the effects of different levels of excitations [15, 18, 21].

In order to determine the optimum coefficients of weighting matrices **Q** and **R**, the response ratio, R_e for peak and root mean square (RMS) values of displacements as well as accelerations of the floor mass are obtained for the symmetric system with $T_y = 1$ s. Initially, the coefficient, r_w of the matrix, **R** is considered as unity and the coefficient, q_w of matrix, **Q** is varied by placing higher weightage on the response quantities. The response are obtained under Imperial Valley, 1940 earthquake and shown in Figure 4(a). From the figure, it can be noticed that initially with the increase in q_w, the ratio, R_e decreases up to certain limit and then increases with further increase in value of q_w. This means there exists an optimum value of coefficient, q_w for which R_e is minimum, which shows the maximum reduction in various responses. Thus, the optimum value for

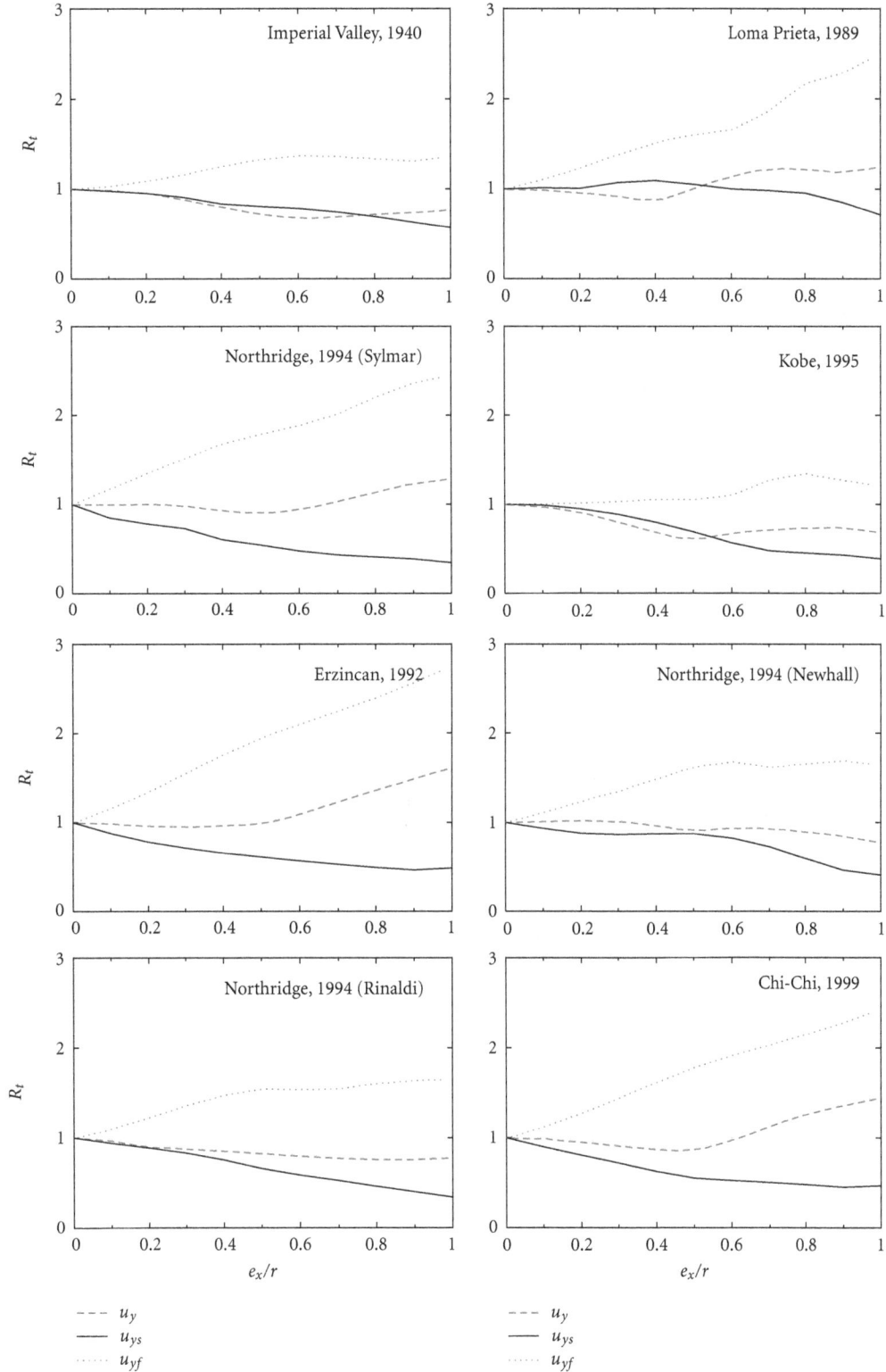

FIGURE 11: Effect of eccentricity on response ratio, R_t for various peak displacements ($T_y = 1$ s, $\Omega_\theta = 1$).

q_w is found to be 10^{11}. Further, trials have been done by keeping q_w as constant (i.e., $q_w = 10^{11}$) and by varying the coefficient, r_w and the results are shown in Figure 4(b). From the figure, it can be noticed that the ratio, R_e is minimum corresponding to the value of $r_w = 1$. Moreover, the above

results are also verified for the systems with different time period, T_y (i.e., varied from 0.1 s to 3 s), and it is found that the optimum values of q_w and r_w remains almost in the similar range such as to have optimum compromise between the reduction in peak and RMS displacement and acceleration responses. Thus, for the study carried out herein, the optimum coefficients for the weighting matrices $\mathbf{Q}$ and $\mathbf{R}$ are considered as $q_w = 10^{11}$ and $r_w = 1$.

The present study is carried out by considering two MR dampers (one on each side of building), each of 200 kN capacity with semiactive clipped-optimal control algorithm with maximum applied voltage of 10 V. The responses are also obtained with passive-off and passive-on cases, with constant zero voltage and with constant maximum voltage, respectively, and compared with the semiactive control case. However, the main focus is to study the behavior of the system equipped with semiactive dampers. Figure 5 shows the time histories of various displacement and acceleration response quantities of the uncontrolled building compared with the corresponding building controlled with semiactive MR dampers with clipped-optimal control algorithm.

The responses are obtained for the system with $T_y = 1$ s, $\Omega_\theta = 1$, and $e_x/r = 0.3$ (intermediate eccentricity) under Imperial Valley, 1940 earthquake. The significant reduction in displacement and acceleration responses at CM, at flexible and stiff edges as well as torsional response can be noticed from the figure. Thus, the semiactive system is quite effective in reducing lateral-torsional responses.

Furthermore, Table 3 depicts the comparison between the three control strategies, namely, passive-off, passive-on, and semiactive clipped-optimal control for the system with $T_y = 1$ s, $\Omega_\theta = 1$, $e_x/r = 0.3$ under Imperial Valley, 1940 earthquake. The ratio, R_e for various responses based on peak and RMS values are summarized herein. In general, ratio R_e comes out be less than unity for all control strategies implying the effectiveness of control systems. It is further observed that passive-on strategy gives higher reduction in peak and RMS displacements and accelerations as compared to passive-off case. Also, it is observed that the reduction in peak displacements and accelerations obtained at CM (u_y and $\ddot{u}_y$), at edges (u_{ys}, u_{yf}, $\ddot{u}_{ys}$, and $\ddot{u}_{yf}$) as well as its torsional components (u_θ and $\ddot{u}_\theta$) for the semiactive case is found as little higher or comparable with those obtained with passive-on case. Moreover, the reduction in RMS of various displacement quantities is slightly higher for passive-on case as compared to corresponding semiactive case, whereas higher reduction in RMS of various acceleration quantities is found for semiactive case as compared to passive-on case. The table also shows the normalized control forces of dampers which are placed at the edges of the building. It is important to note that the RMS control forces required for passive-on case is much more higher than the passive-off case while for the semiactive case it is only little higher than passive-off case. This reflects that, for the semiactive clipped-optimal control algorithm, with little increase in RMS damper forces as compared to passive-off case, the higher or comparable reduction in peak and RMS values of lateral, edge, and torsional responses is obtained than those which is obtained by passive-on case.

Moreover, Figures 6, 7, and 8 represents the hysteresis loops for normalized damper force with displacement and velocity for the considered control strategies, namely,

passive-off, passive-on, and semiactive clipped-optimal control case, respectively, for the system with $T_y = 1$ s, $\Omega_\theta = 1$, $e_x/r = 0.3$ under Imperial Valley, 1940 earthquake. Further, Figure 9 shows the time variation of applied voltage for the semiactive clipped-optimal control case for the dampers located at stiff and flexible edges.

One of the key parameters affecting the lateral-torsional response of buildings is the structural eccentricity. In order to study the effects of eccentricity and torsional coupling on the seismic response of asymmetric systems, the variations of ratio, R_t for lateral and edge displacements are plotted against eccentricity ratio, e_x/r (which is varied from 0 to 1) for eight selected earthquakes. The variations are shown for torsionally flexible ($\Omega_\theta = 0.5$), strongly coupled ($\Omega_\theta = 1$) and torsionally stiff ($\Omega_\theta = 2$) systems with $T_y = 1$ s in Figures 10, 11, and 12, respectively. It is observed that for the systems with $\Omega_\theta = 0.5$, with increase in e_x/r, the ratio, R_t for lateral displacement, u_y decreases and remains less than unity, in general, whereas, the ratio, R_t for u_{ys} increases first, up to an intermediate eccentricity (i.e., $e_x/r < 0.5$) and then decreases with further increase in eccentricity ratio. For flexible edge displacement, u_{yf}, the ratio, R_t first decreases up to an intermediate eccentricity and then increases for higher eccentricity. This shows that for torsionally flexible system with an intermediate eccentricity, u_y and u_{yf} reduces and u_{ys} increases due to torsional coupling as compared to corresponding symmetric system and for the system with higher eccentricity, u_{yf} increases, and u_y and u_{ys} decrease due to torsional coupling except for Erzincan (1992) and Chi-Chi (1999) earthquakes. Hence, for torsionally flexible systems with an intermediate eccentricity, the effectiveness of control system is more for asymmetric system for reducing u_y and u_{yf} as compared to corresponding symmetric systems and hence it will be underestimated by ignoring the eccentricity and modeling the building as 2D. Whereas, the effectiveness of control system is less for asymmetric system for reducing u_{ys} as compared to corresponding symmetric systems and hence it will be overestimated for u_{ys}. Whereas, for torsionally flexible systems with higher eccentricities, effectiveness of control system will be underestimated for reducing u_y and u_{ys} and it will be overestimated for u_{yf}.

Moreover, as shown in Figures 11 and 12 for the systems with $\Omega_\theta = 1$ and 2, the ratio, R_t for u_{ys} decreases and for response, u_{yf}, it increases with increase in eccentricity ratio. This implies that the effectiveness of control system is more for asymmetric system as compared to corresponding symmetric system in reducing u_{ys} and hence it will be underestimated and the effectiveness is less for asymmetric system in reducing u_{yf} and hence it will be underestimated by ignoring the effect of torsional coupling. Also, for the systems with $\Omega_\theta = 1$, the ratio, R_t for u_y remains less that unity except for the systems with higher eccentricities. However, for torsionally stiff systems, the ratio, R_t for lateral and edge displacements, remains very less sensitive to the variation of e_x/r. Thus, the effects of torsional coupling can be ignored for such systems while estimating the effectiveness of semiactive control system for asymmetric system as compared to symmetric system. Further, the difference between edge displacements of asymmetric and

TABLE 3: Response ratio, R_e for Peak and RMS Responses for various control strategies under Imperial Valley, 1940 earthquake ($T_y = 1$, $\Omega_\theta = 1$, $e_x/r = 0.3$).

Response quantities	Control strategy	Passive-off	Passive-on	Semiactive clipped-optimal control
Based on peak response	R_e for u_y	0.56903	0.52736 (**7.32%**)	0.51507 (**9.48%**)
	R_e for u_{ys}	0.65228	0.54286 (**16.77%**)	0.54447 (**16.53%**)
	R_e for u_{yf}	0.71709	0.61866 (**13.73%**)	0.59502 (**17.02%**)
	R_e for u_θ	0.56500	0.36206 (**35.92%**)	0.36790 (**35.89%**)
	R_e for $\ddot{u}_y$	0.61011	0.61266 (**−0.42%**)	0.60387 (**1.02%**)
	R_e for $\ddot{u}_{ys}$	0.67720	0.61226 (**9.59%**)	0.61318 (**9.45%**)
	R_e for $\ddot{u}_{yf}$	0.76390	0.72524 (**5.06%**)	0.71258 (**6.72%**)
	R_e for $\ddot{u}_\theta$	0.53690	0.36741 (**31.57%**)	0.36840 (**31.38%**)
	F_{ds}/W	0.01401	0.02000 (**−42.80%**)	0.02000 (**−42.80%**)
	F_{df}/W	0.01407	0.02000 (**−42.11%**)	0.02000 (**−42.11%**)
Based on RMS response	R_e for u_y	0.61004	0.49222 (**19.31%**)	0.52683 (**13.64%**)
	R_e for u_{ys}	0.52862	0.36608 (**30.75%**)	0.38857 (**26.49%**)
	R_e for u_{yf}	0.50951	0.38461 (**24.51%**)	0.41568 (**19.17%**)
	R_e for u_θ	0.42665	0.24874 (**41.70%**)	0.26326 (**38.30%**)
	R_e for $\ddot{u}_y$	0.68025	0.67366 (**0.97%**)	0.63343 (**6.88%**)
	R_e for $\ddot{u}_{ys}$	0.55606	0.46018 (**17.24%**)	0.44690 (**19.63%**)
	R_e for $\ddot{u}_{yf}$	0.55171	0.53870 (**2.36%**)	0.49283 (**10.67%**)
	R_e for $\ddot{u}_\theta$	0.44723	0.29753 (**33.47%**)	0.29076 (**34.99%**)
	F_{ds}/W	0.00718	0.01619 (**−125.49%**)	0.00969 (**−34.88%**)
	F_{df}/W	0.00734	0.01627 (**−121.68%**)	0.00922 (**−25.63%**)

(Numbers in parentheses indicate percentage reduction as compared to the passive-off case. Positive numbers correspond to a response reduction.)

corresponding symmetric system is higher for strongly coupled and torsionally flexible systems as compared to torsionally stiff systems and the difference increases with the increase in structural eccentricity. Thus, for torsionally flexible and strongly coupled systems, the effects of torsional coupling can not be neglected while estimating the effectiveness of semiactive control system for asymmetric systems as compared to corresponding symmetric systems in controlling edge displacements.

To study the effects of torsional coupling on base shear, V_y, the variations of ratio, R_t for V_y are plotted against e_x/r for three values of frequency ratios (i.e., $\Omega_\theta = 0.5$, 1 and 2) as shown in the Figure 13. From the figure, it can be observed that with increase in e_x/r, ratio, R_t decreases, in general. Further, it is noticed that the ratio, R_t for V_y, is more sensitive to variation of e_x/r for systems with $\Omega_\theta = 1$ followed by the systems with $\Omega_\theta = 0.5$ and $\Omega_\theta = 2$. This indicates that the torsional coupling reduces the base shear for controlled asymmetric systems as compared to corresponding symmetric systems and hence, the effectiveness of semiactive control system in reducing base shear is more for asymmetric systems as compared to corresponding symmetric systems and the effectiveness will be underestimated, if torsional coupling is ignored and building is modeled as 2D. This effect is more pronounced for the systems with higher superstructure eccentricities. Thus, the difference between the base shear of asymmetric and corresponding symmetric system increases with the increase in eccentricity ratio and it is significant for strongly coupled systems followed by torsionally flexible and torsionally stiff systems.

Moreover, the variations of ratio, R_t for lateral and edge accelerations are shown for the systems with $\Omega_\theta = 0.5, 1$, and 2 in Figures 14, 15, and 16, respectively. In general, it is found that the trends for edge accelerations are opposite to those obtained for the edge displacements. It is further observed that for the systems with $\Omega_\theta = 0.5$, with increase in e_x/r, the ratio, R_t for lateral acceleration, $\ddot{u}_y$ and flexible edge acceleration, $\ddot{u}_{yf}$ decreases and that of stiff edge acceleration, $\ddot{u}_{ys}$ increases and remains more than a unity. Thus, for torsionally flexible systems, by ignoring the eccentricity, the effectiveness of semiactive control system will be underestimated for reducing $\ddot{u}_y$ and $\ddot{u}_{yf}$ and overestimated for $\ddot{u}_{ys}$. Similarly, for the system with $\Omega_\theta = 1$, the ratio, R_t for $\ddot{u}_y$ reduces with the increase in e_x/r and for $\ddot{u}_{ys}$, the ratio remains slightly more that unity, in general. Whereas, for $\ddot{u}_{yf}$, the ratio remains less than or close to unity. However, the variation of R_t for $\ddot{u}_{yf}$ remains less sensitive to e_x/r as compared to those of responses, $\ddot{u}_y$ and $\ddot{u}_{ys}$. Also, it is found that the variation of R_t with e_x/r for edge accelerations are more pronounced for systems with $\Omega_\theta = 0.5$ as compared to $\Omega_\theta = 1$. Moreover, as shown in Figure 16, for systems with $\Omega_\theta = 2$, with increase in e_x/r, the ratio, R_t for $\ddot{u}_y$ and $\ddot{u}_{ys}$ slightly decreases and for $\ddot{u}_{yf}$, it increases slightly. However, the rate of change of ratio, R_t with e_x/r is very less sensitive for such systems as compared to systems with $\Omega_\theta = 0.5$ and 1. Thus, the effects of torsional coupling are more pronounced for torsionally flexible systems than strongly coupled systems while estimating the effectiveness of semiactive control system for asymmetric systems in controlling lateral and edge accelerations as compared to corresponding symmetric systems.

Figures 17 and 18 show the variations of response ratio, R_t against e_x/r for three values of Ω_θ, for normalized RMS control forces developed by the dampers which are installed at stiff edge (F_{ds}) and at flexible edge (F_{df}) of the building. It can be observed that for torsionally flexible systems, the ratio, R_t for F_{ds} increases with increase in e_x/r and remains more than unity and for F_{df}, it decreases and remains less than unity except for the systems with higher eccentricities under Erzincan (1992) and Chi-Chi (1999) earthquakes. Hence, by ignoring the effects of torsional coupling for torsionally flexible systems, control forces at stiff edge, F_{ds}, will be underestimated, and at flexible edge, F_{df}, it will be overestimated as compared to corresponding symmetric systems. Further, for strongly coupled systems, the ratio, R_t for F_{ds} decreases with the increase in e_x/r and remains less than or very close to unity, whereas for F_{df}, it increases with

the increase e_x/r and remains more than or close to unity. This implies that for strongly coupled systems, by ignoring the effects of torsional coupling, the control forces at stiff edge will be overestimated and at flexible edge, those will be underestimated as compared to corresponding symmetric systems. Whereas, for torsionally stiff systems, the ratio, R_t for F_{ds} decreases with the increase in e_x/r and for F_{df}, it increases with the increase e_x/r. Thus, by ignoring the effects of torsional coupling, control forces at stiff edge will be overestimated and at flexible edge, those will be underestimated as compared to corresponding symmetric systems. However, the rate of change for ratio, R_t against e_x/r for control forces is more sensitive for systems with $\Omega_\theta = 0.5$ followed by the systems with $\Omega_\theta = 1$ and 2. Thus, the effects of torsional coupling on the estimation of control forces significantly depend on superstructure eccentricity and torsional to lateral frequency ratio.

To study, the effects of torsional coupling on the seismic behavior of laterally stiff and laterally flexible structural systems, the variations of ratio, R_t for edge displacements and accelerations as well as for normalized RMS control forces are plotted against uncoupled lateral time period, T_y (i.e., varied from 0.1 s to 3 s). The results are obtained for an intermediate eccentricity ratio ($e_x/r = 0.3$) and shown in Figure 19 for the systems with $\Omega_\theta = 0.5, 1$ and 2. The response ratios are obtained for eight selected earthquakes and the average trend from those earthquakes are shown in Figure 19. It can be noticed from the figure that for laterally stiff systems with $\Omega_\theta = 0.5$ and 1, the difference between the edge displacements of asymmetric and corresponding symmetric systems is higher as compared to laterally flexible systems and as the flexibility of superstructure increases the difference tends to decreases. Whereas, for torsionally stiff systems, variation in ratio R_t for edge displacement and acceleration response remains insensitive to the change in values of T_y. Further, for laterally stiff systems with $\Omega_\theta = 0.5$, the difference between the flexible edge displacement of asymmetric and corresponding symmetric systems is significantly higher as compared to laterally flexible systems. This means that for laterally stiff systems, the torsional coupling significantly increases the flexible edge displacement for asymmetric system as compared to symmetric system. In addition, for laterally stiff systems with $\Omega_\theta = 0.5$, the values of ratio, R_t for F_{ds} and F_{df} remains more than unity and for laterally flexible systems with $\Omega_\theta = 0.5$, the ratio, R_t for F_{ds} remains more than unity and for F_{df}, it remains less than unity. Thus, by ignoring the effects of torsional couplings for laterally stiff systems with $\Omega_\theta = 0.5$, the control forces at stiff edge and flexible edge damper will be underestimated for asymmetric systems as compared to corresponding symmetric systems. Whereas, for laterally flexible systems with $\Omega_\theta = 0.5$, the control forces at stiff edge damper will be underestimated and at flexible edge damper, it will be overestimated. Similarly, for laterally stiff systems with $\Omega_\theta = 1$, the values of ratio, R_t for F_{ds} remains more than unity and for F_{df}, it remains more than unity and for laterally flexible systems with $\Omega_\theta = 1$, the ratio, R_t for F_{ds} and F_{df} remains close to the unity. Thus, by ignoring the effects of torsional couplings for laterally stiff systems with

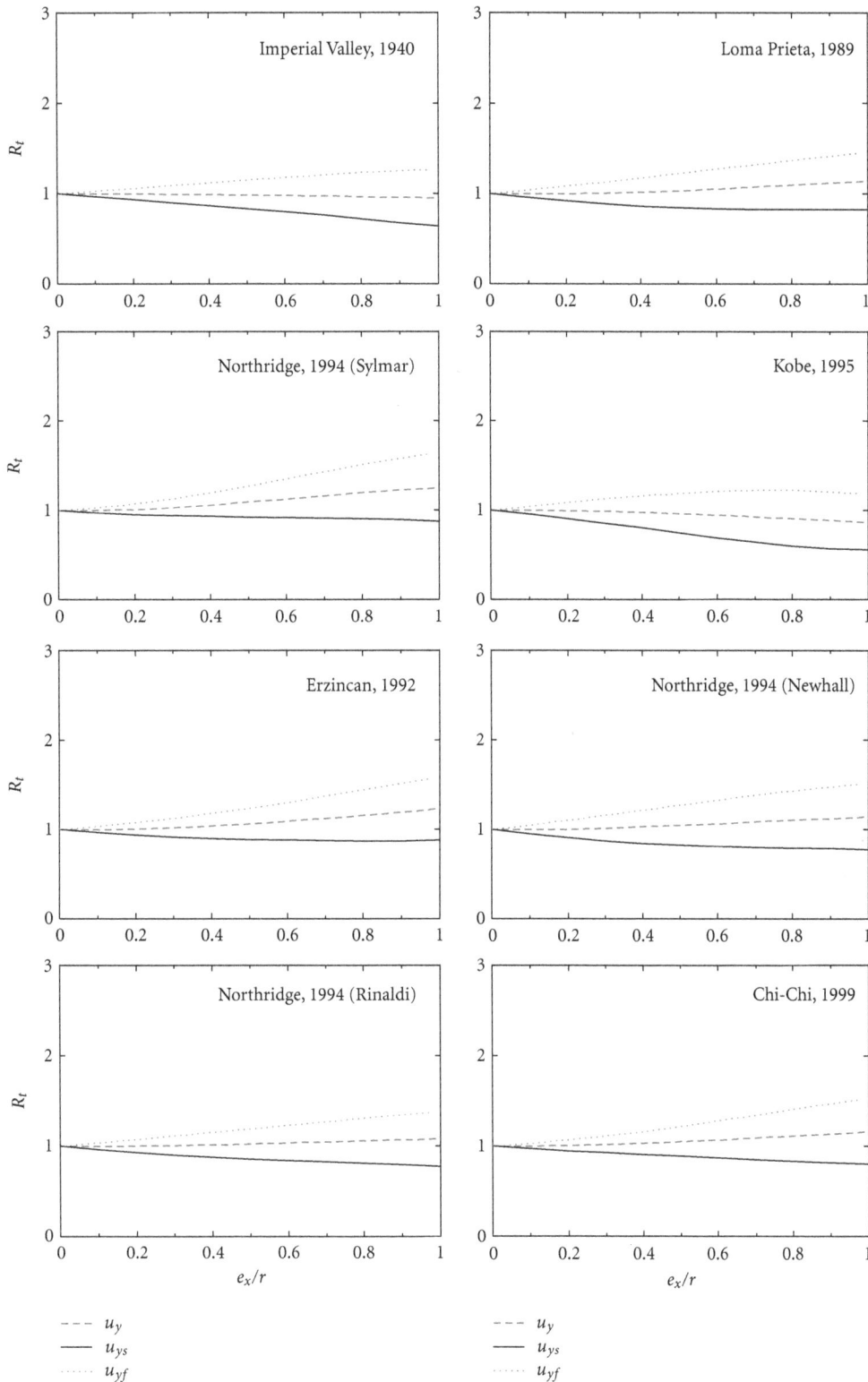

FIGURE 12: Effect of eccentricity on response ratio, R_t for various peak displacements ($T_y = 1$ s, $\Omega_\theta = 2$).

$\Omega_\theta = 1$, the control forces at stiff edge will be overestimated and at flexible edge that will be underestimated. Moreover, for laterally stiff to laterally flexible systems with $\Omega_\theta = 2$, the values of ratio, R_t, for F_{ds} remains slightly less than unity and for F_{df}, it remains slightly more than unity implying that by ignoring the effects of torsional couplings for such systems, the control forces at stiff edge will be overestimated and at flexible edge that will be underestimated. Further, it is important to note that for the systems with $\Omega_\theta = 0.5$, 1 and 2, the variations in values of ratio, R_t, for normalized damper

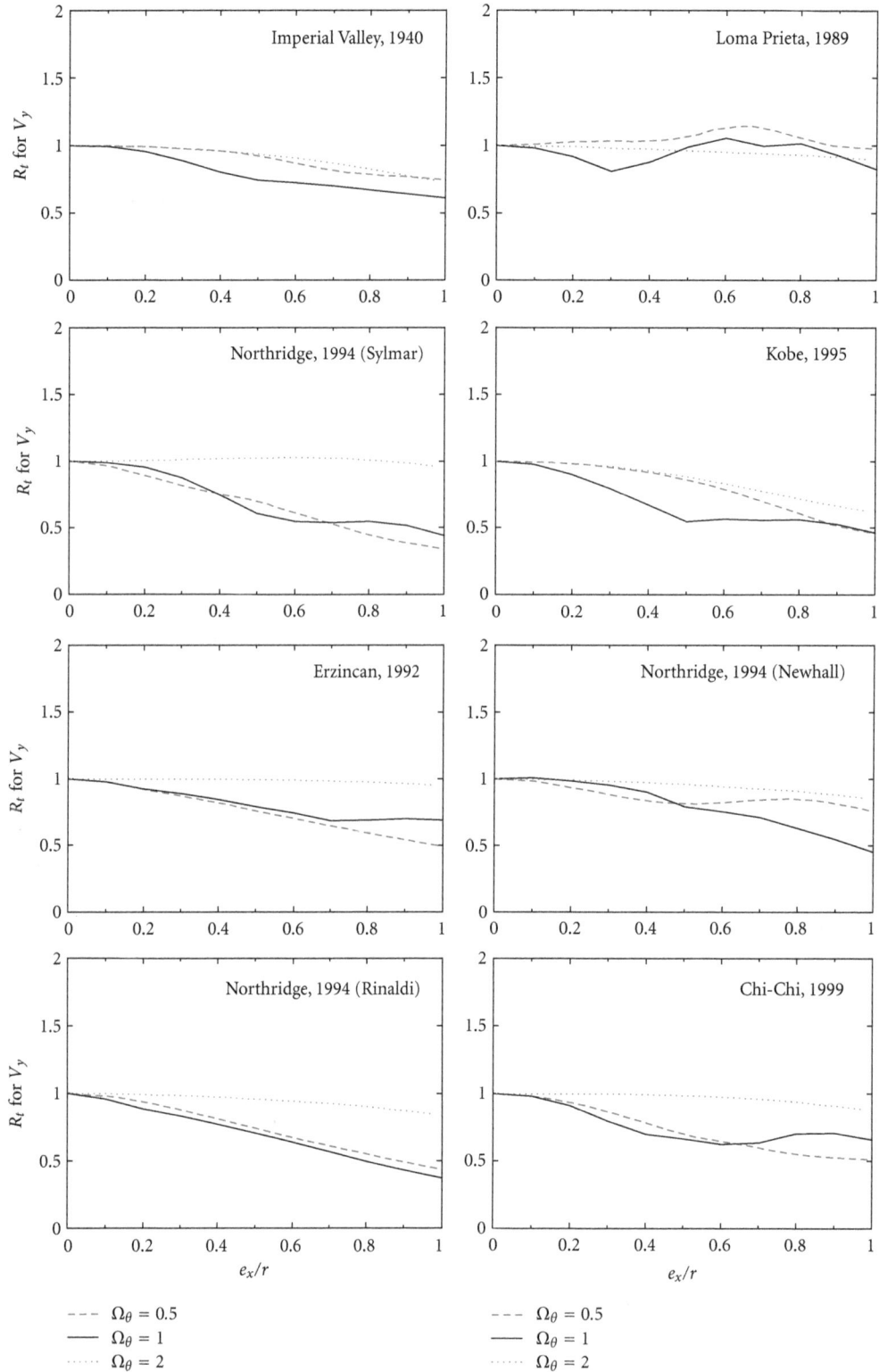

FIGURE 13: Effect of eccentricity on response ratio, R_t for peak base shear, V_y ($T_y = 1$ s).

control forces, F_{ds} and F_{df}, are much more sensitive and significant for laterally stiff systems as compared to laterally flexible systems. Thus, the difference between the damper control forces for asymmetric systems and corresponding symmetric systems is higher for laterally stiff systems as compared to laterally flexible systems and this difference tends to reduce with the increase in lateral flexibility of the superstructure.

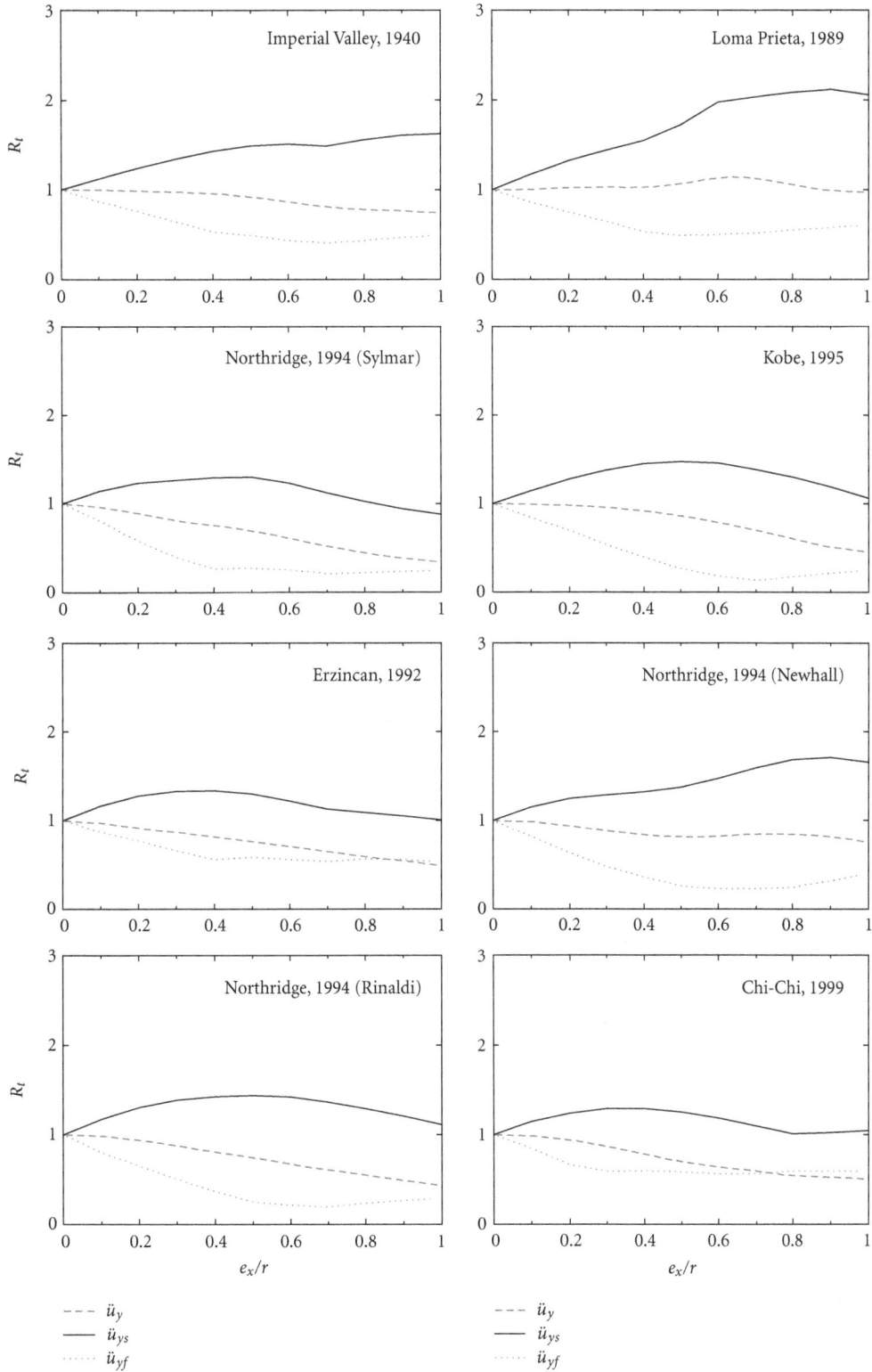

FIGURE 14: Effect of eccentricity on response ratio, R_t for various peak accelerations ($T_y = 1$ s, $\Omega_\theta = 0.5$).

In order, to study the effectiveness of semiactive MR dampers with clipped-optimal control algorithm in reducing the coupled responses, the variations of ratio, R_e against eccentricity ratio, e_x/r for the system with $T_y = 1$ are shown in Figure 20. The response ratios, R_e, for lateral,

torsional, and edge displacements as well as accelerations for different values of Ω_θ for eight selected earthquakes are obtained and the average trends from these earthquakes are shown in Figure 20. It may be recalled that, the ratio, R_e is between the responses of controlled asymmetric system

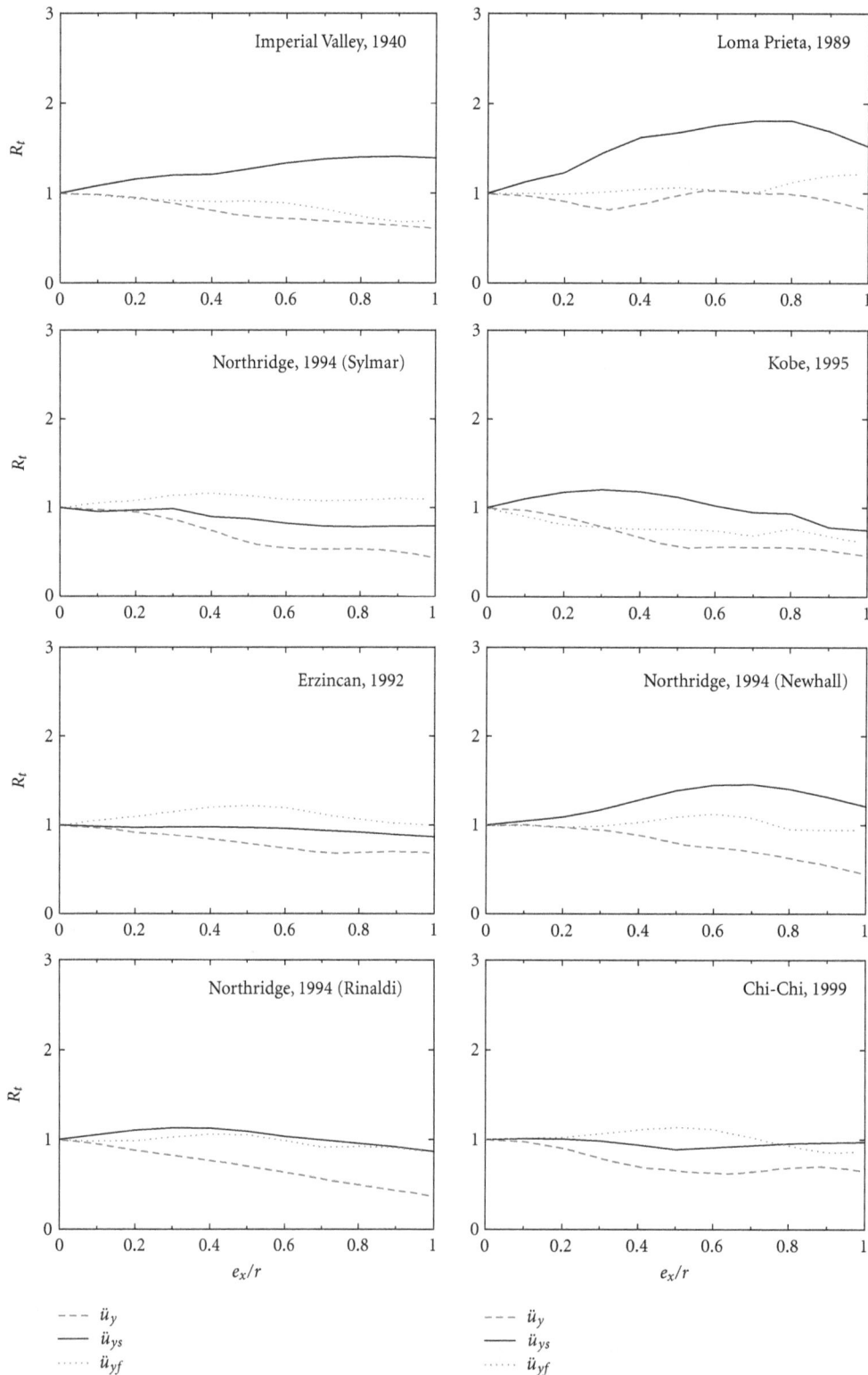

FIGURE 15: Effect of eccentricity on response ratio, R_t for various peak accelerations ($T_y = 1$ s, $\Omega_\theta = 1$).

and corresponding uncontrolled system and its value less than one indicates that the control system is effective in reducing the responses. It can be noticed from the figure that the ratio, R_e, for lateral, edge, and torsional displacements and accelerations comes out to be less than unity for

all considered values of e_x/r implying that the semiactive MR dampers using clipped-optimal algorithm is effective in reducing peak lateral-torsional responses. It is further noticed that for strongly coupled system ($\Omega_\theta = 1$), the values of ratio, R_e, for torsional displacement, u_θ, and

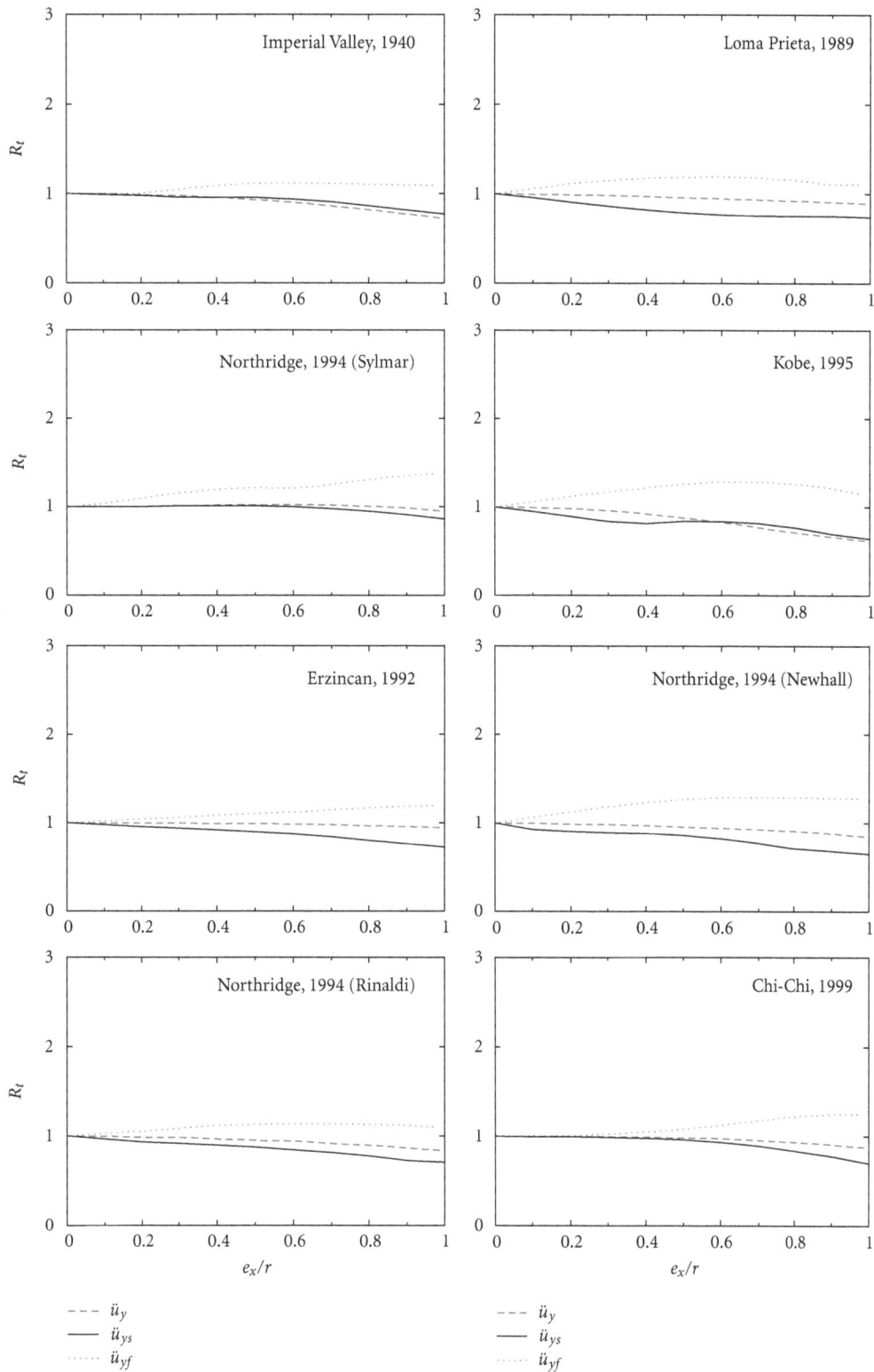

FIGURE 16: Effect of eccentricity on response ratio, R_t for various peak accelerations ($T_y = 1$ s, $\Omega_\theta = 2$).

torsional acceleration, $\ddot{u}_\theta$, increases with increase in e_x/r. This means the effectiveness of control system for controlling u_θ and $\ddot{u}_\theta$ reduces for higher eccentricities for such systems. Moreover, for torsionally stiff systems, the values of ratio, R_e for all responses quantities remains very less sensitive to

the change in e_x/r. However, for such systems with lower eccentricities ratio, the effectiveness of control system is less in reducing torsional acceleration. Furthermore, for strongly coupled system, the values of R_e for torsional displacement and acceleration responses are less than those obtained for

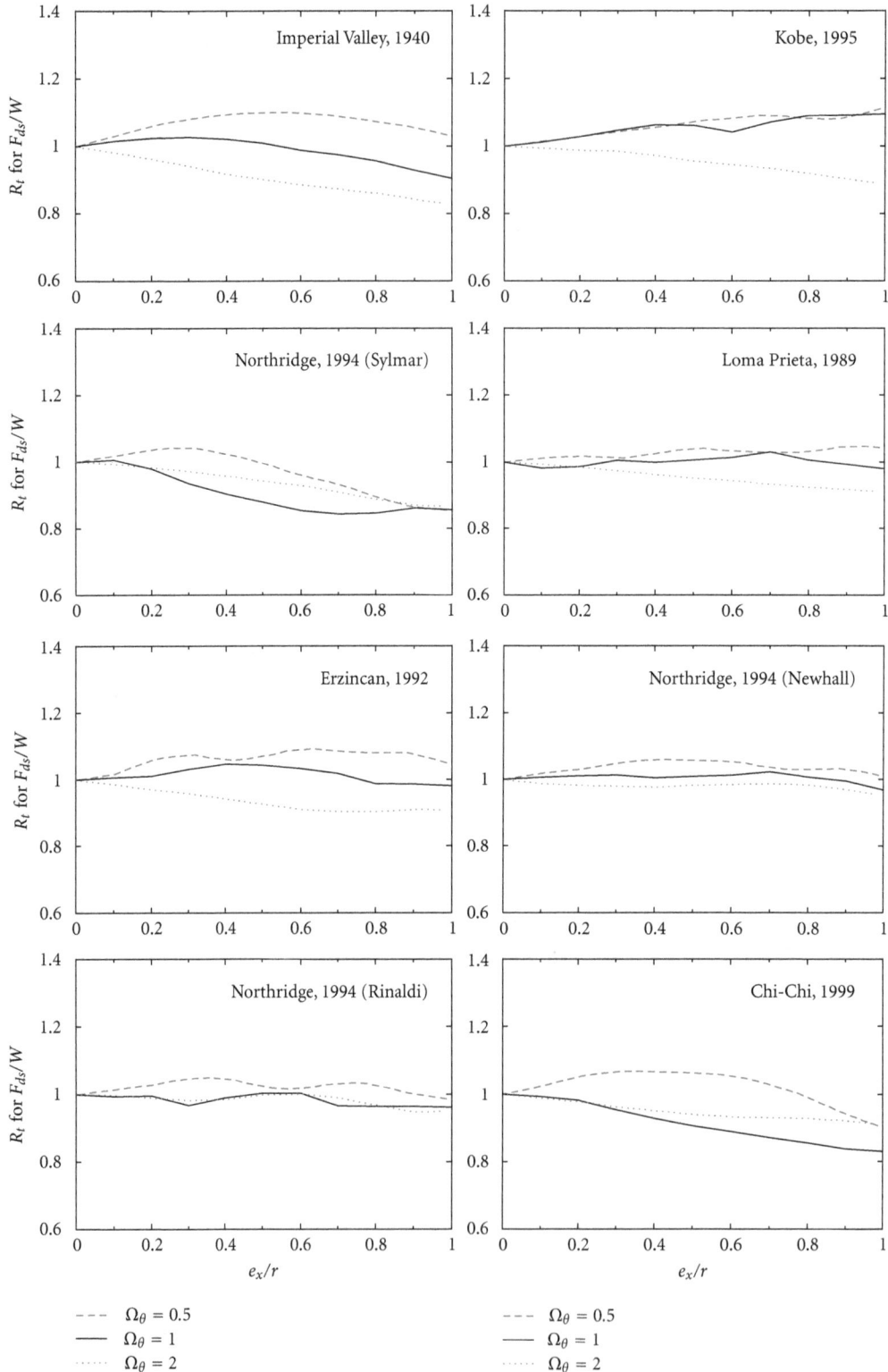

FIGURE 17: Effect of eccentricity on response ratio, R_t for normalized stiff edge RMS damper force ($T_y = 1\,$s).

lateral and edge displacement and acceleration responses for considered range of values of e_x/r, implying that the implemented control system is more effective for reducing the torsional responses than lateral responses of strongly coupled system. Moreover, the variation in values of ratio,

R_e for edge displacements and accelerations strongly depends on values of e_x/r for the systems with $\Omega_\theta = 0.5$ and 1. Thus, the effectiveness of semiactive control system in reducing the various responses has strong dependence on superstructure eccentricity for torsionally flexible and strongly coupled

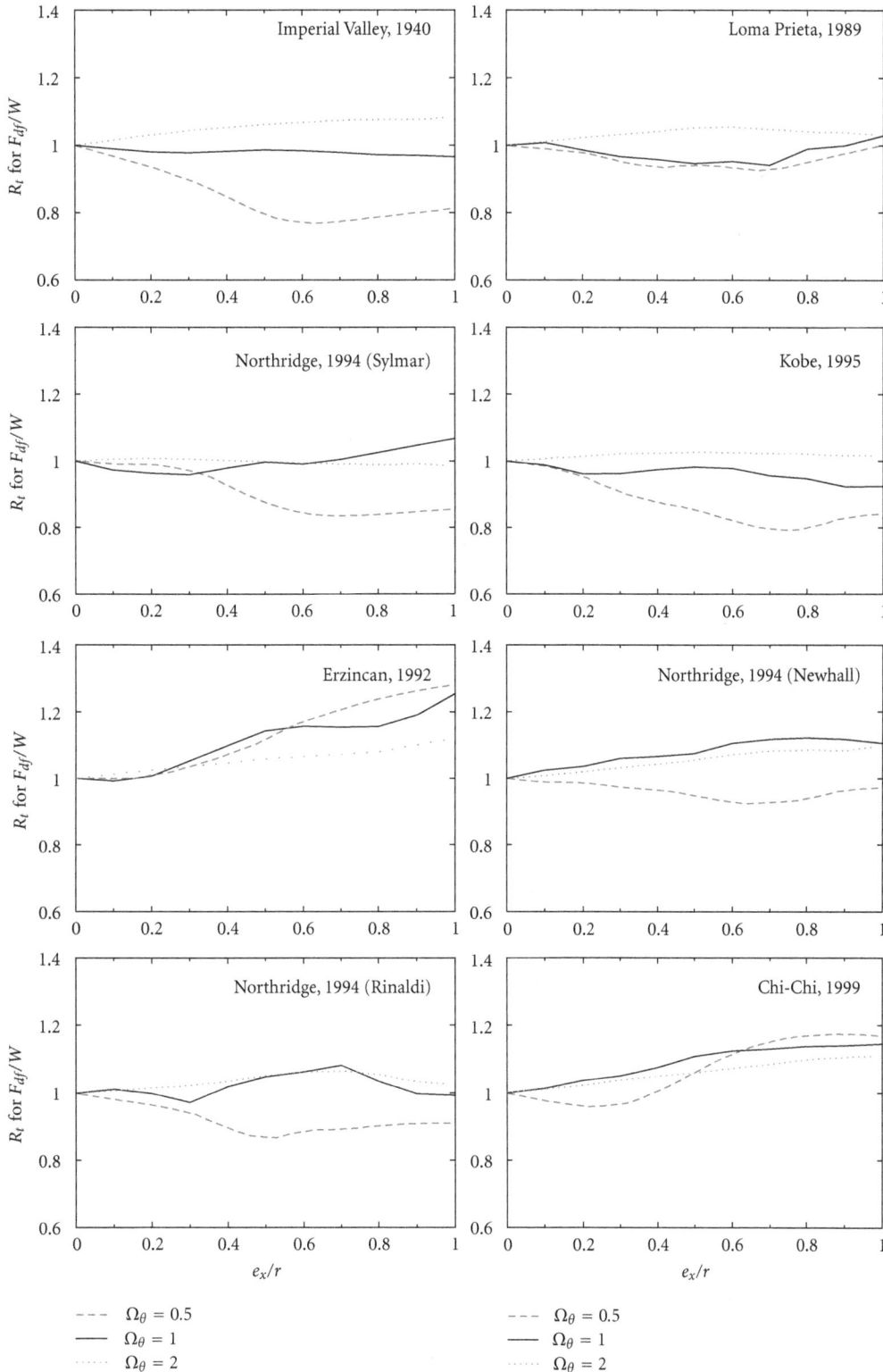

FIGURE 18: Effect of eccentricity on response ratio, R_t for normalized flexible edge RMS damper force ($T_y = 1$ s).

systems and it is very less dependent for torsionally stiff systems.

The investigations carried out in this research paper are based on the idealized single-storey building model with dampers located only at periphery (edges) of the building as it would develop the higher resisting forces. The dampers are assumed to be symmetrically placed in the building. Moreover, from the practical application

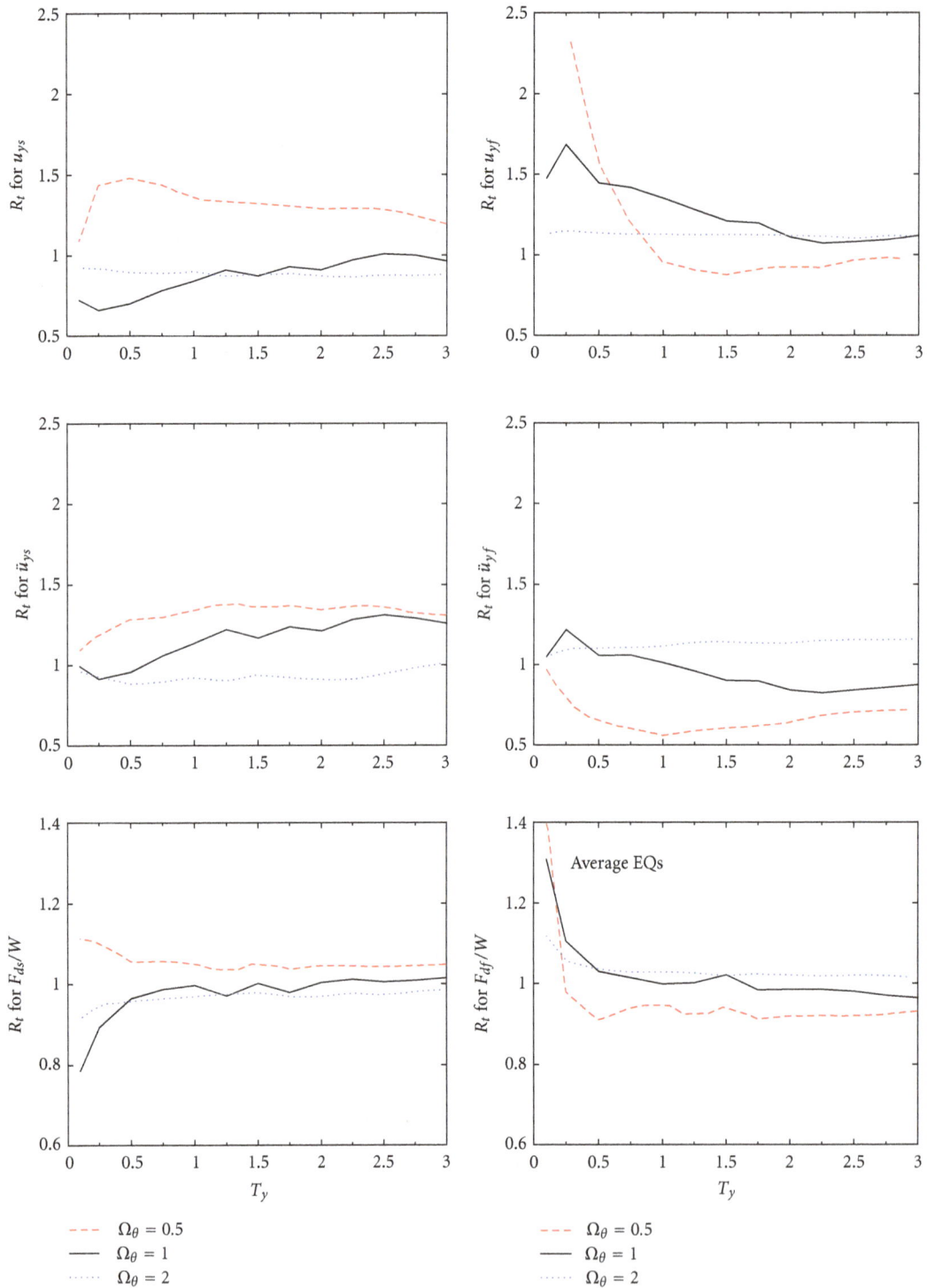

FIGURE 19: Effect of time period, T_y on response ratio, R_t for peak edge displacements and accelerations and RMS damper forces under average earthquakes ($e_x/r = 0.3$).

point of view also, it shall be convenient to install dampers at the edges of the building to satisfy the functional requirements also. However, it shall be further useful to study the effects of torsional coupling for multistory building with different configuration of dampers in plan and/or in elevation.

6. Conclusions

The seismic response of linearly elastic, single-storey, one-way asymmetric building installed with semiactive MR dampers subjected to different earthquake ground motions is investigated. The response is evaluated considering

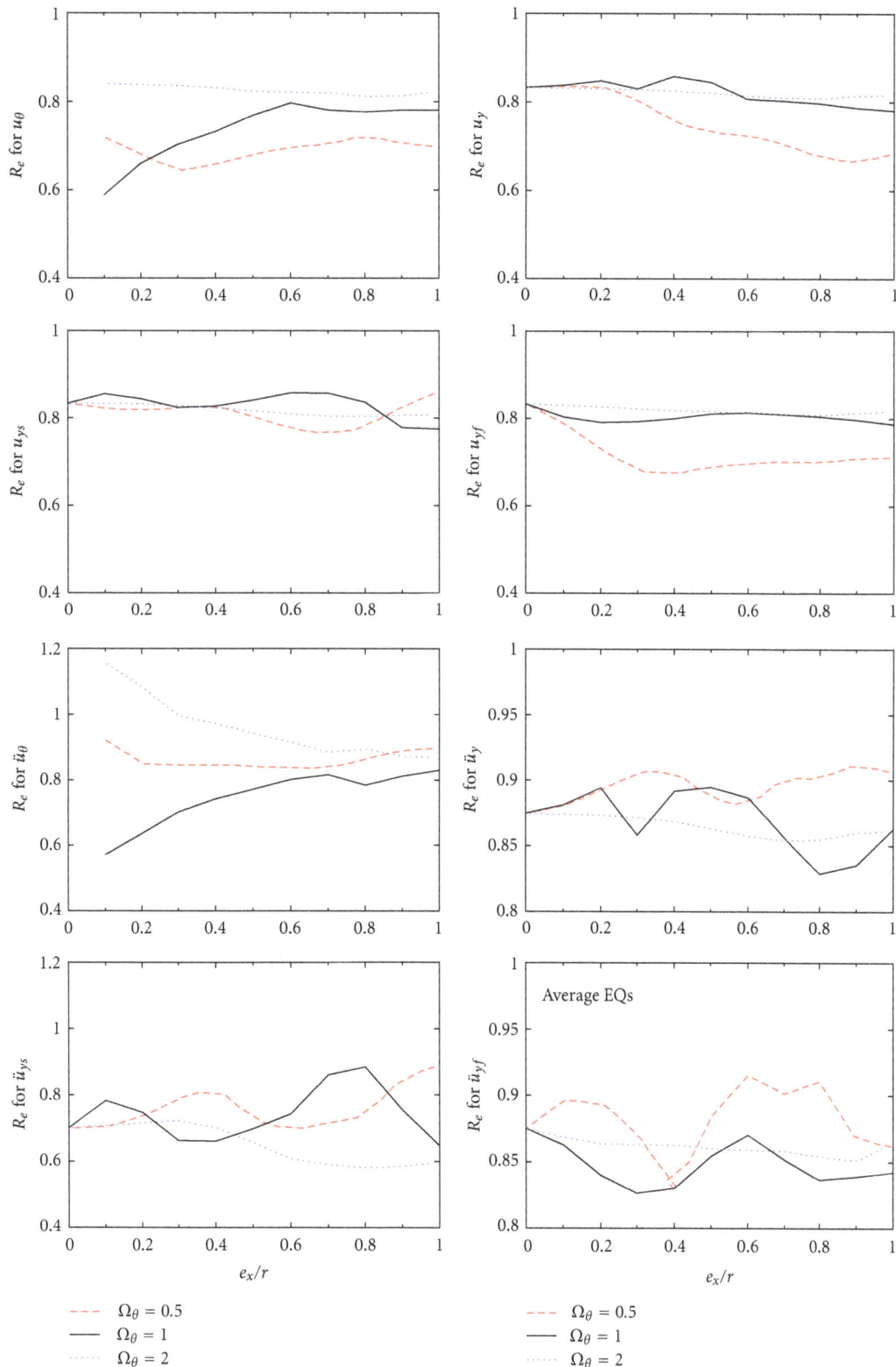

FIGURE 20: Effect of eccentricity on response ratio, R_e for various peak response quantities under average earthquakes ($T_y = 1$).

clipped-optimal control algorithm with parametric variations to study the effectiveness of semiactive control system for torsionally coupled system and the influence of important parameters on the effectiveness of semiactive

control system for asymmetric systems. The important parameters considered are: eccentricity ratio of superstructure, uncoupled lateral time period, and ratio of uncoupled torsional to lateral frequency. From the trend of the results of the present study, the following conclusions can be drawn.

(1) For the semiactive clipped-optimal control algorithm, with little increase in RMS damper forces as compared to passive-off case, the higher or comparable reduction in peak and RMS values of lateral, edge and torsional responses is obtained than those which is obtained by passive-on case.

(2) For torsionally flexible and strongly coupled systems, the effects of torsional coupling can not be neglected while estimating the effectiveness of semiactive control system for asymmetric systems as compared to corresponding symmetric systems in controlling edge displacements.

(3) The difference between the base shear of asymmetric and corresponding symmetric system increases with the increase in eccentricity ratio and it is significant for strongly coupled systems followed by torsionally flexible and torsionally stiff systems.

(4) The effects of torsional coupling are more pronounced for torsionally flexible systems than strongly coupled systems while estimating the effectiveness of semiactive control system for asymmetric systems in controlling lateral and edge accelerations as compared to corresponding symmetric systems.

(5) The effects of torsional coupling on the estimation of control forces significantly depend on superstructure eccentricity and torsional to lateral frequency ratio.

(6) The difference between the damper control forces for asymmetric systems and corresponding symmetric systems is higher for laterally stiff systems as compared to laterally flexible systems and this difference tends to reduce with the increase in lateral flexibility of the superstructure.

(7) The effectiveness of semiactive control system in reducing the various responses has strong dependence on superstructure eccentricity for torsionally flexible and strongly coupled systems and it is very less dependent for torsionally stiff systems.

Notations

a: Plan dimension of building, parallel to the direction of ground motion

a_0 and a_1: Coefficients for Rayleigh's mass and stiffness proportional damping matrix, respectively

$\mathbf{A}$: System matrix

$\mathbf{A_d}$: Discrete-time system matrix

A_m and n: Parameters for MR damper model

b: Plan dimension of building, perpendicular to the direction of ground motion

$\mathbf{B}$: Distribution matrix of control forces

$\mathbf{B_d}$: Discrete-time counterpart of distribution matrix of control forces

c_{0a} and c_{0b}: Parameters for MR damper model

c_{0i} and c_{1i}: Parameters for viscous damping observed at large and low velocities, respectively, for ith MR damper model

c_{1a} and c_{1b}: Parameters for MR damper model

$\mathbf{C}$: Structural damping matrix of the system

$\mathbf{C_m}$ and $\mathbf{D_m}$: Matrices used for measurement equation

e_x: Structural (superstructure) eccentricity between centre of mass (CM) and centre of rigidity (CR) of the building system

$\mathbf{E}$: Distribution matrix of excitation forces

$\mathbf{E_d}$: Discrete-time counterpart of distribution matrix of excitation forces

$\mathbf{f}$: Damper control force vector

f_{ci}: Desired control force for ith damper

$\mathbf{f_c}$: Desired control force vector

f_i: Control force in ith damper

F_{df}: Control force of damper located at flexible edge of building

F_{ds}: Control force of damper located at stiff edge of building

g: Acceleration due to gravity

$H(\cdot)$: Heaviside step function

$\mathbf{I}$: Identity matrix

J: Infinite horizon performance index for the design of controller

k: Time step

k_0: Stiffness at large velocities of MR damper model

k_1: Accumulator stiffness of MR damper model

$\mathbf{K}$: Stiffness matrix of the system

$\mathbf{K_g}$: Gain matrix for the design of controller

K_{xi}: Lateral stiffness of ith column in x-direction

K_y: Total lateral stiffness of the system in y-direction

K_{yi}: Lateral stiffness of ith column in y-direction

$K_{\theta r}$: Torsional stiffness of the system about a vertical axis at the CR

$K_{\theta\theta}$: Torsional stiffness of the system about a vertical axis at the CM

l_{a1}: Coordinate of position of the accelerometer 1, located at flexible edge

l_{a2}: Coordinate of position of the accelerometer 2, located at stiff edge

$\mathbf{L_g}$: Gain matrix for the state estimator

m: Lumped mass of the deck

$\mathbf{M}$: Mass matrix of the system

$\mathbf{P}$: Matrix of algebraic Ricatti equation

q_w: Coefficient for weighting matrix, $\mathbf{Q}$

$\mathbf{Q}$:	Weighting matrix for the vector of regulated/measured responses
r:	Mass radius of gyration about a vertical axis through CM
r_w:	Coefficient for weighting matrix, $\mathbf{R}$
$\mathbf{R}$:	Weighting matrix for the vector of control forces
R_e:	Response ratio to study the effectiveness of control system
R_t:	Response ratio to study the effects of torsional coupling
$\mathbf{S}$:	Matrix of algebraic Ricatti equation
T_y:	Uncoupled lateral time period of system
$\mathbf{u}$ and $\dot{\mathbf{u}}$:	Displacement and velocity vectors, respectively
u_{di}:	Filtered voltage of ith MR damper
u_y:	Lateral displacement at CM, in y-direction
u_{yf}:	Flexible edge displacement
u_{ys}:	Stiff edge displacement
u_θ:	Torsional displacement, in θ-direction
$\ddot{\mathbf{u}}$:	Acceleration vector
$\ddot{\mathbf{u}}_g$:	Ground acceleration vector
$\ddot{u}_{gy}$:	Ground acceleration in y-direction
$\ddot{u}_{mf}$:	Flexible edge acceleration measured by the accelerometer
$\ddot{u}_{ms}$:	Stiff edge acceleration measured by the accelerometer
$\mathbf{v}$:	Measurement noise vector
v_i:	Applied voltage sent to the current driver of ith damper
$V_{\max}$:	Maximum applied voltage
V_y:	Base shear
W:	Weight of deck
x_0:	Initial displacement of spring, k_1
x_{di}:	Displacement of ith MR damper
x_i and y_i:	x-coordinate and y-coordinate distances of ith element from CM
$\dot{x}_{di}$:	Velocity of ith MR damper
y_{di} and $\dot{y}_{di}$:	Displacement and velocity at middle plate of ith MR damper model
$\mathbf{y}_m$:	Vector measured responses
z_i:	Evolutionary variable
$\mathbf{z}$:	State vector
$\hat{\mathbf{z}}$:	Estimated state vector
α_a and α_b:	Parameters for MR damper model
α_i:	Parameter for ith MR damper model
β_m and γ_m:	Parameters for MR damper model
γ_g:	Parameter for statistically independent Gaussian white noise processes
η:	Parameter for MR damper model
Γ:	Influence coefficient vector
Δt:	Time interval
Λ:	Location matrix for control forces
ω_y:	Uncoupled lateral frequency of the system
ω_θ:	Uncoupled torsional frequency of the system
Ω_θ:	Ratio between uncoupled torsional to lateral frequency.

References

[1] R. Hejal and A. K. Chopra, "Lateral-torsional coupling in earthquake response of frame buildings," *Journal of Structural Engineering*, vol. 115, no. 4, pp. 852–867, 1989.

[2] R. S. Jangid and T. K. Datta, "Nonlinear response of torsionally coupled base isolated structure," *Journal of Structural Engineering*, vol. 120, no. 1, pp. 1–22, 1994.

[3] R. S. Jangid and T. K. Datta, "Performance of multiple tuned mass dampers for torsionally coupled system," *Earthquake Engineering and Structural Dynamics*, vol. 26, no. 3, pp. 307–317, 1997.

[4] R. K. Goel, "Effects of supplemental viscous damping on seismic response of asymmetric-plan systems," *Earthquake Engineering and Structural Dynamics*, vol. 27, no. 2, pp. 125–141, 1998.

[5] V. A. Date and R. S. Jangid, "Seismic response of torsionally coupled structures with active control device," *Journal of Structural Control*, vol. 8, no. 1, pp. 5–15, 2001.

[6] W. H. Lin and A. K. Chopra, "Asymmetric one-storey elastic systems with non-linear viscous and viscoelastic dampers: simplified analysis and supplemental damping system design," *Earthquake Engineering and Structural Dynamics*, vol. 32, no. 4, pp. 579–596, 2003.

[7] J. C. de la Llera, J. L. Almazán, and I. J. Vial, "Torsional balance of plan-asymmetric structures with frictional dampers: analytical results," *Earthquake Engineering and Structural Dynamics*, vol. 34, no. 9, pp. 1089–1108, 2005.

[8] L. Petti and M. De Iuliis, "Torsional seismic response control of asymmetric-plan systems by using viscous dampers," *Engineering Structures*, vol. 30, no. 11, pp. 3377–3388, 2008.

[9] Y. Chi, M. K. Sain, K. D. Pham, and B. F. Spencer Jr., "Structural control paradigms for an asymmetric building," in *Proceedings of the 8th ASCE Specialty Conference on Probabilistic Mechanics and Structural Reliability*, University of Notre Dame, July 2000.

[10] O. Yoshida, S. J. Dyke, L. M. Giacosa, and K. Z. Truman, "Experimental verification of torsional response control of asymmetric buildings using MR dampers," *Earthquake Engineering and Structural Dynamics*, vol. 32, no. 13, pp. 2085–2105, 2003.

[11] O. Yoshida and S. J. Dyke, "Response control of full-scale irregular buildings using magnetorheological dampers," *Journal of Structural Engineering*, vol. 131, no. 5, pp. 734–742, 2005.

[12] D. A. Shook, P. N. Roschke, P. Y. Lin, and C. H. Loh, "Semi-active control of a torsionally-responsive structure," *Engineering Structures*, vol. 31, no. 1, pp. 57–68, 2009.

[13] H. N. Li and X. L. Li, "Experiment and analysis of torsional seismic responses for asymmetric structures with semi-active control by MRdampers," *Smart Materials and Structures*, vol. 18, no. 7, Article ID 075007, pp. 1–10, 2009.

[14] B. F. Spencer Jr., S. J. Dyke, M. K. Sain, and J. D. Carlson, "Phenomenological model for magnetorheological dampers," *Journal of Engineering Mechanics*, vol. 123, no. 3, pp. 230–238, 1997.

[15] S. J. Dyke, B. F. Spencer Jr., M. K. Sain, and J. D. Carlson, "Modeling and control of magnetorheological dampers for seismic response reduction," *Smart Materials and Structures*, vol. 5, no. 5, pp. 565–575, 1996.

[16] G. C. Hart and K. Wong, *Structural Dynamics for Structural Engineers*, John Wiley & Sons, New York, NY, USA, 2000.

[17] L. Y. Lu, "Predictive control of seismic structures with semi-active friction dampers," *Earthquake Engineering and Structural Dynamics*, vol. 33, no. 5, pp. 647–668, 2004.

[18] L. M. Jansen and S. J. Dyke, "Semiactive control strategies for MR dampers: comparative study," *Journal of Engineering Mechanics*, vol. 126, no. 8, pp. 795–803, 2000.

[19] H. J. Jung, B. F. Spencer Jr., and I. W. Lee, "Control of seismically excited cable-stayed bridge employing magnetorheological fluid dampers," *Journal of Structural Engineering*, vol. 129, no. 7, pp. 873–883, 2003.

[20] G. Yang, B. F. Spencer Jr., J. D. Carlson, and M. K. Sain, "Large-scale MR fluid dampers: modeling and dynamic performance considerations," *Engineering Structures*, vol. 24, no. 3, pp. 309–323, 2002.

[21] Y. Wang and S. Dyke, "Smart system design for a 3D base-isolated benchmark building," *Structural Control and Health Monitoring*, vol. 15, no. 7, pp. 939–957, 2008.

Permissions

The contributors of this book come from diverse backgrounds, making this book a truly international effort. This book will bring forth new frontiers with its revolutionizing research information and detailed analysis of the nascent developments around the world.

We would like to thank all the contributing authors for lending their expertise to make the book truly unique. They have played a crucial role in the development of this book. Without their invaluable contributions this book wouldn't have been possible. They have made vital efforts to compile up to date information on the varied aspects of this subject to make this book a valuable addition to the collection of many professionals and students.

This book was conceptualized with the vision of imparting up-to-date information and advanced data in this field. To ensure the same, a matchless editorial board was set up. Every individual on the board went through rigorous rounds of assessment to prove their worth. After which they invested a large part of their time researching and compiling the most relevant data for our readers. Conferences and sessions were held from time to time between the editorial board and the contributing authors to present the data in the most comprehensible form. The editorial team has worked tirelessly to provide valuable and valid information to help people across the globe.

Every chapter published in this book has been scrutinized by our experts. Their significance has been extensively debated. The topics covered herein carry significant findings which will fuel the growth of the discipline. They may even be implemented as practical applications or may be referred to as a beginning point for another development. Chapters in this book were first published by Hindawi Publishing Corporation; hereby published with permission under the Creative Commons Attribution License or equivalent.

The editorial board has been involved in producing this book since its inception. They have spent rigorous hours researching and exploring the diverse topics which have resulted in the successful publishing of this book. They have passed on their knowledge of decades through this book. To expedite this challenging task, the publisher supported the team at every step. A small team of assistant editors was also appointed to further simplify the editing procedure and attain best results for the readers.

Our editorial team has been hand-picked from every corner of the world. Their multi-ethnicity adds dynamic inputs to the discussions which result in innovative outcomes. These outcomes are then further discussed with the researchers and contributors who give their valuable feedback and opinion regarding the same. The feedback is then collaborated with the researches and they are edited in a comprehensive manner to aid the understanding of the subject.

Apart from the editorial board, the designing team has also invested a significant amount of their time in understanding the subject and creating the most relevant covers. They scrutinized every image to scout for the most suitable representation of the subject and create an appropriate cover for the book.

The publishing team has been involved in this book since its early stages. They were actively engaged in every process, be it collecting the data, connecting with the contributors or procuring relevant information. The team has been an ardent support to the editorial, designing and production team. Their endless efforts to recruit the best for this project, has resulted in the accomplishment of this book. They are a veteran in the field of academics and their pool of knowledge is as vast as their experience in printing. Their expertise and guidance has proved useful at every step. Their uncompromising quality standards have made this book an exceptional effort. Their encouragement from time to time has been an inspiration for everyone.

The publisher and the editorial board hope that this book will prove to be a valuable piece of knowledge for researchers, students, practitioners and scholars across the globe.

List of Contributors

Ying Wang, Xiangyu Wang and Ping Yung
School of Built Environment, Curtin University of Western Australia, Australia

Jun Wang
School of Construction Management and Real Estate, Chongqing University, China

Guo Jun
CCDI, China

G. Tecchio, M. Grendene and C.Modena
Department of Structural and Transportation Engineering, University of Padova, Via Marzolo 9, 35131 Padova, Italy

S. Alshurafa and H. Alhayek
Department of Civil Engineering, University of Manitoba, 15 Gillson Street Winnipeg, MB, Canada MB R3T 2N2

F. Taheri
Department of Civil and Resource Engineering, Dalhousie University, 1360 Barrington Street, Halifax, NS, Canada

Aleksandra Radlinska, Andrea Welker, Kathryn Greising, Blake Campbell and David Littlewood
Department of Civil and Environmental Engineering, Villanova University, 800 Lancaster Avenue, Villanova, PA 19085, USA

Isamu Yoshitake
Department of Civil and Environmental Engineering, Yamaguchi University, Ube, Yamaguchi 755-8611, Japan

Andrew Scanlon and Farshad Rajabipour
Department of Civil and Environmental Engineering, Pennsylvania State University, University Park, PA 16802, USA

Yoichi Mimura
Department of Civil and Environmental Engineering, Kure National College of Technology, Kure, Hiroshima 737-8506, Japan

Lingyu Yu and Victor Giurgiutiu
Department of Mechanical Engineering, University of South Carolina, Columbia, SC 29208, USA

Sepandarmaz Momeni and Valery Godinez
Mistras Group Inc., 195 Clarksville Road, Princeton Junction, NJ 08550, USA

Paul Ziehl and Jianguo Yu
Department of Civil Engineering, University of South Carolina, Columbia, SC 29208, USA

Andres Lepage
Department of Architectural Engineering, The Pennsylvania State University, 104 Engineering Unit A, University Park, PA 16802, USA

Hooman Tavallali
Leslie E. Robertson Associates, 40 Wall Street, 23rd Floor, New York, NY 10005, USA

Santiago Pujol
School of Civil Engineering, Purdue University, 550 Stadium Mall Drive, West Lafayette, IN 47907, USA

Jeffrey M. Rautenberg
Wiss, Janney, Elstner Associates, 2000 Powell Street, Suite 1650, Emeryville, CA 94608, USA

Adutwum Marfo, Ying Luo and Chen Zhong-an
Faculty of Civil Engineering and Mechanics, Jiangsu University, Zhenjiang 212013, China

Hubo Cai
Division of Construction Engineering and Management, School of Civil Engineering, Purdue University, 550 Stadium Mall Drive, West Lafayette, IN 47907, USA

Osama Abudayyeh, Upul Attanayake and Joseph Barbera
Department of Civil and Construction Engineering, Western Michigan University, Kalamazoo, MI 49008, USA

Eyad Almaita and Ikhlas Abdel-Qader
Department of Electrical and Computer Engineering, Western Michigan University, Kalamazoo, MI 49009, USA

Panagis G. Papadopoulos, Andreas Diamantopoulos, Haris Xenidis and Panos Lazaridis
Department of Civil Engineering, Aristotle University of Thessaloniki, 54124 Thessaloniki, Greece

Aziz Saber and Ashok Reddy Aleti
Department of Civil Engineering, Louisiana Tech University, 600 West Arizona Avenue, Ruston, LA 71272, USA

B. Asgari and S. A. Osman
Department of Civil & Structural Engineering, Faculty of Engineering & Built Environment, National University ofMalaysia (UKM), 43600 Bangi, Selangor, Malaysia

A. Adnan
Department of Civil & Structural Engineering, Faculty of Engineering & Built Environment, UTM University of Malaysia, 81310 Skudai, Johor, Malaysia

Daniel P. Treese and Shirley E. Clark
Environmental Engineering Program, Penn State Harrisburg, Middletown, PA, USA

Katherine H. Baker
Life Sciences Program, Penn State Harrisburg, Middletown, PA, USA

Jejal Reddy Bathi
Department of Civil, Construction and Environmental Engineering, The University of Alabama, Tuscaloosa, AL 35487, USA
Global Solutions International, LLC, P.O. Box 223, Mobile, AL 36652, USA

Robert E. Pitt
Department of Civil, Construction and Environmental Engineering, The University of Alabama, Tuscaloosa, AL 35487, USA

Shirley E. Clark
Penn State Harrisburg, Environmental Engineering Program, Middletown, PA 17057, USA

Junwon Seo
Department of Civil, Construction, Environmental Engineering, Iowa State University, Ames, IA 50011, USA

Daniel G. Linzell
Department of Civil Engineering, The University of Nebraska, Lincoln, NE 68588, USA

Jong Wan Hu
Department of Civil and Environmental Engineering, University of Incheon, Incheon 406-772, Republic of Korea

Eltayeb Mohamedelhassan, Kevin Curtain, Matt Fenos, Kevin Girard, Anthony Provenzano and Wesley Tabaczuk
Department of Civil Engineering, Lakehead University, 955 Oliver Road, Thunder Bay, ON, Canada

Adam Scianna, Zhaoshuo Jiang, Richard Christenson and John De Wolf
Department of Civil and Environmental Engineering, University of Connecticut, Storrs, Connecticut 06269-2037, USA

Andrew N. Daumueller and David V. Jauregui
Department of Civil Engineering, New Mexico State University, Hernandez Hall Box 30001, Las Cruces, NM 88003, USA

Snehal V.Mevada and R. S. Jangid
Department of Civil Engineering, Indian Institute of Technology Bombay, Powai, Mumbai 400 076, India

www.ingramcontent.com/pod-product-compliance
Lightning Source LLC
Chambersburg PA
CBHW080658200326
41458CB00013B/4906